AF773430

2015 Annual Competitiveness Analysis, Agricultural Productivity and Development Vision for ASEAN-10

NUS National University of Singapore

LKY Lee Kuan Yew School of Public Policy
ASIA COMPETITIVENESS INSTITUTE

In 2016 ACI was ranked 13th globally, 2nd in Asia and 1st in Singapore amongst 90 think-tanks world-wide under the "Best University Affiliated Think-Tank" category by Think-Tanks & Civil Societies Program at The University of Pennsylvania, USA

About ACI

The Asia Competitiveness Institute (ACI) was established in August 2006 as a research centre at the Lee Kuan Yew School of Public Policy (LKYSPP), National University of Singapore (NUS). It aims to build the intellectual leadership and network for understanding and developing competitiveness in the Asia region. ACI seeks to contribute to the enhancement of inclusive growth, living standards and institutional governance through competitiveness research on sub-national economies in Asia. It identifies mitigating issues and challenges for potential public policy interventions through close collaboration with regional governments, business corporations, policy think-tanks and academics. ACI's three key research pillars include (i) sub-national economies level competitiveness analysis and city-level liveability analysis, (ii) firm-level competitiveness analysis in 16 Asia economies, and (iii) Singapore's long-term growth strategies and public policy analysis.

ACI's value propositions may be encapsulated in its acronym:

Analytical inputs to initiate policies for policy-makers and business leaders in Asia

Capacity building to enable others through improvement in productivity and efficiency

Intellectual leadership to create pragmatic models of competitiveness and inclusive growth

The institute's core research competencies can also be encapsulated in this acronym describing our evidence-based assessments conducted on public policies for ASEAN in the context of the rise of China and India.

Vision and Mission

- ACI's over-arching vision is to build up its research credibility with policy impact, contributing as a professional, world-class think-tank.
- ACI's mission is to establish our niche as a leading policy think-tank by identifying competitiveness trends, opportunities and challenges, as well as promoting competition and synergising complementarities amongst Asian economies and business corporations.
- ACI endeavours to articulate sound recommendations, entice discourse, and shape agenda in the arena of public policy amongst Asian governments.
- ACI undertakes evidence-based analysis of public policy issues and decisions, in order to provide assessment of their effectiveness as well as economic and societal impact.

Research Initiatives and Collaborations

I. Identify trends of competitiveness and policy analysis on trade and investment of ASEAN, within the regional context of competition and complementarities with China and India.

II. Identify competitive strengths and conduct policy analysis on Singapore within the context of regional economies with international benchmarking.

III. We are regularly releasing three indices on liveability ranking including 64 Global Cities, 100 Greater China Cities and 17 Shandong Cities.

IV. We have established an Ease of Doing Business Index on Attractiveness to Investors, Business Friendliness and Competitive Policies for 21 states and federal territories of India and 33 provinces of Indonesia.

V. We have signed Memoranda of Understanding with The World Bank (2016 and 2015), SPRING Singapore (2014) and European Central Bank (2014).

VI. We have signed Memoranda of Understanding with various institutions in Greater China economies, including Shanghai Academy of Social Sciences (2016), Shandong Academy of Social Sciences (2015), Chongqing Municipal People's Government (2015), China Institute for Reform and Development, Haikou (2015), Counsellors' Office of the People's Government of Guangdong Province (LOI, 2014) and Chung-Hua Institution for Economic Research (2015), Taiwan.

VII. We have signed Memoranda of Understanding with six Chief Ministers' Offices in Andhra Pradesh (2015), Bihar (2015), Chhattisgarh (2015), Madhya Pradesh (2015), Odisha (2015) and Punjab (2015).

VIII. We have signed Memoranda of Understanding with Indonesia Investment Coordinating Board (2014), Indonesian Agency for Agricultural Research and Development at Ministry of Agriculture (2014), Indonesian President's Delivery Unit for Developing Monitoring and Oversight (2013) and Employer's Association of Indonesia (2013).

IX. We have signed Memoranda of Understanding with ASEAN think-tanks, including Centre for Strategic and International Studies, Indonesia (2015), Institute of Strategic & International Studies, Malaysia (2015), Philippine Institute for Development Studies (2015), Thailand Development Research Institute (2015) and Central Institute for Economic Management, Vietnam (2015).

About ACI's Research Pillars

ACI has consciously engaged in economic research that has significant relevance to Singapore and the Asian region. Over the years, ACI has focused on our expertise in quantitative competitiveness analysis and simulation, spinning off volumes of research output in the applications of our methodology and regional insight. At this stage, ACI has identified three core research pillars that will guide and define its research efforts moving forward. The three research pillars are as follows:

Pillar I. Sub-national Economies Competitiveness Analysis

ACI engages in systematic and methodical competitiveness analyses of the *sub-national economies* by using an evidence-based, empirical approach involving a comprehensive list of relevant indicators, which are categorised under multiple layers called the 'environments'. A number of our projects, including competitiveness analyses of Greater China, ASEAN-10, India, and Indonesia, as well as other thematic research projects, were conducted by adopting this common methodology, with potential variations in the specific environments and indicators. ACI's competitiveness analysis of Asian economies goes beyond the usual ranking to offer constructive policy recommendations on how individual member states can improve their rankings vis-à-vis their sub-national or regional peers through the application of the 'what-if' simulation, which provides the projected improvements of each state's ranking whereby the bottom one-fifth of its indicators are enhanced.

Pillar II. Micro-based Firm Level Competitiveness Analysis

Micro-based firm level competitiveness analysis is ACI's second research pillar that has been developed in view of the intrinsic importance of firm-level competitiveness in terms of productivity, efficiency, and governance. In the context of globalisation, mobility of economic activities, and blurring of borders, an understanding of the determinants and dynamics of firm-level competitiveness is paramount, in order for policy makers to make adjustments and prepare their industries for an increasingly competitive economic landscape. To this end, ACI has partnered with the European Central Bank to carry out research in this area, possessing a solid foundation in firm-level productivity research in the form of European Competitiveness Network database and methodology. ACI envisaged the expansion of CompNet into Asia, thereby pioneering the Asia CompNet research network encompassing 16 economies in Asia.

Pillar III. Singapore's Long-term Economic Growth Strategies and Public Policies Analysis

The third and imminent research pillar focuses on Singapore's long-term economic growth strategies in the context of changing circumstances, future trends, and emerging opportunities for Singapore in the decades ahead. In particular, ACI will take on the task of critically examining Singapore's public policy strengths and areas of improvement by assessing policy successes of the past, identifying new issues to address and the current policy gaps, through a systematic and evidence-based research inquiry using quantitative methodology and empirical data, leveraging on our network of policy experts.

Founding Patron and International Advisory Panel

Founding Patron: **Mr George Yeo**
Visiting Scholar, Lee Kuan Yew School of Public Policy, National University of Singapore & Former Minister of Foreign Affairs, Singapore

International Advisory Panel

Co-Chairs:

Professor Michael Porter (2006–2010)
Bishop William Lawrence University Professor

Ms Marjorie Yang (2006–2012)
Chairman, Esquel Group

Members:

Professor Kishore Mahbubani
Dean, Lee Kuan Yew School of Public Policy, National University of Singapore

Dr Kuntoro Mangkusubroto
Head, President's Delivery Unit for Developing Monitoring and Oversight (UKP4), Indonesia

Mr Narayana Murthy
Chairman, Infosys Technologies Limited

Mr Philip Yeo
Chairman, SPRING Singapore

Mrs Ow Foong Pheng
Permanent Secretary, Ministry of Trade and Industry

Ms Yong Ying-I
Permanent Secretary, Public Service Division

Professor Chan Kam Leung Alan
Dean, College of Humanities, Arts and Social Sciences,
Nanyang Technological University

Asia Competitiveness Institute

Co-Directors:

Associate Professor Tan Khee Giap (2011 till now)
Professor Tan Kong Yam (2011 till now)

Asia Competitiveness Institute - World Scientific Series

ISSN: 2424-9912

Series Editor: Tan Khee Giap (*Lee Kuan Yew School of Public Policy, NUS, Singapore)*

2014

2014 Annual Competitiveness Ranking and Simulation Study for ASEAN-10 and Development Strategies to Enhance Asia Economic Connectivity
by Tan Khee Giap, Sangiita Yoong Wei Cher, Sasidaran Gopalan & Nguyen Le Phuong Anh (Lee Kuan Yew School of Public Policy, NUS, Singapore)

2014 Annual Provincial and Regional Competitiveness Analysis for Greater China Economies: Development Strategies under the New Normal
by Tan Khee Giap, Yuan Randong & Sangiita Yoong Wei Cher (Lee Kuan Yew School of Public Policy, NUS, Singapore)

2014 Regional Competitiveness Analysis and a Master Plan on Regional Development Strategies for India: Annual Competitiveness Update and Evidence on Economic Development Model for Selected States of India
by Tan Khee Giap, Linda Low, Kartik Rao & Tan Kong Yam (Lee Kuan Yew School of Public Policy, NUS, Singapore)

2014 Provincial and Inaugural Regional Competitiveness Analysis: Safeguarding Indonesia's Growth Momentum
by Tan Khee Giap, Nurina Merdikawati, Mulya Amri & Tan Kong Yam (Lee Kuan Yew School of Public Policy, NUS, Singapore)

2014 Annual Competitiveness Analysis and Development Strategies for Indonesian Provinces
by Tan Khee Giap, Nurina Merdikawati, Mulya Amri & Blake Harley Berger (Lee Kuan Yew School of Public Policy, NUS, Singapore)

2015

2015 Annual Competitiveness Analysis, Agricultural Productivity and Development Vision for ASEAN-10
by Tan Khee Giap, Nguyen Le Phuong Anh, Sasidaran Gopalan & Luu Nguyen Trieu Duong (Lee Kuan Yew School of Public Policy, NUS, Singapore)

2015 Estimation of Agricultural Productivity and Annual Update of Competitiveness Analysis for Greater China: Optimising Agricultural Productivity and Promoting Innovation Driven Growth
by Tan Khee Giap, Yuan Randong & Xie Teleixi (Lee Kuan Yew School of Public Policy, NUS, Singapore)

2015 Agricultural Total Factor Productivity and Competitiveness Analysis for States and Federal Territories and Five Regions of India: Annual Competitiveness Update and Evidence on the Agricultural Development Models for Selected Indian States
by Tan Khee Giap, Sasidaran Gopalan & Anuja Tandon (Lee Kuan Yew School of Public Policy, NUS, Singapore)

2015 Agricultural Productivity, Decentralisation, and Competitiveness Analysis for Provinces and Regions of Indonesia
by Tan Khee Giap, Nurina Merdikawati, Mulya Amri & Tan Kong Yam (Lee Kuan Yew School of Public Policy, NUS, Singapore)

2015 Annual Competitiveness Analysis and Development Strategies for Indonesian Provinces
by Tan Khee Giap, Mulya Amri, Nurina Merdikawati & Nursyahida Binte Ahmad (Lee Kuan Yew School of Public Policy, NUS, Singapore)

The complete list of the volumes in the series published since 2013 can be found at
http://www.worldscientific.com/series/aciwss

Asia Competitiveness Institute - World Scientific Series

2015 Annual Competitiveness Analysis, Agricultural Productivity and Development Vision for ASEAN-10

Tan Khee Giap

Co-Director, Asia Competitiveness Institute
Lee Kuan Yew School of Public Policy, National University of Singapore

Nguyen Le Phuong Anh

Research Fellow & Deputy Director (Administration), Asia Competitiveness Institute
Lee Kuan Yew School of Public Policy, National University of Singapore

Sasidaran Gopalan

Research Fellow, Asia Competitiveness Institute
Lee Kuan Yew School of Public Policy, National University of Singapore

Luu Nguyen Trieu Duong

Research Assistant, Asia Competitiveness Institute
Lee Kuan Yew School of Public Policy, National University of Singapore

Reviewers

Nguyen Cam Tu
Deputy Minister
Ministry of Industry and Trade, Vietnam

Bustanul Arifin
Professor
University of Lampung and
Board of Founder/Senior Economist
Institute for Development of Economics and Finance, Indonesia

Roehlano Briones
Senior Research Fellow
Philippine Institute for Development Studies, Philippines

Nguyen Dinh Cung
President
Central Institute for Economic Management, Vietnam

Somhatai Panichewa
Chief Executive Officer
Amata VN Public Company Limited, Thailand

Chartchai Panichewa
President
ThaiSri Insurance Public Company Limited, Thailand

Waleerat Suphannachart
Assistant Professor
Department of Agricultural and Resource Economics
Kasetsart University, Thailand

Lee Kuan Yew School of Public Policy
Asia Competitiveness Institute
NUS National University of Singapore
World Scientific

Published by

World Scientific Publishing Co. Pte. Ltd.
5 Toh Tuck Link, Singapore 596224
USA office: 27 Warren Street, Suite 401-402, Hackensack, NJ 07601
UK office: 57 Shelton Street, Covent Garden, London WC2H 9HE

British Library Cataloguing-in-Publication Data
A catalogue record for this book is available from the British Library.

Asia Competitiveness Institute - World Scientific Series
2015 ANNUAL COMPETITIVENESS ANALYSIS, AGRICULTURAL PRODUCTIVITY AND DEVELOPMENT VISION FOR ASEAN-10

Copyright © 2017 by World Scientific Publishing Co. Pte. Ltd.

All rights reserved. This book, or parts thereof, may not be reproduced in any form or by any means, electronic or mechanical, including photocopying, recording or any information storage and retrieval system now known or to be invented, without written permission from the publisher.

For photocopying of material in this volume, please pay a copying fee through the Copyright Clearance Center, Inc., 222 Rosewood Drive, Danvers, MA 01923, USA. In this case permission to photocopy is not required from the publisher.

ISBN 978-981-3147-44-7

Desk Editor: Jiang Yulin

Typeset by Stallion Press
Email: enquiries@stallionpress.com

Printed in Singapore

Contents

About ACI iii
Foreword xiii
Preface xv
Acknowledgements xvii
About the Authors xix
List of Abbreviations xxi
List of Figures and Tables xxv

Chapter 1 The ASEAN Economic Community by 2016? 1
1.1 Introductory Notes 1
1.2 Theoretical Literature on Economic Integration 4
1.3 The Evolution of Economic Integration in ASEAN 5
1.3.1 ASEAN Economic Integration 1970s–2000s 5
1.3.2 The ASEAN Economic Community 8
1.4 The ASEAN Economic Community by 2016? 12
1.4.1 Achievements and Progress 12
Tariff Reduction and Elimination 12
Trade Facilitation 14
1.4.2 Obstacles and Challenges 17
Low Utilisation of Preferential Tariffs 17
Non-Tariff Barriers to Trade 19
ASEAN Single Window and ASEAN Trade Repository 20
Trade in Services 22
Skilled Labour Mobility 25
The ASEAN Single Aviation Market (ASAM) 26
1.4.3 Can the ASEAN Economic Community be Completed by 2016? 28
1.5 A Post-2015 Vision for ASEAN 30
1.5.1 Trans-Pacific Partnership (TPP) 34
1.5.2 Regional Comprehensive Economic Partnership (RCEP) 35
1.5.3 Implications for Post-2015 ASEAN 39
Toward the Free Trade Area of the Asia-Pacific (FTAAP) 39
The Role of Indonesia 43
Domestic Reforms 46

1.6 Concluding Remarks 50
References 53

Discussant Notes 1.1 Mr Nguyen Cam Tu, Deputy Minister, Ministry of Industry and Trade, Vietnam 60
Discussant Notes 1.2 Dr Nguyen Dinh Cung, President, Central Institute for Economic Management (CIEM), Vietnam 64

Chapter 2 2015 Update on Annual Competitiveness Analysis for ASEAN-10 67
2.1 Introduction 67
2.2 ACI Competitiveness Framework and Methodology 68
2.2.1 Literature Review on Competitiveness 68
2.2.2 ACI Competitiveness Framework 69
2.2.3 Data Sources 71
2.2.4 Computation of Standardised Scores 71
2.2.5 *What-if* Competitiveness Simulation Analysis 72
2.3 Competitiveness Analysis for ASEAN-10 Economies, 2000–2012 72
2.3.1 Overall Competitiveness 72
2.3.2 Macroeconomic Stability 74
2.3.3 Government and Institutional Setting 77
2.3.4 Financial, Businesses and Manpower Conditions 80
2.3.5 Quality of Life and Infrastructure Development 83
2.3.6 Competitiveness Over Time Analysis 86
2.3.7 Maximising Strengths and Minimising Weaknesses 93
Top 20% Strongest and Weakest Indicators 95
Median Competitiveness Web Analysis 104
Maximum Competitiveness Web Analysis 110
2.3.8 Upgrading Competitiveness through *What-if* Simulation Analysis 116
2.4 Concluding Remarks and Policy Implications 119
References 121

Discussant Notes 2.1 Mrs Somhatai Panichewa, Chief Executive Officer, Amata VN Public Company Limited, Thailand and Mr Chartchai Panichewa, President, ThaiSri Insurance Public Company Limited, Thailand 122

Chapter 3 How Productive is the Agricultural Sector in ASEAN? 125
3.1 Introduction 125
3.2 Agriculture in ASEAN: Trends, Opportunities and Challenges 128
3.2.1 Overview of the Agricultural Sectors in ASEAN-8 132
3.2.2 Opportunities and Challenges from a Regional Perspective 136
3.3 Literature Review 139

3.4 Methodology and Data 142
3.4.1 Growth Accounting Methodology 143
3.4.2 Malmquist Index Data Envelopment Analysis (DEA) Methodology 143
3.4.3 Data 145
3.5 Empirical Findings 146
3.6 Case Study: Agricultural Development in Malaysia 148
3.6.1 Key Features of Malaysia's Agricultural Sector 149
3.6.2 Malaysia's Agricultural Strategies and Policies 152
3.6.3 Challenges and Prospects 157
3.7 Conclusion and Policy Implications 159
References 162

Discussant Notes 3.1 Dr Roehlano Briones, Senior Research Fellow, Philippine Institute for Development Studies, Philippines 167
Discussant Notes 3.2 Dr Bustanul Arifin, Professor, University of Lampung and Board of Founder/Senior Economist, Institute for Development of Economics and Finance, Indonesia 170
Discussant Notes 3.3 Dr Waleerat Suphannachart, Assistant Professor, Department of Agricultural and Resource Economics, Kasetsart University, Thailand 172

Chapter 4 Drivers of Growth in the Travel and Tourism Industry in Malaysia, Thailand and Singapore: A Geweke Causality Analysis 175
4.1 Introduction 175
4.2 Background of the Travel and Tourism Industry in Malaysia, Thailand and Singapore 178
4.2.1 Malaysia 178
4.2.2 Thailand 180
4.2.3 Singapore 182
4.3 Literature Review and Hypotheses 184
4.4 Methodology and Data 185
4.4.1 The Geweke Causality Analysis 185
4.4.2 Data 190
4.5 Empirical Findings 192
4.5.1 Geweke Causality Analysis for Malaysia 192
4.5.2 Geweke Causality Analysis for Thailand 195
4.5.3 Geweke Causality Analysis for Singapore 198
4.6 Conclusion and Policy Implications 202
References 203

Chapter 5 Concluding Remarks and Future Research Agenda 207
5.1 Summary of Key Findings 207
5.2 Development Vision and Strategies for ASEAN-10 211
5.3 Future Research Agenda 221
5.3.1 Impacts of Exchange Rate Volatility on Foreign Direct Investment and Trade 221
5.3.2 Multi-country Global Vector Autoregressive Model on Economic Interdependencies 222
5.3.3 Tracking Firm-Level Productivity, Efficiency and Competitiveness for the 16 Asian Economies 223
References 223

Appendix 1 Policy Brief: Make or Break for Vietnam in the TPP Era? 225
Appendix 2 Competitiveness Analysis of ASEAN-10 Economies, 2000–2012 — List of Indicators 233
Appendix 3 Competitiveness Analysis — The Algorithm 239

Foreword

The year 2015 marks a monumental juncture in the 30 years of Vietnam's *Doi Moi* (renovation) reform aiming to transform Vietnam from a centrally planned to a prosperous, inclusive and sustainable market economy. Firstly, the ASEAN Community was officially established on 31st December 2015, in which Vietnam has been an active member. Secondly, on 5th October 2015, Vietnam, together with 11 Pacific Rim nations including the United States, successfully concluded the high-standard and wide-ranging Trans-Pacific Partnership (TPP), which is projected to bring about an additional 13% GDP growth for Vietnam by 2025.

Nonetheless, studies undertaken by the Central Institute for Economic Management (CIEM) have shown that the Vietnam economy continues to be afflicted by various market and state failures arising from the fundamental imbalance between ineffectual market institutions and rising competition from increasing economic integration with the world, which has resulted in eroded national competitiveness. Indeed, Vietnam's average GDP growth rate from 2008 up to now is about 6% — 1.6 to 1.8 percentage point lower than the average growth of 7.6% between 1991 and 2007.

In this light, the annual competitiveness analysis and simulation study for ASEAN-10 conducted by the Asia Competitiveness Institute (ACI) at the Lee Kuan Yew School of Public Policy (LKYSPP), National University of Singapore (NUS), is extremely timely. What I find most impressive about the ACI's competitiveness framework is that it not only provides the ranking of countries but also identifies areas for improvement with concrete policy recommendations for individual countries in a rigorous, eclectic and systematic manner. The estimation of agricultural productivity growth conducted in this publication is particularly useful for Vietnam where over 60% of the labour force are engaged in agricultural production. Furthermore, the ACI's comprehensive assessment of the ASEAN Economic Community will serve as an instructive roadmap for the formulation and implementation of the ASEAN agenda in years to come. In this respect, the extensive research work undertaken by the ACI provides an important and insightful complement to the research done by the CIEM, contributing toward evidence-based and astute policy formulation in ASEAN in general, and in Vietnam in particular.

I hope the ACI will continue its excellent work as well as look into exploring various themes, particularly those that are pertinent to Vietnam. I extend wholehearted support for the ACI's research projects and look forward to collaborating with the ACI in the future.

Dr Nguyen Dinh Cung
President
Central Institute for Economic Management (CIEM), Vietnam

Preface

The Asia Competitiveness Institute (ACI) at the Lee Kuan Yew School of Public Policy (LKYSPP), National University of Singapore (NUS), has been at the forefront of competitiveness studies with a special focus on ASEAN in the context of rising China and India. Its extensive research work and policy impact have been widely recognised — for example, The Think Tanks and Civil Society Programme at the University of Pennsylvania ranked ACI as the 13th "Best University Affiliated Think Tank" in its 2015 Global Go To Think Tank Index.

The phenomenal rise of ASEAN as the seventh-largest economy in the world by 2014 — in spite of the global financial tsunami of 2008–2009 — has established the ten-member bloc as one of the most cohesive and enduring regional associations in the developing world today. 2015 marks a significant milestone in the history of ASEAN, as the ASEAN Community — which comprises the ASEAN Political-Security Community, the ASEAN Economic Community and the ASEAN Socio-Cultural Community — was officially launched on 31st December. However, only about 80% of the measures listed in the ASEAN Economic Community (AEC) Blueprint were implemented by December 2015. Against this backdrop, this annual compendium, now in its third edition, extends beyond competitiveness analysis to provide an in-depth assessment as to whether the establishment of the AEC could feasibly be completed by the end of 2016.

This book also presents the latest findings on the relative competitiveness of the ASEAN-10 economies. The framework adopted by ACI is one of a kind, going beyond ranking to incorporate a *what-if* policy simulation on how each economy can improve its competitiveness. A customised approach such as this, rather than a one-size-fits-all approach, is vital for a region as diverse as ASEAN. In light of the importance of agriculture in the majority of the ASEAN member states, an empirical estimation of the agricultural Total Factor Productivity (TFP) growth in eight ASEAN economies was conducted. This offers insights into how policy-makers in the region could foster agricultural growth for inclusive and balanced economic development. Finally, the Geweke causality analysis was employed to examine the drivers of growth in the travel and tourism industry in Malaysia, Singapore and Thailand. Extensive sets of data — compiled from reputable sources and on-the-ground surveys and supported by rigorous evidence-based quantitative methodologies — serve as the basis for this ambitious blueprint for ASEAN leaders, policy-makers, managers, thinkers and others who are interested in the region.

With this publication on ASEAN, as well as similar companion volumes on Greater China, India and Indonesia, ACI has taken the lead in understanding and developing competitiveness in Asia. We hope this study augments the ongoing policy discourse in ASEAN and helps the region to fulfil its potential and to transform the lives of its people.

Professor Kishore Mahbubani
Dean
Lee Kuan Yew School of Public Policy
National University of Singapore

Acknowledgements

We from the Asia Competitiveness Institute (ACI) at the Lee Kuan Yew School of Public Policy (LKYSPP), National University of Singapore (NUS), have combined leading-edge research methodologies with extensive data collection efforts to evaluate and track economic competitiveness since 2000 at both national and sub-national levels for over 100 Asian economies including Greater China, India, Indonesia and ASEAN-10. Building upon our previous works, in 2015 we extended our study to include various volumes on *2015 Estimation of Agricultural Productivity and Annual Update of Competitiveness Analysis for Greater China, 2015 Agricultural Total Factor Productivity and Competitiveness Analysis for States and Federal Territories and Five Regions of India, 2015 Agricultural Productivity, Decentralisation, and Competitiveness Analysis for Provinces and Regions of Indonesia*, as well as for ASEAN-10 as in this volume.

In this book on *2015 Annual Competitiveness Analysis, Agricultural Productivity and Development Vision for ASEAN-10*, we assess the possibility of completing the ASEAN Economic Community by the year 2016, in light of the challenges encountered in the last final mile with only about 80% of the stipulated requirements implemented by the formal launch on 31st December 2015. The competitiveness analysis for ASEAN-10 economies is also improved and updated. Holistically, ACI offers evidence-based assessment of both economic, institutional and social indicators at the macro level and sectoral indicators at the micro level for competitiveness evaluation. Given the significance of the agricultural sector in the majority of the ASEAN member states, we also conduct an empirical estimation of annual agricultural Total Factor Productivity (TFP) growth in eight ASEAN countries. Malaysia's success in agricultural development, as evident in its agricultural TFP growth being the highest in the region, is also explored in an in-depth case study. Finally, the Geweke causality analysis was undertaken to examine the drivers of growth in the travel and tourism industry in Malaysia, Singapore and Thailand.

The publication of this book would not have been possible without the tremendous support from various country experts from the public and private sector who attended the two ASEAN Review Seminars organised by ACI at LKYSPP, NUS on 26th August 2014 and 31st August 2015. We wish to record our heartfelt gratitude for the time and effort invested by the following reviewers who gave their constructive comments and offered instructive insights on the respective chapters as incorporated in this book:

Mr Nguyen Cam Tu
Deputy Minister, Ministry of Industry and Trade, Vietnam

Dr Bustanul Arifin
Professor, University of Lampung and Board of Founder/Senior Economist, Institute for Development of Economics and Finance, Indonesia

Dr Roehlano Briones
Senior Research Fellow, Philippine Institute for Development Studies, Philippines

Dr Nguyen Dinh Cung
President, Central Institute for Economic Management (CIEM), Vietnam

Mrs Somhatai Panichewa
Chief Executive Officer, Amata VN Public Company Limited, Thailand

Mr Chartchai Panichewa
President, ThaiSri Insurance Public Company Limited, Thailand

Dr Waleerat Suphannachart
Assistant Professor, Department of Agricultural and Resource Economics, Kasetsart University, Thailand

We would also like to place on record our appreciation for the support and encouragement from Dean Kishore Mahbubani; Associate Professor Eduardo Araral, Vice Dean (Research); Mr Donald Low, Associate Dean (Research and Executive Education); and other colleagues at LKYSPP. Finally, this book would not have been possible without the dedicated effort of all research and administrative staff at ACI. Thus we note with tremendous appreciation the high team spirit of all ACI staff including Dr Ramkishen S. Rajan, Xie Teleixi, Yap Xin Yi, Tracy Cai Jiao and Nurliyana Binte Yusoff.

Tan Khee Giap
Nguyen Le Phuong Anh
Sasidaran Gopalan
Luu Nguyen Trieu Duong
Asia Competitiveness Institute
Lee Kuan Yew School of Public Policy
National University of Singapore

About the Authors

Tan Khee Giap is a Co-Director of the Asia Competitiveness Institute (ACI) and Associate Professor at the Lee Kuan Yew School of Public Policy, National University of Singapore. He is also the Chairman of the Singapore National Committee for Pacific Economic Cooperation. Upon graduating with a PhD from University of East Anglia, England, in 1987 under the Overseas Research Scheme awarded by the Committee of Vice-Chancellors and Principals of the Universities of the United Kingdom, he joined the banking sector as a treasury manager. Thereafter, he taught at the Department of Economics and Statistics, National University of Singapore, 1990–1993. Dr Tan joined Nanyang Technological University in 1993 and was Associate Dean, Graduate Studies Office, 2007–2009. Dr Tan has consulted extensively with the various government ministries, statutory boards and government-linked companies of the Singapore government. He has also served as a consultant to international agencies such as the Asian Development Bank, Asian Development Bank Institute, United Nations Industrial Development Group, World Gold Council, ASEAN Secretariat, Central Policy Unit of Hong Kong, Kerzner International, Las Vegas Sands and Marina Bay Sands. Dr Tan is the lead author of twenty books, serves as journal editors and published widely in international refereed journals. His current research interests include econometric forecasting, cost of living index, global liveable cities index, ease of doing business index and competitiveness analysis on sub-national economies of China, India, Indonesia and ASEAN. Dr Tan was Deputy President of the Singapore Economic Society, 2004. He served in the 2002 Economic Review Committee (ERC), served as the Deputy Chairman of the IPS Forum for Economic Restructuring (IFER) in 2003 and served as a member of the Resource Panel of the Government Parliamentary Committee for Transport and Government Parliamentary Committee for Finance and Trade since 2007. Dr Tan has extensively advised and guided multinational corporations leading to public listings, especially those companies from Mainland China and Taiwan. He is also currently an Independent Director of the publicly listed Artivision, BreadTalk Group, Boustead Projects and TEE Land.

Nguyen Le Phuong Anh is a Research Fellow and Deputy Director (Administration) at the Asia Competitiveness Institute (ACI) at the Lee Kuan Yew School of Public Policy (LKYSPP), National University of Singapore (NUS). Phuong Anh received her Bachelor (Honours) degree in Economics from Nanyang Technological University (NTU), Singapore and a Master of Science degree in Development Studies from London School of Economics and Political Science (LSE), United Kingdom. Prior to joining ACI, she gained extensive experience working in the financial sector with OCBC, DBS and Bloomberg L.P. in Singapore, and additionally held a research internship position at the Mekong Institute in Khon Kaen, Thailand. At ACI, she co-authored two books titled *2014 Annual Indices for Expatriates and Ordinary Residents on Cost of Living, Wages and Purchasing Power for World's Major Cities* and *2014 Annual Competitiveness Ranking and Simulation Study for ASEAN-10 and Development Strategies to Enhance Asia Economic Connectivity*, and an editorial opinion on Vietnam for the Strait Times. She is currently coordinating the annual competitiveness study of the ASEAN-10 economies. Phuong Anh's research interests include economic development policy, international political economy, ASEAN economies and applied econometrics.

Sasidaran Gopalan is a Research Fellow at the Asia Competitiveness Institute (ACI) at the Lee Kuan Yew School of Public Policy (LKYSPP), National University of Singapore (NUS). Prior to joining ACI, he was based in Hong Kong as a Post-Doctoral Fellow at the HKUST Institute for Emerging Market Studies and HKUST Jockey Club Institute for Advanced Study at the Hong Kong University of Science and Technology (HKUST). Dr Gopalan completed his PhD in Public Policy, specialising in International Finance and Policy at the School of Policy, Government and International Affairs (SPGIA), George Mason University, USA. He obtained his Bachelors and Masters' degrees in Economics from Loyola College (Chennai, India) and Madras School of Economics (Chennai, India), respectively. After spending a year at the Indian Institute of Management Bangalore (IIMB) post-graduation as a Research Associate, he was based at the Institute of South Asian Studies (ISAS) at National University of Singapore (NUS) for two years doing research work on international economics and policy issues concerning South Asia. He has published two co-authored books with leading publishing houses, several academic articles in refereed international journals, book chapters in edited volumes, a number of policy briefs as well as op-eds. for leading global financial dailies. Dr Gopalan's research interests span the broad fields of financing for development and international macroeconomic and financial policies with respect to emerging and developing economies in Asia. More specific topics include an empirical analysis of financial openness and international capital flows, financial development, financial inclusion, foreign aid, remittances as well as international trade and migration issues. His research has a heavy empirical and policy flavour. At ACI, he is the lead coordinator for the flagship projects on India, specifically relating to competitiveness analysis and ease of doing business. He also supervises and contributes to the formulation as well as implementation of several empirical research projects, targeted towards journal publications and broader policy outreach.

Luu Nguyen Trieu Duong is a Research Assistant at the Asia Competitiveness Institute (ACI), Lee Kuan Yew School of Public Policy (LKYSPP), National University of Singapore (NUS). Luu graduated from NUS in June 2015 with a second upper class honours Bachelor degree in Economics. Luu was a recipient of the ASEAN Undergraduate Scholarship and was placed on the Dean's List and Dean's Scholar List on separate occasions during his course of study. At ACI, he focuses on projects related to ASEAN economies, research on cost of living for expatriates and ordinary citizens in major cities around the world, growth slowdown analysis by income thresholds for Asian economies and multi-country econometric modelling. Luu's research interests include growth and development, ASEAN economies, applied econometrics and economics of aging.

List of Abbreviations

ABAC	APEC Business Advisory Council
ABMI	Asian Bond Markets Initiative
ACI	Asia Competitiveness Institute
ACIA	ASEAN Comprehensive Investment Agreement
ACTS	ASEAN Customs Transit System
ADB	Asian Development Bank
AEC	ASEAN Economic Community
AECV2030	Asia Economic Connectivity Vision 2030
AF	Airport Facilities
AFAS	ASEAN Framework Agreement on Services
AFC	Asian Financial Crisis
AFTA	ASEAN Free Trade Area
AHN	ASEAN Highway Network
AHTN	ASEAN Harmonised Tariff Nomenclature
AIA	ASEAN Investment Area
AIFS	ASEAN Integrated Food Security
AIIB	Asian Infrastructure Investment Bank
AMRO	ASEAN+3 Macroeconomic Research Office
APEC	Asia Pacific Economic Cooperation
APTA	ASEAN Preferential Trading Agreement
APTERR	ASEAN Plus Three Emergency Rice Reserve
AQ	Air Quality
ASAM	ASEAN Single Aviation Market
ASEAN	Association of Southeast Asian Nations
ASW	ASEAN Single Window
ATIGA	ASEAN Trade in Goods Agreement
BPM	Bank Pertanian Malaysia
B&R	Belt & Road
CEP	Comprehensive Economic Partnership
CEPEA	Comprehensive Economic Partnership for East Asia
CEPT	Common Effective Preferential Tariff

CGE	Computable General Equilibrium
CI	Capital Investment on Tourism
CJK FTA	China–Japan–South Korea Tripartite Free Trade Agreement
CLMV	Cambodia, Laos, Myanmar and Vietnam
CMI	Chiang Mai Initiative
CMIM	Chiang Mai Initiative Multilateralisation
COT	Competitiveness Over Time
CPI	Consumer Price Index
DEA	Data Envelopment Analysis
EAS	East Asia Summit
EEE	Electrical and Electronic Equipment
EIE	Environment and Infrastructure Environment
EIU	Economist Intelligence Unit
EPI	Environment Performance Index
ER	Exchange Rate
ETP	Economic Transformation Programme
EU	European Union
FAMA	Federal Agricultural Marketing Authority
FAO	Food and Agriculture Organisation
FBMC	Financial, Businesses and Manpower Conditions
FDI	Foreign Direct Investment
FELCRA	Federal Land Consolidation and Rehabilitation Authority
FELDA	Federal Land Development Authority
FTA	Free Trade Agreement
FTAAP	Free Trade Area of the Asia-Pacific
G2G	Government to Government
GATS	General Agreement on Trade in Services
GCR	Global Competitiveness Report
GDP	Gross Domestic Product
GET	Government Expenditure on Tourism
GFC	Global Financial Crisis
GIS	Government and Institutional Setting
GMID	Global Market Information Database
GPE	Government Policy Environment
GVAR	Global Vector Autoregressive
IAI	Initiative for ASEAN Integration
IADPs	Integrated Agriculture Development Projects
ICT	Information and Communication Technology
IMF	International Monetary Fund

IPRs	Intellectual Property Rights
ISEAS	Institute of Southeast Asian Studies
ITA	International Tourism Arrivals
IU	Internet Users
LKYSPP	Lee Kuan Yew School of Public Policy
LGUs	Local Government Units
MAAS	Multilateral Agreement on Air Services
MAFC	Malaysian Agriculture Food Corporation
MAFLPAS	Multilateral Agreement for Full Liberalisation of Passenger Air Services
MAFLAFS	Multilateral Agreement for Full Liberalisation of Air Freight Services
MARDI	Malaysian Agricultural Research and Development Institute
MFN	Most Favoured Nation
MNCs	Multinational Corporations
MNP	Movement of Natural Persons
MPAC	Master Plan on ASEAN Connectivity
MPOB	Malaysian Palm Oil Board
MPOPC	Malaysian Palm Oil Promotion Council
MRAs	Mutual Recognition Arrangements
MS	Macroeconomic Stability
MTTP	Malaysian Tourism Transformation Programme
NAFTA	North American Free Trade Agreement
NAP1	First National Agricultural Policy
NAP2	Second National Agricultural Policy
NAP3	Third National Agricultural Policy
NAP4	National Agro-food Policy
NEM	New Economic Model
NEP	National Ecotourism Plan
NIEs	Newly Industrialised Economies
NKEA	National Key Economic Area
NSWs	National Single Windows
NTBs	Non-Tariff Barriers
NTMs	Non-Tariff Measures
NTP	National Tourism Policy
NUS	National University of Singapore
ODI	Overseas Development Institute
OECD	Organisation for Economic Cooperation and Development
PECC	Pacific Economic Cooperation Council
PORIM	Palm Oil Research Institute of Malaysia
PORLA	Palm Oil Regulatory and Licensing Association

PPE	Purchasing Power Environment
PPPs	Public-Private Partnerships
PSS	Public Security and Safety
QLID	Quality of Life and Infrastructure Development
R&D	Research and Development
RCEP	Regional Comprehensive Economic Partnership
RDAs	Regional Development Authorities
RIATS	Roadmap for Integration of Air Travel Sector
RISDA	Rubber Industry Smallholder Development Authority
ROOs	Rules of Origin
RSVI	Ranked Standardised Value of Indicator
RU	Religious Unrest
SD	Standard Deviation
SMEs	Small and Medium-sized Enterprises
SOEs	State-Owned Enterprises
SRI	Strategic Reform Initiative
SSC	State Securities Commission of Vietnam
SSE	Safety and Security Environment
STRI	Services Trade Restrictiveness Index
STB	Singapore Tourism Board
SVAR	Structural Vector Autoregressive
SVI	Standardised Value of Indicator
TAC	Treaty of Amity and Cooperation
TAT	Tourism Authority of Thailand
TC	Technological Change
TDC	Tourism Development Corporation of Malaysia
TEC	Technical Efficiency Change
TFP	Total Factor Productivity
TFPC	Total Factor Productivity Change
TPP	Trans-Pacific Partnership
UN	United Nations
UNCTAD	United Nations Conference on Trade and Development
UNDP	United Nations Development Programme
UNWTO	United Nations World Tourism Organisation
US	United States
USDA	US Department of Agriculture
WCY	World Competitiveness Yearbook
WEF	World Economic Forum
WTO	World Trade Organisation
WTTC	World Travel & Tourism Council

List of Figures and Tables

Chapter 1

Table 1.1	ASEAN's Diversity in Size, Level of Development and Economic Openness, 2013	2
Table 1.2	Real Income Gains of the ASEAN Economic Community (AEC) *vis-à-vis* the ASEAN Free Trade Area (AFTA)	11
Table 1.3	Documents Required to Import and Export in ASEAN Countries, 2005–2014	15
Table 1.4	Time Taken to Import and Export in ASEAN Countries, 2005–2014	16
Table 1.5	Margins Between the Average Common Effective Preferential Tariff (CEPT) Rates and Average Applied Most Favoured Nation (MFN) Rates in ASEAN (%), 2007–2014	18
Table 1.6	Foreign Ownership Allowed in the Air Transport Sector in ASEAN Countries (%)	23
Table 1.7	Tariff Elimination Coverage under ASEAN+1 Free Trade Agreements (Percentage of Total Tariff Lines)	32
Table 1.8	Liberalisation Timeframe for ASEAN+1 Free Trade Agreements	33
Table 1.9	Real Income Effects of the Trans-Pacific Partnership (TPP)	35
Table 1.10	Real Income Effects of the Regional Comprehensive Economic Partnership (RCEP)	36
Table 1.11	Real Income Effects of the Trans-Pacific Partnership (TPP), the East Asian Track and the Free Trade Area of the Asia-Pacific (FTAAP)	41
Figure 1.1	ASEAN's Real GDP and GDP per Capita Growth Rates, 1990–2014	2
Figure 1.2	Four Pillars and 17 Core Elements of the ASEAN Economic Community (AEC)	10
Figure 1.3	Average Common Effective Preferential Tariff (CEPT) Rates in ASEAN, 2007–2014	13
Figure 1.4	Restrictiveness of Services Trade Policies in Cambodia, Indonesia, Malaysia, Philippines, Thailand and Vietnam, 2008 and 2012	24
Figure 1.5	Restrictiveness of Services Trade Policies in Selected ASEAN Economies in 2012 and Restrictiveness of the AEC Blueprint Services Liberalisation Targets	25

Chapter 2

Table 2.1	ASEAN-10 in a Snapshot — 2013	68
Table 2.2	Overall Competitiveness Rankings and Standardised Scores, 2000–2012	73
Table 2.3	Macroeconomic Stability Rankings and Standardised Scores, 2000–2012	75
Table 2.4	Government and Institutional Setting Rankings and Standardised Scores, 2000–2012	78
Table 2.5	Financial, Businesses and Manpower Conditions Rankings and Standardised Scores, 2000–2012	81
Table 2.6	Quality of Life and Infrastructure Development Rankings and Standardised Scores, 2000–2012	84
Table 2.7	Top 20% Strongest Indicators List for Brunei, 2012	95
Table 2.8	Top 20% Weakest Indicators List for Brunei, 2012	95
Table 2.9	Top 20% Strongest Indicators List for Cambodia, 2012	96
Table 2.10	Top 20% Weakest Indicators List for Cambodia, 2012	96
Table 2.11	Top 20% Strongest Indicators List for Indonesia, 2012	97
Table 2.12	Top 20% Weakest Indicators List for Indonesia, 2012	97
Table 2.13	Top 20% Strongest Indicators List for Laos, 2012	98
Table 2.14	Top 20% Weakest Indicators List for Laos, 2012	98
Table 2.15	Top 20% Strongest Indicators List for Malaysia, 2012	99
Table 2.16	Top 20% Weakest Indicators List for Malaysia, 2012	99
Table 2.17	Top 20% Strongest Indicators List for Myanmar, 2012	100
Table 2.18	Top 20% Weakest Indicators List for Myanmar, 2012	100
Table 2.19	Top 20% Strongest Indicators List for Philippines, 2012	101
Table 2.20	Top 20% Weakest Indicators List for Philippines, 2012	101
Table 2.21	Top 20% Strongest Indicators List for Singapore, 2012	102
Table 2.22	Top 20% Weakest Indicators List for Singapore, 2012	102
Table 2.23	Top 20% Strongest Indicators List for Thailand, 2012	103
Table 2.24	Top 20% Weakest Indicators List for Thailand, 2012	103
Table 2.25	Top 20% Strongest Indicators List for Vietnam, 2012	104
Table 2.26	Top 20% Weakest Indicators List for Vietnam, 2012	104
Table 2.27	*What-if* Competitiveness Simulation Analysis on Overall Competitiveness, 2012	117
Table 2.28	*What-if* Competitiveness Simulation Analysis on Macroeconomic Stability, 2012	117
Table 2.29	*What-if* Competitiveness Simulation Analysis on Government and Institutional Setting, 2012	118
Table 2.30	*What-if* Competitiveness Simulation Analysis on Financial, Businesses and Manpower Conditions, 2012	118
Table 2.31	*What-if* Competitiveness Simulation Analysis on Quality of Life and Infrastructure Development, 2012	119

Figure 2.1	ACI Competitiveness Framework	70
Figure 2.2	Overall Competitiveness — 2000, 2011 and 2012	74
Figure 2.3	Macroeconomic Stability — 2000, 2011 and 2012	76
Figure 2.4	Macroeconomic Stability (Economic Vibrancy) — 2000, 2011 and 2012	76
Figure 2.5	Macroeconomic Stability (Openness to Trade and Services) — 2000, 2011 and 2012	77
Figure 2.6	Macroeconomic Stability (Attractiveness to Foreign Investors) — 2000, 2011 and 2012	77
Figure 2.7	Government and Institutional Setting — 2000, 2011 and 2012	79
Figure 2.8	Government and Institutional Setting (Government Policies and Fiscal Sustainability) — 2000, 2011 and 2012	79
Figure 2.9	Government and Institutional Setting (Institutions, Governance and Leadership) — 2000, 2011 and 2012	80
Figure 2.10	Government and Institutional Setting (Competition, Regulatory Standards and Rule of Law) — 2000, 2011 and 2012	80
Figure 2.11	Financial, Businesses and Manpower Conditions — 2000, 2011 and 2012	82
Figure 2.12	Financial, Businesses and Manpower Conditions (Financial Deepening and Business Efficiency) — 2000, 2011 and 2012	82
Figure 2.13	Financial, Businesses and Manpower Conditions (Labour Market Flexibility) — 2000, 2011 and 2012	83
Figure 2.14	Financial, Businesses and Manpower Conditions (Productivity Performance) — 2000, 2011 and 2012	83
Figure 2.15	Quality of Life and Infrastructure Development — 2000, 2011 and 2012	85
Figure 2.16	Quality of Life and Infrastructure Development (Physical Infrastructure) — 2000, 2011 and 2012	85
Figure 2.17	Quality of Life and Infrastructure Development (Technological Infrastructure) — 2000, 2011 and 2012	86
Figure 2.18	Quality of Life and Infrastructure Development (Standard of Living, Education and Social Stability) — 2000, 2011 and 2012	86
Figure 2.19	Competitiveness Over Time Analysis for Overall Competitiveness, Period B (2007–2012) Against Period A (2000–2006)	87
Figure 2.20	Competitiveness Over Time Analysis for Economic Vibrancy, Period B (2007–2012) Against Period A (2000–2006)	88
Figure 2.21	Competitiveness Over Time Analysis for Openness to Trade and Services, Period B (2007–2012) Against Period A (2000–2006)	89
Figure 2.22	Competitiveness Over Time Analysis for Attractiveness to Foreign Investors, Period B (2007–2012) Against Period A (2000–2006)	89
Figure 2.23	Competitiveness Over Time Analysis for Government Policies and Fiscal Sustainability, Period B (2007–2012) Against Period A (2000–2006)	90

Figure 2.24	Competitiveness Over Time Analysis for Institutions, Governance and Leadership, Period B (2007–2012) Against Period A (2000–2006)	90
Figure 2.25	Competitiveness Over Time Analysis for Competition, Regulatory Standards and Rule of Law, Period B (2007–2012) Against Period A (2000–2006)	91
Figure 2.26	Competitiveness Over Time Analysis for Financial Deepening and Business Efficiency, Period B (2007–2012) Against Period A (2000–2006)	91
Figure 2.27	Competitiveness Over Time Analysis for Labour Market Flexibility, Period B (2007–2012) Against Period A (2000–2006)	92
Figure 2.28	Competitiveness Over Time Analysis for Productivity Performance, Period B (2007–2012) Against Period A (2000–2006)	92
Figure 2.29	Competitiveness Over Time Analysis for Physical Infrastructure, Period B (2007–2012) Against Period A (2000–2006)	93
Figure 2.30	Competitiveness Over Time Analysis for Technological Infrastructure, Period B (2007–2012) Against Period A (2000–2006)	94
Figure 2.31	Competitiveness Over Time Analysis for Standard of Living, Education and Social Stability, Period B (2007–2012) Against Period A (2000–2006)	94
Figure 2.32	Median Competitiveness Web Analysis for Brunei, 2012	105
Figure 2.33	Median Competitiveness Web Analysis for Cambodia, 2012	106
Figure 2.34	Median Competitiveness Web Analysis for Indonesia, 2012	106
Figure 2.35	Median Competitiveness Web Analysis for Laos, 2012	107
Figure 2.36	Median Competitiveness Web Analysis for Malaysia, 2012	107
Figure 2.37	Median Competitiveness Web Analysis for Myanmar, 2012	108
Figure 2.38	Median Competitiveness Web Analysis for Philippines, 2012	108
Figure 2.39	Median Competitiveness Web Analysis for Singapore, 2012	109
Figure 2.40	Median Competitiveness Web Analysis for Thailand, 2012	110
Figure 2.41	Median Competitiveness Web Analysis for Vietnam, 2012	110
Figure 2.42	Maximum Competitiveness Web Analysis for Brunei, 2012	111
Figure 2.43	Maximum Competitiveness Web Analysis for Cambodia, 2012	111
Figure 2.44	Maximum Competitiveness Web Analysis for Indonesia, 2012	112
Figure 2.45	Maximum Competitiveness Web Analysis for Laos, 2012	113
Figure 2.46	Maximum Competitiveness Web Analysis for Malaysia, 2012	113
Figure 2.47	Maximum Competitiveness Web Analysis for Myanmar, 2012	114
Figure 2.48	Maximum Competitiveness Web Analysis for Philippines, 2012	114
Figure 2.49	Maximum Competitiveness Web Analysis for Singapore, 2012	115
Figure 2.50	Maximum Competitiveness Web Analysis for Thailand, 2012	115
Figure 2.51	Maximum Competitiveness Web Analysis for Vietnam, 2012	116

Chapter 3

Table 3.1	Share of Agriculture in GDP and Employment in ASEAN, 2000 and 2011	126
Table 3.2	Annual Growth of Labour Productivity and Non-Labour Inputs in ASEAN-8, 2000–2011	132
Table 3.3	Average Annual Yield and Yield Growth of Selected Crops in ASEAN, China and India, 2000–2011	139
Table 3.4	Existing Literature on Agricultural Total Factor Productivity (TFP) in ASEAN	140
Table 3.5	Total Factor Productivity (TFP) Growth in ASEAN-8 Using Growth Accounting Methodology, 2000–2011	147
Table 3.6	Total Factor Productivity (TFP) Growth in ASEAN-8 Using Malmquist Index DEA Methodology, 2000–2011	147
Table 3.7	Average Total Factor Productivity Change (TFPC) and Its Decomposition in ASEAN-8, 2000–2011	147
Table 3.8	Malmquist Index Robustness Check for ASEAN-7 (Myanmar Excluded), 2000–2011	148
Table 3.9	Average Annual Growth Rates of Malaysia's Agricultural Trade, 1981–2011	151
Table 3.10	Trade Performance of Major Subsectors of Malaysian Agriculture (in Thousands US$), 1980–2011	151

Figure 3.1	Rural Poverty Incidence in ASEAN, 2012	126
Figure 3.2	Arable Land as a Percentage of Total Land Area in ASEAN-8, 2000 and 2011	129
Figure 3.3	Share of Major Crops in ASEAN's Gross Agricultural Production by Value (US$), 2000 and 2011	130
Figure 3.4	Share of Major Crops in ASEAN's Agricultural Exports, 2000 and 2011	130
Figure 3.5	Annual Agricultural Output Growth Rates of ASEAN-8, China and India, 2000–2011	131
Figure 3.6	ASEAN-8 Countries' Shares of Regional Agricultural Output, 2000 and 2011	131
Figure 3.7	Share of Major Crops in Malaysia's Gross Agricultural Production by Value (US$), 1980–2011	150
Figure 3.8	Malaysia's Agricultural Exports and Imports, 1980–2011	150

Discussant Notes 3.3

Table 1	Public Agricultural R&D Intensity Ratios in ASEAN	174

Chapter 4

Table 4.1 Potential Drivers of Growth in the Travel and Tourism Industry — Four Environments and Nine Indicators 190

Table 4.2 Summary of Linear Stepwise Regression Analysis on Tourism Development in Malaysia 192

Table 4.3 Estimated Measures of Bi-Directional Feedback between International Tourism Arrivals (ITA), Capital Investment on Tourism (% of GDP) (CI), Government Expenditure on Tourism (% of GDP) (GET), Airport Facilities (AF) and Religious Unrest (RU) for Malaysia, 2000–2012 193

Table 4.4 Estimated Measures of Multi-Directional Feedback between International Tourism Arrivals (ITA), Capital Investment on Tourism (% of GDP) (CI), Government Expenditure on Tourism (% of GDP) (GET), Airport Facilities (AF) and Religious Unrest (RU) for Malaysia, 2000–2012 193

Table 4.5 Estimated Measures of Bi-Directional Feedback between International Tourism Arrivals (ITA), Purchasing Power Environment (PPE) and Government Policy Environment (GPE) and Safety and Security Environment (SSE) for Malaysia, 2000–2012 194

Table 4.6 Estimated Measures of Multi-Directional Feedback between International Tourism Arrivals (ITA), Purchasing Power Environment (PPE), Government Policy Environment (GPE) and Safety and Security Environment (SSE) for Malaysia, 2000–2012 195

Table 4.7 Summary of Linear Stepwise Regression Analysis on Tourism Development in Thailand 195

Table 4.8 Estimated Measures of Bi-Directional Feedback between International Tourism Arrivals (ITA), Consumer Price Index (CPI) and Religious Unrest (RU) for Thailand, 2000–2012 196

Table 4.9 Estimated Measures of Multi-Directional Feedback between International Tourism Arrivals (ITA), Consumer Price Index (CPI) and Religious Unrest (RU) for Thailand, 2000–2012 196

Table 4.10 Estimated Measures of Bi-Directional Feedback between International Tourism Arrivals (ITA), Purchasing Power Environment (PPE), Government Policy Environment (GPE), Safety and Security Environment (SSE) and Environment and Infrastructure Environment (EIE) for Thailand, 2000–2012 197

Table 4.11 Estimated Measures of Multi-Directional Feedback between International Tourism Arrivals (ITA), Purchasing Power Environment (PPE), Government Policy Environment (GPE), Safety and Security Environment (SSE) and Environment and Infrastructure Environment (EIE) for Thailand, 2000–2012 198

Table 4.12 Summary of Linear Stepwise Regression Analysis on Tourism Development in Singapore 199

Table 4.13 Estimated Measures of Bi-Directional Feedback between International Tourism Arrivals (ITA), Government Expenditure on Tourism (% of GDP) (GET), Capital Investment on Tourism (% of GDP) (CI), Consumer Price Index (CPI), Airport Facilities (AF) and Religious Unrest (RU) for Singapore, 2000–2012 199

Table 4.14 Estimated Measures of Multi-Directional Feedback between International Tourism Arrivals (ITA), Government Expenditure on Tourism (% of GDP) (GET), Capital Investment on Tourism (% of GDP) (CI), Consumer Price Index (CPI), Airport Facilities (AF) and Religious Unrest (RU) for Singapore, 2000–2012 200

Table 4.15 Estimated Measures of Bi-Directional Feedback between International Tourism Arrivals (ITA), Purchasing Power Environment (PPE), Government Policy Environment (GPE), Environment and Infrastructure Environment (EIE) and Safety and Security Environment (SSE) for Singapore, 2000–2012 201

Table 4.16 Estimated Measures of Multi-Directional Feedback between International Tourism Arrivals (ITA), Purchasing Power Environment (PPE), Government Policy Environment (GPE), Environment and Infrastructure Environment (EIE) and Safety and Security Environment (SSE) for Singapore, 2000–2012 201

Figure 4.1 Growth of Visitor Arrivals in Singapore and Thailand, 2007–2013 177

Figure 4.2 Direct Contribution of Tourism and Travel as a Percentage of GDP in Malaysia, Singapore and Thailand, 2005–2013 180

Figure 4.3 Number of International Tourist Arrivals (in Millions) to East Asia and Pacific, Malaysia, Singapore and Thailand, 2000–2013 181

Figure 4.4 International Tourism Receipts as a Percentage of Total Exports in East Asia and Pacific, Malaysia, Singapore and Thailand, 2006–2013 181

Figure 4.5 International Tourism Receipts (in Current US$ Billion) in East Asia and Pacific, Malaysia, Singapore and Thailand, 2000–2013 182

Chapter 5

Figure 5.1 ACI's Development Vision for ASEAN-10 212

Appendix 1

Figure 1 Overall Competitiveness Scores and Rankings of ASEAN-10, 2000–2013 227

Chapter 1

The ASEAN Economic Community by 2016?

1.1 INTRODUCTORY NOTES

In the tradition of the Asia Competitiveness Institute (ACI) at the Lee Kuan Yew School of Public Policy (LKYSPP), National University of Singapore (NUS), our research publications make a distinctive contribution to the literature by focusing on pertinent regional and national economic issues of the Association of Southeast Asian Nations (ASEAN) as one of our flagship projects.[1] Now in its third edition,[2] this book on *2015 Annual Competitiveness Analysis, Agricultural Productivity and Development Vision for ASEAN-10* presents our latest competitiveness ranking and simulation study for the ten ASEAN member states (Chapter 2) and in-depth analysis on other policy topics that are of crucial importance to the region. These are: an extensive assessment of the ASEAN Economic Community (AEC) which was formally launched in December 2015 (this Chapter), an empirical estimation of agricultural productivity in individual ASEAN-8 countries (Brunei and Singapore excluded) (chapter 3), and a Geweke causality analysis on drivers of growth in the travel and tourism industry in Malaysia, Singapore and Thailand (Chapter 4). With this volume, we seek to provide evidence-based empirical findings and policy analysis to aid policy formulation for enhanced competitiveness in the region.

Established in 1967 initially as a security grouping in the midst of the Vietnam War, the presently ten-member ASEAN[3] has evolved to become one of the most enduring and cohesive regional associations in the developing world today. Having braved through multiple domestic, regional and global economic crises, particularly the Asian Financial Crisis (AFC) of 1997–1998 and the global financial crisis (GFC) of 2007–2008, ASEAN as a whole has proven its increasing resilience in the face of an increasingly uncertain global economic landscape (see Figure 1.1) to become the seventh largest economy in the world with a combined Gross Domestic Product (GDP) of US$2.4 trillion by 2014.[4] For the last 48 years, the regional bloc has also played an integral part in driving economic transformation in Asia, notwithstanding its remarkable heterogeneity in size, resource endowment, economic openness and national income (see Table 1.1). 2015 marks another momentous milestone in the history of the regional grouping as the ASEAN

[1]See Tan *et al.* (2015b, 2015c, 2015d, 2015e) for other flagship research projects on Greater China, India and Indonesia at the sub-national level.

[2]See Tan *et al.* (2013a, 2015a) for the first and second edition of ACI's publications on ASEAN.

[3]ASEAN was established in 1967 in Thailand with the signing of the ASEAN Declaration (Bangkok Declaration) by Indonesia, Malaysia, the Philippines, Singapore and Thailand. The group was subsequently expanded with the inclusion of five new members — Brunei (joined in 1984), Vietnam (1995), Laos and Myanmar (1997) and finally Cambodia (1999) — commonly referred to as ASEAN-10.

[4]See McKinsey Global Institute (2014).

Figure 1.1: ASEAN's Real GDP and GDP per Capita Growth Rates, 1990–2014

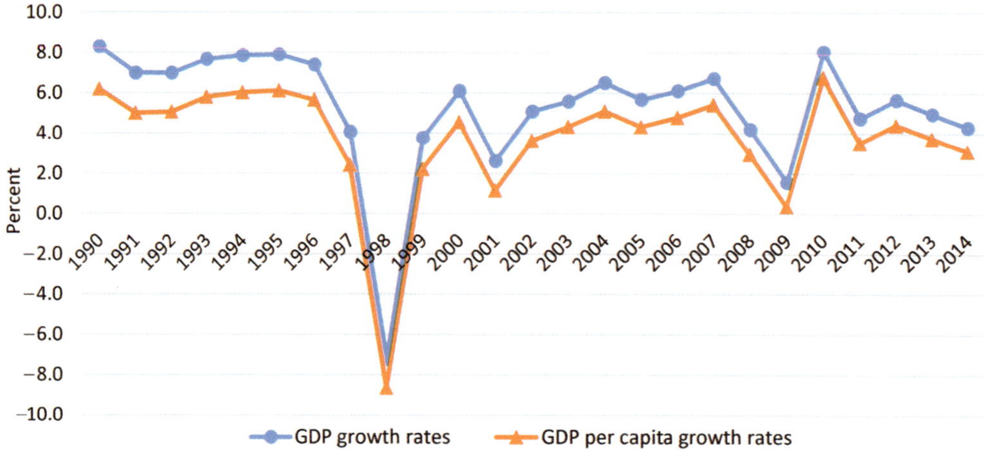

Note: The annual growth rates of GDP and GDP per capita are derived on the basis of constant price series.
Source: United Nations Conference on Trade and Development (UNCTAD).

Table 1.1: ASEAN's Diversity in Size, Level of Development and Economic Openness, 2013

Country	Land Area (Thousand Square Kilometres)	Population (Million)	GDP (Constant Billion)	GDP per Capita (Constant US$ 2000)	Trade (Percentage US$ 2000 of GDP)	Foreign Direct Investment (FDI) Net Inflows (Percentage US$ 2000 of GDP)
Brunei	5.3	0.4	9.7	23245.5	90.8	4.8
Cambodia	176.5	15.1	9.3	615.7	100.0	8.7
Indonesia	1811.6	249.9	291.1	1165.1	45.0	2.6
Laos	230.8	6.8	4.1	610.2	48.8	3.8
Malaysia	328.6	29.7	205.9	6928.3	144.4	3.5
Myanmar	653.1	53.3	10.0	187.7	28.2	3.8
Philippines	298.2	98.4	159.4	1619.9	49.8	1.4
Singapore	0.7	5.4	266.3	49314.1	298.3	21.4
Thailand	510.9	67.0	268.4	4005.3	132.7	3.8
Vietnam	310.1	89.7	47.9	534.3	161.9	5.2

Source: World Bank, International Monetary Fund (IMF) and ASEAN Statistics.

Community was officially established on 31 December, taking regional integration to a greater height as "a single market and production base."

The ASEAN Community comprises three pillars, namely economic cooperation (AEC), political stability and security (ASEAN Political-Security Community) and socio-cultural development (ASEAN Socio-Cultural Community). With demographic dividends of a youthful population with less than 15% aged 55 and above, the formation of the ASEAN Community could not be more timely, which, if successful, will augment regional competitiveness with more balanced economic development amongst the member economies, projected to grow by 2025 an aggregate output of 7% per annum by creating 14 million new jobs, to be better plugged into the global economy and to potentially become the 4th largest economy in the world by 2030.[5]

The AEC, in particular, would see lower tariffs and non-tariff barriers, harmonised custom procedures and higher trade and investment flows within ASEAN through making it more attractive to foreign investors and stimulating business activities for small- and medium-sized enterprises (SMEs). Professional groupings including tour operators, engineers, dentists, architects, accountants, nurses and doctors are going be in high demand. More affluent middle class will be emerging, well exceeding 100 million by 2025, with rising household disposable income driving consumption of goods and services, as lifestyle is being tilted toward urban living with intensified migration from rural areas to bigger cities. The AEC also envisages a narrowing development gap between the older, more mature ASEAN-6 and the less developed newer members of Cambodia, Laos, Myanmar and Vietnam. Enhanced extra-regional economic integration also forms one of the principal pillars of the AEC with negotiations already under way for the ASEAN-led Regional Comprehensive Economic Partnership (RCEP) which involves long-standing and emerging economic powers (Japan, China, India) as well as other key trading partners (South Korea, Australia, New Zealand).

Nevertheless, while the AEC was formally launched on 31st December 2015, multiple agenda remain work in progress with several member countries opting out of certain liberalisation measures such as Indonesia. With only about 80% of the stipulated requirements for the AEC completed by 31st December 2015, what are the difficulties for the last final mile? Can the AEC be completed by 2016? In meeting the target of the AEC by 2016, could ASEAN members stay united or would the bloc become disorganised with economic agenda being disrupted by geopolitical tussles between foreign major powers? What are the challenges ahead for ASEAN beyond the 2015 milestone?

This chapter provides an in-depth analysis of the foregoing pertinent questions that concern not only the completion of the AEC *per se* but also the future of regional integration in ASEAN and the region-wide aspiration for sustainable and equitable economic development. Following this Introduction, Section 1.2 discusses some of the major theoretical works on economic integration. Section 1.3 summarises the evolution of economic integration in ASEAN from the 1970s to the 2000s and presents a brief overview of the AEC. Section 1.4 provides a detailed assessment of the progress of the AEC and analyses whether it can be practically completed by 2016. Section 1.5 presents the post-2015 vision for ASEAN with important implications from the ongoing liberalisation efforts through the Trans-Pacific Partnership (TPP) and the RCEP as well as the potential formation of the Free Trade Area of the Asia-Pacific (FTAAP). Section 1.6 concludes.

---

[5] See Heng (2015).

1.2 THEORETICAL LITERATURE ON ECONOMIC INTEGRATION

The discourse on economic integration dates back to the 1950s when political initiatives to unify Europe following the Second World War gained traction, notably with the establishment of the European Coal and Steel Community in 1952 which ultimately led to the founding of the European Union (EU). While there are many definitions and interpretations of economic integration, Balassa (1961) is widely recognised as the most seminal and first comprehensive theoretical framework on the topic. According to Balassa (1961), economic integration, defined as "the abolition of discrimination within an area," takes four forms which varies in the degree of integration:

(i) Free trade area, which is the first level of economic integration where member countries remove all trade barriers while retaining their own barriers with non-members.
(ii) Customs union, where, apart from being a free trade area, members' trade policies toward non-members are also harmonised, for instance, through the adoption of common external tariff.
(iii) Common market, which is the 3rd level of integration allowing free movement of labour and capital between member states.
(iv) Economic union, where members of a common market unify their monetary and fiscal policies, reflecting the most complex level of economic integration.

Prior to Balassa (1961), Viner (1950) pioneered the first theory on customs union, which argues that the establishment of a customs union has a two-sided quality, liberalising commerce among members while discriminating against third parties. In particular, the abolition of tariffs would simultaneously lead to an increase in trade between the two trading partners (trade creation) and a diversion of trade from the third state which is outside the customs union (trade diversion). While trade creation produces a positive welfare effect arising from both lower tariffs and increased trade volume, trade diversion has a negative welfare effect as it does the opposite. Viner's theory provides the point of departure for a large body of subsequent literature which explores the welfare implications of various preferential trade agreements.

Drawing on Viner (1950), Johnson (1965), Cooper and Massell (1965) and Bhagwati (1968) found that states could reduce the costs of achieving any given level of import-competing industrialisation by forming a preferential trade agreement capable of exploiting economies of scale and discriminating against goods emanating from outside sources. Krugman (1991) examined how scale economies within regional trade arrangements can foster greater specialisation as the location of production is shifted among members.

Another strand of literature specifically looks at whether regional trade agreements bolster multilateral liberalisation. Kemp and Wan (1976) demonstrated that it is possible for any group of countries to form a free trade area without reducing the welfare of third parties, and that incentives exist for the trade bloc to expand until it includes all states. Krugman (1993) argued that regional institutions reduce the number of actors engaged in multilateral negotiations, thereby mitigating problems of bargaining and other actions that can hamper such negotiations. Summers (1991) and Lawrence (1996) asserted that regional trade arrangements can induce members to undertake and consolidate economic reforms and such reforms are likely to promote multilateral openness.

Several recent studies have focused on "new regionalism," which refers to the latest phase of economic integration starting from the early 1990s. According to Mansfield and Milner (1999), "new regionalism" has been primarily driven by governments and the nature of domestic institutions. They argued that certain governments had opted to enter into a regional trade agreement because doing so seemed likely to facilitate liberal economic and political reforms that could dilute the political efficacy of societal groups that opposed such changes. Mittelman (2000) and Grugel (2004) viewed "new regionalism" as a response to rising global competition. Mittelman (2000) contended that competition in attracting foreign capital to support production had led states to collaborate in order to attain a larger market space. Grugel (2004) described the latest wave of regionalism as a "route through which states mediate the range of economic and social pressures generated by globalisation."

It has thus been argued that economic integration within ASEAN is part of "new regionalism," motivated not just by economic reasons but also by political and strategic imperatives that force the ten small economies to collectively manage their vulnerabilities.[6] For instance, Buszynski (1997) noted that the ASEAN Free Trade Area (AFTA) was instituted in the early 1990s to provide new political purpose to the regional grouping after the end of the Cold War and the Cambodian crisis. Hew (2007) pointed out that the formation of the AEC was driven by several global forces, most important of which were: the AFC of 1997–1998, China's rapid growth and the proliferation of regional trading arrangements outside Asia, which will be discussed in greater detail in the next section.

1.3 THE EVOLUTION OF ECONOMIC INTEGRATION IN ASEAN

1.3.1 ASEAN Economic Integration 1970s–2000s

Notable for its distinct economic and institutional diversity, ASEAN has emerged as one of the most cohesive regional associations in the developing world. When it was established in 1967 by Indonesia, Malaysia, the Philippines, Thailand and Singapore, ASEAN did not aspire toward a deeply integrated regional grouping with a complex institutional structure *a la* the EU model. It set for itself, in the 1967 Bangkok Declaration, seven modest objectives which included, among others: "to accelerate the economic growth, social progress and cultural development in the region," "to promote regional peace and stability," and "to promote active collaboration and mutual assistance on matters of common interest."[7] In essence, ASEAN consciously chose the path of economic cooperation to foster regional harmony amidst a highly uncertain geopolitical environment at the time.

---

[6]See, for instance, Buszynski (1997), Hew (2007) and Tham and Basu Das (2015).

[7]The Bangkok Declaration, or ASEAN Declaration, is ASEAN's founding document signed in Bangkok on 8th August 1967 by the Foreign Ministers of the five ASEAN founding members, namely Indonesia, Malaysia, Philippines, Singapore and Thailand. The document is accessible from: http://cil.nus.edu.sg/rp/pdf/ 1967%20 ASEAN%20Declaration-pdf.pdf.

In fact, the *raison d'être* for the establishment of ASEAN was primarily political and driven by the collective desire for regional security among the prospective member states. It was during the height of the Cold War when Indonesia just renounced its *konfrontasi* (confrontation) against Malaysia and Singapore.[8] Malaysia and Singapore had separated after a brief union mired with race riots. Malaysia and Philippines were in territorial dispute over Sabah. Communist insurgencies in Vietnam, Laos and Cambodia were at their peak. Against such backdrop, regime change was under way in Indonesia in 1965 and 1966 with the new administration under President Suharto signalling its intention to restore peace and focus on economic development. Regional reconciliation and restoration of relations thus provided the impetus for the formation of ASEAN.

With memories of bilateral tensions still fresh in the leaders' minds, it was agreed upon the advent of ASEAN that regional cooperation must be predicated upon mutual respect for national sovereignty and should be non-legally binding, with priority given to issues of domestic importance such as internal security and economic development. The so-called "ASEAN Way," which stresses informality and consultation, consensus-based decision-making and non-interference in member states' internal affairs, was thus adopted as the guiding principle of all ASEAN's initiatives.[9] Such approach allows member states to focus on the more pressing task of nation-building while maintaining national autonomy in all regional efforts.[10] ASEAN's institutional structure was therefore very minimal with foreign ministers of member states responsible for regional policymaking and coordination of cooperation activities.

It was not until February 1976 at the first leaders' summit in Bali, Indonesia, that the scope of cooperation in ASEAN was extended to the economic sphere. By then, the regional consensus was that import substitution had failed, and hence the original ASEAN-5 embarked on an export-led economic policy with special emphasis on attracting foreign direct investment (FDI). The summit also marked the beginning of increasing formalisation of regional cooperation initiatives with the signing of the ASEAN Preferential Trading Agreement (APTA), the ASEAN Industrial Projects, the ASEAN Industrial Complementation and the ASEAN Industrial Joint Ventures. APTA, the most significant among the four agreements, sought to promote intra-ASEAN trade through the adoption of preferential tariffs and non-tariff measures on a product-by-product basis.[11] The remaining three agreements were aimed at establishing inter-governmental industrial projects in each member country and facilitating resource pooling and specialisation of complementary products. However, the four cooperation programs had very limited impacts on intra-regional trade and investment.[12]

---

[8] *Konfrontasi* was a policy of hostility launched by Indonesia's President Sukarno from 1963 to 1966 in opposition to the formation of the Federation of Malaysia, of which Singapore was a part between 1963 and 1965. While there was no declared war, numerous raids were organised by Indonesian volunteers in the border area between Indonesia and East Malaysia on the island of Borneo. The conflict ended in 1966 following the ouster of President Sukarno.

[9] See Acharya (2011) for details on the emergence of the "ASEAN Way."

[10] See Sukma (2014, 5).

[11] The APTA agreement is accessible from: http://cil.nus.edu.sg/rp/pdf/1977%20Agreement%20on% 20ASEAN%20 Preferential%20Trading%20Arrangements-pdf.pdf.

[12] See Hill and Menon (2010, 4).

China's rapid rise as an important FDI destination since the early 1990s posed serious challenges for ASEAN to maintain competitiveness. Meanwhile, various global and regional trade initiatives were gathering momentum, making deeper economic integration a pressing task for ASEAN. Notable were the establishment of the Asia–Pacific Economic Cooperation (APEC) in 1989 and the imminent accession of Mexico, a middle-income competitor in the US export market, into the North America Free Trade Agreement (NAFTA). At the Singapore Summit in 1992, the leaders of the original ASEAN-5 and Brunei, acceded in 1984, launched the AFTA. For the first time in the region's cooperation history, there was a clear implementation timetable. The six member states agreed to reduce the Common Effective Preferential Tariff (CEPT) rates on products in their Inclusion Lists to 0–5% by 2008, with an interim target of 20% by 1998–2000. This deadline was later brought forward to 2003. A target of 90% tariff lines was also set by 2011. AFTA was subsequently extended to the new members of ASEAN, namely Vietnam (joined in 1995), Laos and Myanmar (1997), and Cambodia (1999). Vietnam was given until 2006 to bring down tariffs on products in its Inclusion List to no more than 5%. For Laos and Myanmar, it was 2008 while Cambodia had until 2010.

In line with global trends, particularly the conclusion of the Uruguay Round in 1994, ASEAN embarked on deepening integration in services trade and investment. The ASEAN Framework Agreement on Services (AFAS) was signed in December 1995 with an aim to progressively liberalise regional trade in services beyond the scope and depth set by the General Agreement on Trade in Services (GATS). Given the importance of FDI to the region, the bloc also launched the ASEAN Investment Area (AIA) in October 1998, which was subsequently expanded into the ASEAN Comprehensive Investment Agreement (ACIA) in February 2009, making ASEAN one of the first regional blocs in the developing world to formalise commitment to promote intra-regional investment.

In mid-1997, the AFC erupted, affecting Thailand, Indonesia, Malaysia, the Philippines, Brunei and Singapore, which saw an average 8% contraction of GDP.[13] In addition to losing attractiveness as a host of FDI, ASEAN was also seen by many as ineffective and fragile owing to its inability to respond decisively to the crisis. The AFC thus led to a collective rethink of the future of regional economic cooperation, which culminated with the adoption of the ASEAN Vision 2020 in December 1997. The Vision did not only reinstate the member states' commitment to fully implement AFTA and accelerate AFAS and AIA but also expanded regional cooperation agenda to "closer consultations in macroeconomic and financial policies." The Vision also enunciates the bloc's goal to "create a stable, prosperous and highly competitive ASEAN Economic Region in which there is a free flow of goods, services and investments, a freer flow of capital, equitable economic development and reduced poverty and socio-economic disparities."[14]

Two other notable developments took place since the AFC broke out. First, ASEAN forged closer economic ties with countries outside the region by initiating a host of cooperation initiatives and free trade agreements (FTAs). In December 1997 in the midst of the AFC, ASEAN

[13] See Asian Development Bank Institute (2014, 7).

[14] See ASEAN Secretariat (1997).

invited the leaders of Japan, South Korea and China to an informal meeting, which kick-started a host of ASEAN+3 initiatives. The two most important projects were the creation of the ASEAN+3 Chiang Mai Initiative (CMI)[15] in 2000 and the Asian Bond Markets Initiative (ABMI)[16] in 2003, both aiming to bolster regional financial safety nets. In December 2005, ASEAN convened the first East Asia Summit (EAS) with the participation of the Plus Three countries, India, Australia and New Zealand, which together form the ASEAN+6 grouping. The United States and Russia subsequently acceded to the group. Between 2004 and 2009, ASEAN concluded bilateral ASEAN+1 FTAs with all the Plus Six countries.

Second, against the backdrop of a fast-changing and increasingly uncertain global economic environment, ASEAN sought to re-define itself as well as strengthen internal institutions. The ASEAN Charter adopted in December 2007 turns ASEAN into a legal entity and provides the bloc with a new institutional framework. The semi-annual ASEAN Summit was identified as the highest policy-making body which comprises the heads of state of all member countries. The Charter also expands the roles and responsibilities of the ASEAN Secretariat and the ASEAN Secretary-General, particularly with the mandate to facilitate and monitor the implementation of ASEAN's agreements and commitments. ASEAN's objectives were extended to include promoting an ASEAN identity and maintaining the centrality of ASEAN as the primary driving force in regional cooperation.[17]

1.3.2 The ASEAN Economic Community

By the end of the 2000s, ASEAN had largely completed the easy phase of trade liberalisation. Significant progress had been made in tariff elimination and reduction under the AFTA–CEPT scheme, which began in 1993. By 2010, the share of all tariff lines with 0% tariff rate was around 90% for ASEAN-6, while the share of tariff lines with 0–5% tariff rates was 99% for Vietnam, Cambodia and Myanmar and 95% for Laos.[18] Nonetheless, intra-ASEAN trade remained far from liberalised.

First, non-tariff barriers (NTBs) in the form of non-automatic licensing, technical regulations and quality standards are widely prevalent. Second, infrastructural constraints have not been effectively mitigated, thus hampering intra-regional trade flows. Third, large development gaps have resulted in limited trade creation effects from tariff reduction for the newer, less developed

[15] Launched at the ASEAN+3 Finance Ministers' Meeting in Chiang Mai in 2000, the CMI was originally a currency swap arrangement between ASEAN, China, Japan and South Korea. The CMI was later multilateralised in 2009 to become a self-managed US dollar liquidity pool from which a member state can swap its local currency for US dollars in the event of a balance of payments or liquidity crisis. At the 15th ASEAN+3 Finance Ministers and Central Bank's Governors' Meeting in May 2012 in Manila, member countries agreed to expand the CMI Multilateralisation (CMIM) Agreement from US$120 billion at the time of signing in 2009 to US$240 billion. The ASEAN+3 Macroeconomic Research Office was established in 2011 to support CMIM implementation as an independent surveillance unit as well as to monitor and detect emerging vulnerabilities. See Kawai *et al.* (2014, 83–92).

[16] The ABMI aims to develop efficient and liquid bond markets in Asia to better utilise Asian savings for Asian investments. The initiative would also contribute to the mitigation of currency and maturity mismatches in financing. See Kurihara (2012).

[17] See ASEAN Secretariat (2007).

[18] See Okabe and Urata (2013, 3).

member states compared to the older members.[19] Furthermore, while FDI flows into the bloc increased remarkably in the 2000s, they went mostly to the more developed members, particularly Singapore.[20]

Meanwhile, China continued to gain dominance in global trade as it became a member of the World Trade Organisation (WTO) in 2001. In fact, a significant diversion of trade away from ASEAN and into China has been under way since the 1990s.[21] This has created tremendous concerns among ASEAN leaders to deepen economic integration and provide economies of scale to foreign investors. Concurrently, China's market size and increasing purchasing power present immense opportunities for exports that ASEAN could benefit from.

As ASEAN made concerted efforts to increase integration with countries outside the region following the AFC, it was acutely aware of its limited ability in mitigating its vulnerabilities to and preventing economic shocks from wreaking havoc in the member economies, which was evident from the lack of a coordinated and cohesive response to the crisis. ASEAN was therefore in need of a more effective collective mechanism for averting and managing crises, which can only be realised through intensifying regional economic cooperation. Strategically, the issue of maintaining the centrality of ASEAN in the plethora of global and regional initiatives has also become increasingly prominent. While ASEAN had spearheaded a number of cooperative mechanisms with a diverse group of international partners, advancing its own interests in such big forums, with the presence of long-standing as well as emerging powers, requires higher degrees of cohesiveness and solidarity.

In light of the foregoing trends and issues, in October 2003 ASEAN decided to further deepen and broaden regional integration, drawing on the ASEAN Vision 2020, by establishing the ASEAN Community by 2020, which consists of the ASEAN Political and Security Community, the ASEAN Socio-Cultural Community and the AEC. At the 12th ASEAN Summit in January 2007 which coincided with the bloc's 40th anniversary held in Singapore, the leaders of ASEAN member states agreed to advance the establishment of the AEC by 2015. In November 2007, the AEC Blueprint outlining various measures and a specific "strategic schedule" was adopted.

Since its conception, the AEC is widely regarded as a highly ambitious project which aims to transform ASEAN into (i) a single market and production base, (ii) a highly competitive economic region, (iii) a region of equitable economic development and (iv) a region fully integrated into the global economy.[22] Correspondingly, the AEC Blueprint lays out four pillars and 17 core elements as shown in Figure 1.2. The creation of a "single market and production base," with enhanced connectivity, common policy frameworks and deeper integration into the global economy, would allow ASEAN member states to complement each other in an intra-regional production network leveraging on their diverse levels of economic development, exploit economies of scale and thus enhance efficiency in production processes. Furthermore, the formation of the AEC would facilitate easier access to the regional market of more than 600 million people which is evidently attractive to both local and international investors.

---

[19] See Okabe and Urata (2013) for an empirical assessment of the trade creation effects from AFTA.

[20] See Basu Das (2015, 5).

[21] See Hew *et al.* (2005).

[22] See ASEAN Secretariat (2008).

Figure 1.2: Four Pillars and 17 Core Elements of the ASEAN Economic Community (AEC)

Source: ASEAN Secretariat (2015).

In particular, the first pillar envisages ASEAN as a more liberalised market in which its population have greater opportunities to trade, invest and do business within the region, with reduced trade costs and an improved investment regime that make ASEAN a more attractive host of investment for both international and local investors. This pillar encompasses five core elements, namely (i) free flow of goods, (ii) free flow of services, (iii) free flow of investment, (iv) free flow of capital, and (v) free flow of skilled labour, which facilitates the development of regional production networks and enhance ASEAN's capacity as a production centre in the global supply chain.

Going deeper than elimination of trade tariffs, the first pillar of the AEC stipulates the removal of non-tariffs barriers (NTBs), integration of customs procedures through the establishment of the ASEAN Single Window (ASW) as well as harmonising rules of origin (ROOs), technical standards and conformance procedures. Free flow of trade in services requires no restriction on services suppliers in ASEAN in providing services and in setting up companies across national borders in the region. Strengthening capital market integration and development and allowing for greater capital mobility are two components under the first pillar. As for skilled labour, the AEC allows for managed mobility of professionals engaged in cross-border trade and investment related activities.

While integration across all economic sectors is desirable, the AEC specifies 12 priority sectors for accelerated integration. The 12 priority integration sectors are: agro-based products, air transport, automotive, e-commerce, electronics, fisheries, healthcare, rubber-based products, textiles and apparel, tourism, wood-based products and logistics.

The second pillar seeks to create a business-friendly environment conducive for innovation, which is achievable through the adoption of common frameworks and standards across areas such as competition policy, intellectual property rights and consumer protection. It also aims to enhance multi-modal connectivity including maritime, land and air transport, and other infrastructure to facilitate cross-border transportation and reduce the costs of doing business. This pillar also stipulates collaboration measures such as the Trans-ASEAN Gas Pipeline and the ASEAN Power Grid for the optimisation of the region's energy resources to enhance energy security in the region.

The third pillar aims to attain sustainable and balanced economic growth through developing and integrating small- and medium-sized enterprises (SMEs) into regional and global value chains. Narrowing the development divide is also central in this pillar with special emphasis on building the capacity of the newer member states to ensure their effective integration into the AEC. The implementation of this core element is in accordance with the ASEAN Policy Blueprint for SME Development 2004–2014 and the Initiative for ASEAN Integration (IAI) strategic framework launched in 2000.[23] The IAI specifies the following priority areas for capacity building and technical assistance: infrastructure, human resource development, information and communications technology, energy, investment climate, tourism, poverty reduction and improvement in the quality of life.

The 4th and final pillar envisages an increasingly integrated ASEAN into the global economy through a coherent approach which ensures the bloc's collective interests are effectively advanced. Enhanced participation in global supply networks is also a core element under this pillar, particularly for the newer member states. Principal measures comprise maintaining ASEAN centrality in its negotiations for free trade and comprehensive economic partnership agreements and developing a comprehensive package of technical assistance for the less developed member countries to upgrade their industrial capability and productivity and enhance their participation in regional and global integration initiatives.

With such an ambitious set of deliverables, the AEC is projected to generate substantial economic benefits to ASEAN as a whole. According to a Computable General Equilibrium (CGE) model computed by Petri, Plummer and Zhai (2012), the economic gains induced by AEC would exceed those resulting from AFTA by more than six times (see Table 1.2). On average, ASEAN would see a 5.3%, or US$69.4 billion, increase in their aggregate GDP relative to the baseline,

Table 1.2: Real Income Gains of the ASEAN Economic Community (AEC) *vis-à-vis* the ASEAN Free Trade Area (AFTA)

	AFTA	AEC	AFTA	AEC
	US Dollar billion (2004 Prices)		**Percentage of Baseline GDP**	
Brunei	0.2	0.5	2.6	7.0
Cambodia	0.3	0.6	2.7	6.3
Indonesia	1.0	27.6	0.2	6.2
Laos	0.0	0.2	0.6	3.6
Malaysia	2.7	5.7	1.4	3.0
Myanmar	0.0	0.6	0.3	4.4
Philippines	0.9	4.5	0.6	3.2
Singapore	2.6	15.1	1.6	9.7
Thailand	1.6	12.2	0.6	4.9
Vietnam	0.9	2.4	1.1	2.8
ASEAN-10	10.1	69.4	0.8	5.3

Source: Petri, Plummer and Zhai (2012, 107).

[23] Details of the IAI framework is accessible from: http://www.asean.org/wp-content/uploads/images/archive/ 22325.pdf.

as compared to merely 0.8% (US$10.1 billion) under AFTA. All ten member states stand to benefit more from AEC than from AFTA both in absolute term and as a percentage of baseline GDP, although some may benefit more than others. The model assumes complete elimination of tariffs and reduction of NTBs on goods trade, the liberalisation of five services sectors (utilities, construction, transport, private services and government services), increased FDI inflows and a 5% reduction in trade costs.

1.4 THE ASEAN ECONOMIC COMMUNITY BY 2016?

The establishment of the AEC marks a significant milestone in the history of economic integration in ASEAN. While the AEC was officially launched on 31st December 2015, according to an assessment by the ASEAN Secretariat, only 486 out of 611 integration measures (79.5%) have been completed.[24] Nonetheless, a closer look at the actual progress of the AEC points to various obstacles and challenges, particularly in the so-called deep integration issues including NTBs, trade facilitation and trade in services.

It is thus timely for policy-makers in the region to take stock of the progress as well as obstacles in the final mile toward the full formation of the AEC. This section provides a comprehensive assessment of the AEC through detailing the achievements and key challenges which have been hindering the full implementation of the AEC blueprint. Drawing on this analysis, the likelihood that the AEC will be fully realised by the end of 2016 will also be explored.

1.4.1 Achievements and Progress

Notwithstanding the diverse levels of economic development within the ten-member regional grouping, substantial progress have been made in terms of tariff reduction and elimination as well as trade facilitation, which are the two critical core elements under the AEC blueprint.

Tariff Reduction and Elimination

since the signing of the Agreement on the CEPT Scheme under AFTA in 1992, ASEAN members have reduced intra-regional tariffs in successive rounds. The commitments and schedules stipulated in the CEPT–AFTA and the relevant Amendments and Protocols serve as the basis for tariff elimination under the AEC blueprint. Accordingly, tariffs on products not considered "general exceptions" or originally placed on the Sensitive List or Highly Sensitive List are to be eliminated. This is to be completed by 2010 for ASEAN-6 and by 2015 for Cambodia, Laos, Myanmar and Vietnam (CLMV), with flexibilities on some sensitive products until 2018. Implementation of the commitments was given an extra boost with the signing of the ASEAN Trade in Goods Agreement (ATIGA) in 2009. The ATIGA consolidates and streamlines provisions in the CEPT–AFTA while formalising related ministerial decisions under the AEC blueprint as a single legal instrument.

Significant progress in tariff removal has been made, especially for CLMV. As shown in Figure 1.3, the average CEPT rate for all ASEAN economies fell from 2.58% in 2007 to 0.54%

[24] See ASEAN Secretariat (2015a, 10)

Figure 1.3: Average Common Effective Preferential Tariff (CEPT) Rates in ASEAN, 2007–2014

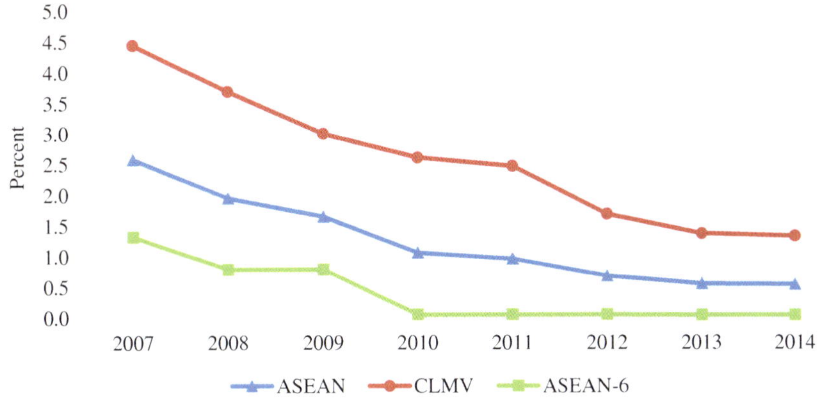

Source: ASEAN Secretariat (2015b, 9).

in 2014. For ASEAN-6 in particular, the average CEPT rate has become negligible since 2010, standing at a mere 0.04% in 2014. Meanwhile, the average rate for CLMV has also steadily decreased over the years from 4.44% in 2007 to reach 1.33% in 2014.

Major progress in tariff elimination is also reflected in the share of tariff lines at zero-percent preferential duty in total tariff lines. By 2015, the share of zero-percent tariff lines for all ten ASEAN economies was 96.0%, up from 80.3% in 2011. This increase was mainly driven by tariff elimination in CLMV as the share of zero-percent tariff lines for these countries rose from 49.3% in 2011 to 90.8% in 2015. The share of zero-percent tariff lines for ASEAN-6 has remained constant at the remarkable 99.2% since 2012.[25]

As noted, not all products traded among ASEAN members are slated for eventual tariff elimination. While products on the Sensitive List[26] are to be phased into the CEPT scheme, member states are only required to reduce tariffs on these products to up to 5%. This phasing in and tariff reduction process is to be completed by 2010 for ASEAN-6, 2013 for Vietnam, 2015 for Laos and Myanmar and 2017 for Cambodia.

In addition, Malaysia, Indonesia and Philippines are allowed to place certain rice products on the Highly Sensitive List.[27] Although these products are also to be phased in to the CEPT scheme by 2010, Malaysia, Indonesia and Philippines retain the right to set the tariff rates on them subjected to maximum caps as specified in Annex 3 of the Protocol on the Special Arrangement for Sensitive and Highly Sensitive Products.[28]

---

[25] Data on share of zero-percent tariff lines in total tariff schedule is from the ASEAN Integration Report 2015. See ASEAN Secretariat (2015b, 10).

[26] See Annex 2 of the Protocol on the Special Arrangement for Sensitive and Highly Sensitive Products which is accessible at http://www.asean.org/?static_post=annex-2-list-of-sensitive-products.

[27] See Annex 1 of the Protocol on the Special Arrangement for Sensitive and Highly Sensitive Products which is accessible at http://www.asean.org/?static_post=annex-1-list-of-highly-sensitive-products.

[28] The ending tariff rates for products on the Highly Sensitive List after they are phased into the CEPT scheme for Malaysia and Indonesia shall be 20% while the rates for Philippines shall be determined in accordance with the CEPT framework. See http://www.asean.org/?static_post=annex-3-tariff-reduction-for-highly-sensitive-products.

Finally, ASEAN members reserve the freedom to apply protectionist measures necessary to protect national security, public morals, human, animal or plant life and health and articles of artistic, historic or archaeological values. These are considered the "general exceptions" to the CEPT scheme.[29]

Trade Facilitation

Aside from tariff reduction and elimination, the AEC blueprint also aspires to simplify, harmonise and standardise trade and customs processes, procedures and related information flows for reduction of transaction costs in intra-ASEAN cross-border trade. Accordingly, its trade facilitation agenda focuses largely on harmonisation of standards, conformity assessment procedures and technical regulations; customs integrations as well as the development of the ASW. In general, the progress of implementing trade facilitation measures has been positive but uneven.

Harmonisation of standards and technical requirements is guided by the ASEAN Policy Guideline on Standards and Conformance. One important initiative under this Guideline is the development of sectoral Mutual Recognition Arrangements (MRAs) on conformity assessment in accordance with the ASEAN Framework Agreement on Mutual Recognition. MRAs are agreements made between two or more parties to recognise some or all aspects in each other's conformity assessment results.[30] To date, MRAs have been concluded for three sectors: electrical and electronic equipment (EEE), cosmetics and medicinal products. In addition, agreements to establish harmonised regulatory frameworks for EEE and cosmetics have been signed and enforced across all ASEAN member states while the agreement for medical devices was signed in 2014 and is pending ratification by individual member states.[31] Going forward, ASEAN is looking to expand the scope of the ASEAN Framework Agreement on Mutual Recognition and develop MRAs for prepared foodstuffs, automotive and building and construction sectors.[32]

For customs integration, progress in some areas is better than others. For example, all ASEAN members have adopted the ASEAN Harmonised Tariff Nomenclature (AHTN) for intra-regional trade since 2012. The AHTN facilitates tariff calculations, trade negotiations and collection of trade data. However, the agenda for establishing the ASEAN Customs Transit System (ACTS) has only seen modest progress. While the goal had been expounded in the AEC blueprint in 2007, the protocol which provides for the establishment of ACTS (Protocol 7 of the ASEAN

---

[29] An example of products which qualify as "general exceptions" to the CEPT scheme is "sin" products such as alcohol and tobacco. ASEAN members retain the right to apply high tariffs and other prohibitive measures in order to reduce the imports of these products.

[30] See ASEAN Secretariat (2015b, 20).

[31] The ASEAN Harmonised EEE Regulatory Regime was signed in 2005 and ratified by all ASEAN members in 2012. The ASEAN Harmonised Cosmetics Regulatory Scheme was signed in 2003 and came into force immediately for all member states as no ratification was required for this agreement. On the other hand, the ASEAN Agreement on Medical Device Directive was signed on 21st November 2014 and though it officially came into force in January 2015, it is only in force in countries which have ratified it. Refer to the List of Instruments published by the ASEAN Secretariat at http://agreement.asean.org/search/by_pillar/2.html.

[32] See ASEAN Secretariat (2015b, 20).

Framework Agreement on the Facilitation of Goods in Transit) was only signed in February 2015. To date, it has not been ratified by all member states.[33] In addition, the piloting phase of ACTS is scheduled to start in April 2016 and currently involves only three ASEAN members, namely Malaysia, Singapore and Thailand. Subsequent expansion to include the rest of ASEAN will only come at a later stage.[34]

On balance, data from the World Bank shows notable efficiency gains in terms of time taken and procedures required to import and export across all ASEAN countries, especially CLMV, over the 2005–2014 period. As seen in Table 1.3, Cambodia and Laos required fewer documents

Table 1.3: Documents Required to Import and Export in ASEAN Countries, 2005–2014

	2005	2006	2007	2008	2009	2010	2011	2012	2013	2014
Documents required to import (number)										
Brunei	NA	5	5	5	5	5	5	5	5	5
Cambodia	11	10	10	10	10	9	9	9	9	9
Indonesia	8	8	8	8	8	8	8	8	8	8
Laos	15	15	10	10	10	10	10	10	10	10
Malaysia	4	4	4	4	4	4	4	4	4	4
Myanmar	NA	NA	NA	NA	NA	NA	NA	9	9	8
Philippines	8	8	8	8	8	7	7	7	7	7
Singapore	3	3	3	3	3	3	3	3	3	3
Thailand	12	12	9	5	5	5	5	5	5	5
Vietnam	8	8	8	8	8	8	8	8	8	8
Documents required to export (number)										
Brunei	NA	5	5	5	5	5	5	5	5	5
Cambodia	6	9	9	9	9	8	8	8	8	8
Indonesia	4	4	4	4	4	4	4	4	4	4
Laos	12	12	10	10	10	10	10	10	10	10
Malaysia	4	4	4	4	4	4	4	4	4	4
Myanmar	NA	NA	NA	NA	NA	NA	NA	9	9	8
Philippines	6	6	6	6	6	6	6	6	6	6
Singapore	3	3	3	3	3	3	3	3	3	3
Thailand	9	9	7	5	5	5	5	5	5	5
Vietnam	5	5	5	5	5	5	5	5	5	5

Note: NA indicates data is not available.
Source: World Bank.

[33] See the list of legal instrument published by the ASEAN Secretariat at http://agreement.asean.org/search/by_pillar/2.html.

[34] See http://arise.asean.org/asean-customs-transit-system-acts-2/.

to import and export in 2014 as compared to 2006. Likewise, the number of documents to import and export for Myanmar have also fallen over the period of 2012–2014. For ASEAN-6, between 2005 and 2014 procedural efficiency has improved most remarkably in Thailand as the number of documents required to import and export dropped significantly from 12 to 5. The number of documents required to import in the Philippines was also reduced.

Meanwhile, Table 1.4 shows that the number of days taken to import and export have decreased over the years for most ASEAN countries. The most significant decreases were in Laos and Cambodia. As the lion's share of the time taken to import and export is filled by document preparation,[35] Table 1.4 reflects marked improvement in customs processes and substantial reduction of red tape burdens in ASEAN. In sum, the downward trends in the number of

Table 1.4: Time Taken to Import and Export in ASEAN Countries, 2005–2014

	2005	2006	2007	2008	2009	2010	2011	2012	2013	2014
Time taken to import (days)										
Brunei	NA	19	19	19	19	20	15	15	15	15
Cambodia	54	45	45	29	29	26	26	26	24	24
Indonesia	27	27	27	27	27	27	27	23	23	26
Laos	65	65	37	37	37	37	33	26	26	26
Malaysia	10	10	10	10	10	10	10	8	8	8
Myanmar	NA	NA	NA	NA	NA	NA	NA	27	27	22
Philippines	18	18	18	16	16	14	14	14	14	15
Singapore	4	4	4	4	4	4	4	4	4	4
Thailand	22	22	14	13	13	13	13	13	13	13
Vietnam	23	23	23	23	21	21	21	21	21	21
Time taken to export (days)										
Brunei	NA	27	27	27	27	25	19	19	19	19
Cambodia	43	37	37	22	22	22	22	22	22	22
Indonesia	22	22	18	18	18	17	17	17	17	17
Laos	55	55	39	38	38	36	32	25	23	23
Malaysia	13	13	13	13	13	13	13	11	11	11
Myanmar	NA	NA	NA	NA	NA	NA	NA	25	25	20
Philippines	17	17	17	16	16	15	15	15	15	15
Singapore	6	6	6	6	6	6	6	6	6	6
Thailand	24	24	17	14	14	14	14	14	14	14
Vietnam	24	24	24	24	22	22	22	21	21	21

Note: NA indicates data is not available.
Source: World Bank.

[35]See ASEAN Secretariat and World Bank (2015, 70).

documents required and time taken to import and export are encouraging signs pointing to the positive progress of the trade facilitation agenda under the AEC.

However, the data compiled by the World Bank also indicates that ASEAN countries are some distance away from the regional best practice as exemplified by Singapore, signalling the need for further improvement. In fact, two major initiatives which would have greatly enhanced trade facilitation have not been completed, namely the ASW and the ASEAN Trade Repository, which will be discussed in detail in the Section on ASEAN Single Window and ASEAN Trade Repository.

1.4.2 Obstacles and Challenges

While substantial progress has been made in trade in goods, important challenges remain for various other key areas. These include: (i) low utilisation of preferential tariffs, (ii) non-tariff barriers to trade, (iii) the formation of the ASW and ASEAN Trade Repository, (iv) trade in services, (v) skilled labour mobility and (vi) the establishment of the ASEAN Single Aviation Market (ASAM).

Low Utilisation of Preferential Tariffs

As described in the Section on Tariff Reduction and Elimination, ASEAN's efforts to reduce and eliminate tariffs on intra-regional trade in goods is highly commendable. However, actual utilisation of the CEPT preferential tariffs among ASEAN members has been modest. According to a 2013 report jointly published by the ASEAN Secretariat and the World Bank, utilisation rates have been generally low. Only imports by Cambodia and Philippines have shown significant utilisation rates.[36] This is a cause for concern as it implies that efforts to eliminate tariffs have not yielded meaningful benefits for the business community in ASEAN.

There are several possible reasons for low utilisation of the CEPT tariff scheme. In particular, the phenomenon may be due to low preference margins between the CEPT rates and the Most Favoured Nation (MFN) rates.[37] The applied MFN tariffs adopted in ASEAN are generally low and have also been reduced in conjunction with the implementation of the CEPT commitments. As a result, the margins between CEPT and MFN rates are small, offering little incentive for businesses in ASEAN to take advantage of the CEPT preferential rates. As shown in Table 1.5, the average margin between the average CEPT rates and the average applied MFN rates for the period of 2007–2014 was 7.34% points for CLMV, 5.23% points for ASEAN-6 and around 6% points for ASEAN as a whole. In comparison, it has been suggested that CEPT preferential

[36]See ASEAN Secretariat and World Bank (2013, 6).

[37]MFN tariffs are what a country promises to impose on imports from other members of the WTO with which it does not have a preferential trade agreement. In other words, MFN rates are the most restrictive that WTO members charge one another in practice. See http://wits.worldbank.org/wits/wits/ witshelp/Content/Data_Retrieval/P/Intro/C2.Types_of_Tariffs.htm.

Table 1.5: Margins Between the Average Common Effective Preferential Tariff (CEPT) Rates and Average Applied Most Favoured Nation (MFN) Rates in ASEAN (%), 2007–2014

Year	ASEAN	CLMV	ASEAN-6
2007	5.57	7.14	4.55
2008	6.18	7.89	5.04
2009	5.78	7.10	4.86
2010	6.31	7.44	5.53
2011	6.07	6.76	5.52
2012	6.11	7.24	5.32
2013	6.35	7.56	5.51
2014	6.36	7.60	5.51
Average	6.09	7.34	5.23

Note: Average reported is simple average.
Source: Authors' calculation based on data from ASEAN Secretariat (2015b, 9).

tariffs would only affect intra-regional imports favourably if preference margins were much higher at around 25% points.[38]

Business surveys conducted in ASEAN have also revealed that low utilisation of preferential tariffs may be due to: (i) the prevalence of NTBs, (ii) cost effectiveness of the ROOs and their related administrative procedures, and most crucially, (iii) lack of information among the business community about the integration initiatives under the AEC.[39] In particular, in a 2013 survey conducted by the Asian Development Bank (ADB) and the Institute of Southeast Asian Studies (ISEAS), overall 55% of the respondents in ASEAN said they were not aware of the AEC.[40] This seems to show that integration efforts under the AEC have thus far followed a top-down state-led approach with insufficient engagement of private sector stakeholders.

Country-specific data and studies further affirm the prevailing lack of private sector participation in the establishment of the AEC. For instance, the percentage of Vietnamese respondents in the aforementioned survey indicating they were not aware of the AEC was 76%, much higher than regional average of 55%. Indeed, the resources allocated by the Vietnamese government to raise awareness among local enterprises, professionals and the general populace about the AEC and its benefits have been inadequate.[41] This certainly poses a concern in the longer term as limited utilisation of the AEC initiatives may erode domestic support for further integration.

[38] See Manchin and Pelkmans-Balaoing (2007).

[39] See Kawai and Wignaraja (2011) and Hu (2013) as quoted in Basu Das (2016).

[40] See Hu (2013).

[41] See Vo (2015).

Non-Tariff Barriers to Trade

The proliferation of NTBs is one of the most prevalent challenges to achieving intra-regional free flow of goods. The AEC blueprint distinguishes NTBs from other non-tariff measures (NTMs) by virtue of their trade-limiting impacts and aims to eliminate NTBs. The blueprint simultaneously calls for standstill and roll-back on NTBs and greater transparency on the remaining NTMs. It should be noted that the difference between NTBs and NTMs is not always clear-cut in practice because a measure's restrictiveness is primarily dependent on its application on the ground. For example, a seemingly mild sanitary or phytosanitary measure may be highly trade-inhibiting if the country requires every shipment to be inspected rather than using risk profiling to minimise the number of inspections.[42]

Nevertheless, elimination of NTBs has been the focus of the AEC integration agenda. The AEC blueprint stipulates that all NTBs are to be removed by 2010 for Brunei, Indonesia, Malaysia, Singapore and Thailand; by 2012 for Philippines and by 2015 with flexibilities to 2018 for CLMV in accordance with the Work Programme on NTBs Elimination. The ASEAN Secretariat has also attempted to highlight the most pervasive classes of NTBs including: customs surcharges, additional charges, single channel for imports, state-trading administration, technical measures, product characteristic requirements, marketing requirements and technical regulations.[43] The onus is on member states to submit their lists of specific NTMs (inclusive of NTBs) to the ASEAN Secretariat, which compiles these lists into a regional NTMs database. Each member state is then required to develop specific work programmes to remove its NTBs. This is referred to as the "notify-negotiate-eliminate" approach.

There has been a growing consensus amongst pundits and policy-makers in the region that little progress has been made in this specific agenda. For example, according to the ASEAN Integration Monitoring Report jointly published by the World Bank and the ASEAN Secretariat in 2013, "there has been little real progress in the elimination by ASEAN Member States of their NTBs."[44] Echoing this sentiment, the Singapore's Minister for Trade and Industry, Mr Lim Hng Kiang, noted in 2015 that "non-tariff barriers in ASEAN are many," and that businesses have to "grapple with custom processes, documentation requirements and a complex regulatory environment within ASEAN."[45]

Furthermore, the regional NTMs database central to the entire process to eliminate NTBs is highly problematic. The lists of NTMs submitted by member states are not in a standardised format and the information is outdated. The country which updated its list most recently is Laos in 2012 while information provided by other ASEAN members can be dated as far back as 2007.[46] Furthermore, there is no mechanism for fellow members or the ASEAN Secretariat to

---

[42] See ASEAN Secretariat and World Bank (2013, 9).

[43] See http://www.asean.org/?static_post=non-tariff-barriers.

[44] See ASEAN Secretariat, and World Bank (2013, 10).

[45] See Lim (2015).

[46] The Non-Tariff Measures Database containing information on individual ASEAN member states is accessible at http://www.asean.org/?static_post=non-tariff-measures-database.

verify these lists. Most tellingly, the ASEAN Secretariat admitted that it is not feasible to calculate the number of NTMs identified as NTBs or the number of NTBs eliminated with the available data.[47] Without proper mechanisms to track the situations on the ground, the NTBs elimination agenda is unlikely to make credible progress.

In fact, the current approach which relies on member states to voluntarily disclose information on their NTMs has been criticised for suffering from several fundamental problems. Firstly, there is a clear conflict of interest as member governments have an incentive to hoard information about their NTMs to preserve their bargaining power in negotiations about elimination of NTBs as well as in future trade negotiations with partners within and outside of ASEAN. Secondly, government agencies in charge of setting and enforcing NTMs may also have the incentive to under-report their regulatory activities for fears of being admonished in case open disclosure leads to a drop in the country's rank in international competitiveness or ease of doing business rankings. Finally, even when there is no such incentive misalignment, institutional weaknesses may still hamper member states' efforts to compile information on their NTMs. As NTMs usually fall under the purview of various government departments and agencies, the task requires effective coordination mechanisms, which are often lacking in member states with low institutional capacity.[48]

Aside from the main "notify-negotiate-eliminate" approach, ASEAN has also initiated another programme known as the Matrix of Actual Cases on NTMs/NTBs. Under this, actual cases of grievances against NTMs are flagged so that member states can work to reduce or eliminate them.[49] While this programme has had a measure of success,[50] the overall prospect of NTMs/NTBs reform under the AEC remains grim in light of the current progress.

ASEAN Single Window and ASEAN Trade Repository

Establishing the ASW and ASEAN Trade Repository are two other important initiatives under the AEC which have not been fully completed. The ASEAN Trade Repository is an online portal for trade-related information on individual ASEAN member states including: tariff nomenclature, MFN tariffs, intra-ASEAN preferential tariffs and tariffs offered in trade agreements of ASEAN with its dialogue partners, ROOs, NTMs, national trade and customs laws and rules, procedures and documentary requirements, administrative rulings, best practices in trade facilitation applied by each member state and lists of authorised traders in individual member states. As a regional interface providing a single point of access, the ASEAN Trade Repository is linked to the respective National Trade Repositories which contain the actual contents. The ASEAN Trade Repository serves to improve regulatory transparency and information accessibility for the

[47] See ASEAN Secretariat (2015b, 6).
[48] See Cadot, Munadi and Ing (2013).
[49] The Matrix is published by the ASEAN Secretariat and accessible at http://www.asean.org/storage/images/2015/July/Matrix_of_Actual_Cases/Matrix%20of%20Actual%20Cases%20on%20NTMs_Resolved%20Cases%20as%20of%2018th%20CCA.PDF.
[50] See Intal *et al.* (2014, 126).

private sector. It also facilitates NTBs elimination as the site is supposed to host the database on NTMs in ASEAN.

As of March 2016, Brunei and Myanmar are yet to launch their respective National Trade Repositories. Furthermore, although the ASEAN Trade Repository has been operationalised (accessible at http://atr.asean.org/), its functionality leaves much room for improvement. While the Repository does provide one-stop access to individual National Trade Repositories except for those of Brunei and Myanmar, it does not link to specific trade-related information for some members. For example, a search query about "Rules of Origin" on the ASEAN Trade Repository website yields the relevant link to access the information for Cambodia but does not do so for Philippines and Vietnam despite the fact that information for Philippines and Vietnam are available on their respective National Trade Repository websites.

Meanwhile, the progress of the ASW is even slower. The ASW is an environment which connects and integrates all members' National Single Windows (NSWs). The NSWs enable single submission of data, single synchronous processing of data and single decision-making for customs clearance and cargo in individual ASEAN member states. As such, the ASW, comprising ten NSWs, is envisaged to expedite customs clearance, reduce transaction costs and thus facilitate a more seamless intra-regional flow of goods. The AEC blueprint originally proposes an ambitious timeline for ASEAN-6 to operationalise their NSWs by 2008 and by 2012 for CLMV. Subsequently, these deadlines were postponed to 2012 and 2015 respectively.[51]

To date, the NSWs are in operation for ASEAN-6 and Vietnam, while they are still under development for Cambodia, Laos and Myanmar.[52] With regard to the ASW, the Protocol on the Legal Framework for implementation was only signed in September 2015 and is currently pending ratification by member states. The project itself currently remains in the pilot phase. The full-fledged ASW Pilot Project Component 2 has been carried out in stages since April 2015 with initial participation by Indonesia, Malaysia, Thailand, Singapore and Vietnam. Brunei and Philippines have planned to join live operation of the system by the second quarter of 2016.[53]

Slow implementation of the ASW agenda is primarily due to disparities in capacity amongst the ASEAN member states. In this regard, it should be noted that a functional ASW entails not only connections of the various NSWs to a single platform but also seamless connections of different government agencies in individual member states to the system. Given the different degrees that e-government is implemented by ASEAN members, it is perhaps unsurprising that the progress of the ASW is slower than hoped for.

Apart from technical difficulties, some ASEAN members need to address existing legal gaps pertaining to e-banking, e-commerce, digital signature, validity or proof or digital support documents, archives of digital documents and so on before they can embark on establishing their respective NSWs.[54] Issuing new legislations in these areas can be time consuming. In this light, it is doubtful if a fully functional ASW can be achieved by the end of 2016.

[51] See Kow and Mowerman (2013, 8).

[52] See http://asw.asean.org/nsw/filter-by-category/general-information.

[53] See http://asw.asean.org/about-asw.

[54] See Kow and Mowerman (2013, 12).

Trade in Services

Achieving free flow of trade in services is slated as an important element in realising the AEC.[55] Services have played an increasingly critical role in sustaining economic growth in ASEAN. As income per capita rises, the demand for services in the region will increase, creating opportunities for growth. Moreover, a competitive services sector is crucial for gainful participation in the global value chains as services act as the enabling interface for production to take place in different geographical locations. However, while ASEAN as a whole has recognised the importance of services integration, actual progress in this direction can only be described as modest at best.

ASEAN has embarked on liberalising trade in services since 1995 with the signing of the AFAS. The AFAS requires ASEAN members to undertake negotiations on a sector-by-sector basis, provides the legal basis for MRAs between member states to facilitate licensing or certification of service suppliers and most importantly, targets for ASEAN members to achieve liberalisation beyond their commitments under the WTO's General Agreement on Trade in Services (GATS).[56] Provisions in the AFAS provide the overall architecture, disciplines and procedures for ASEAN's service integration efforts.

The AEC blueprint reiterates the objectives of the AFAS and sets out clear measures for implementation. It aims to substantially remove all restrictions on trade in services by 2015, with particular priority for air transport, e-ASEAN, healthcare and tourism by 2010 and for logistic services by 2013. It mandates ASEAN members to undertake liberalisation in consecutive biennial rounds and sets out threshold targets on the number of new sub-sectors to be committed in each round. Furthermore, in line with the AFAS, the AEC blueprint specifies liberalisation agenda for four modes of supply: cross-border supply (Mode 1), consumption abroad (Mode 2), commercial presence (Mode 3) and movement of natural persons (Mode 4).

Despite the ambitious objective to achieve free flow of trade in services, the service liberalisation agenda under the AEC leaves much to be desired. Firstly, some targets set out in the AEC blueprint are far from ambitious. For instance, Mode 3 liberalisation only envisages a maximum of 70% of foreign equity participation.[57] In comparison, the TPP agreement mandates its signatories to allow for full foreign ownership in all sectors which the signatories do not take exception.[58] Meanwhile, Mode 4 liberalisation is confined to skilled labour mobility through the implementation of MRAs. As the discussion in section on Skilled Labour Mobility will

[55]See ASEAN Secretariat (2008, 10).

[56]The full text of AFAS is accessible at: http://cil.nus.edu.sg/rp/pdf/1995%20ASEAN%20Framework% 20Agreement%20on%20Services-pdf.pdf.

[57]The AEC Blueprint allows for foreign (ASEAN) equity participation of not less than 51% by 2008 and 70% by 2010 for air transport, e-ASEAN, healthcare and tourism; not less than 49% by 2008, 51% by 2010 and 70% by 2013 for logistic services; and not less than 49% by 2008, 51% by 2010 and 70% by 2015. See ASEAN Secretariat (2008, 11).

[58]TPP Parties adopt a "negative-list" basis, meaning that their markets are fully open to foreign investors, except where they have taken an exception (non-conforming measure) in one of two country-specific annexes: (1) current measures on which a Party accepts an obligation not to make its measures more restrictive in the future and to bind any future liberalization, and (2) measures and policies on which a Party retains full discretion in the future. See https://ustr.gov/about-us/policy-offices/press-office/press-releases/2015/october/summary-trans-pacific-partnership.

illustrate, the MRAs would only have limited impacts on labour mobility and therefore do not contribute much to services liberalisation.

Secondly, flexibilities provided in the AEC blueprint can hamper the integration effort. In particular, the ASEAN-minus-x formula applied in negotiating and implementing liberalisation initiatives allows members to undertake commitments according to their readiness. However, there is neither objective assessment of whether a member is "ready," nor clear requirements for the excluded members to eventually meet the AEC targets within a particular time period. Thus, the vaguely worded formula leaves much room for discretion. Furthermore, in financial services, the blueprint does not specify the scope for liberalisation as the sector is deemed sensitive. The blueprint further states that liberalisation of financial services can only take place "with due respect for national policy objectives and the level of economic and financial sector development of the individual members" in accordance with the ASEAN-minus-x formula. As a result, member states can carve out sub-sectors which they have no desire to liberalise.[59]

Meanwhile, existing goals stated in the AEC blueprint have not always been met. For example, ASEAN members have difficulties in achieving the targets for foreign equity participation limits, modest as these targets are. Air transport, a priority sector no less, is the case in point. As stipulated in the blueprint, the limit on foreign ownership in the air transport sector should have been increased to 51% by 2008 and 70% by 2010. However, according to the ASEAN Services Integration Report jointly published by the ASEAN Secretariat and the World Bank in 2015, except for Singapore, the maximum foreign ownership allowed in ASEAN countries is only 49% (see Table 1.6).

Table 1.6: Foreign Ownership Allowed in the Air Transport Sector in ASEAN Countries (%)

Country	Foreign Ownership Allowed in a Subsidiary	Foreign Ownership Allowed in Acquisition of a Local Company	Foreign Ownership Allowed in Acquisition of a Local State-Owned Company
Cambodia	49	49	49
Indonesia	49	49	49
Laos	49	49	49
Malaysia	49	49	49
Myanmar	0	49	0
Philippines	40	40	0
Singapore	100	100	100
Thailand	49	49	49
Vietnam	49	49	49

Note: Data for Brunei is not available.
Source: ASEAN Secretariat and World Bank (2015).

[59]See Nikomborirak and Jitdumrong (2013) and Menon and Melendez (2015).

Figure 1.4: Restrictiveness of Services Trade Policies in Cambodia, Indonesia, Malaysia, Philippines, Thailand and Vietnam, 2008 and 2012

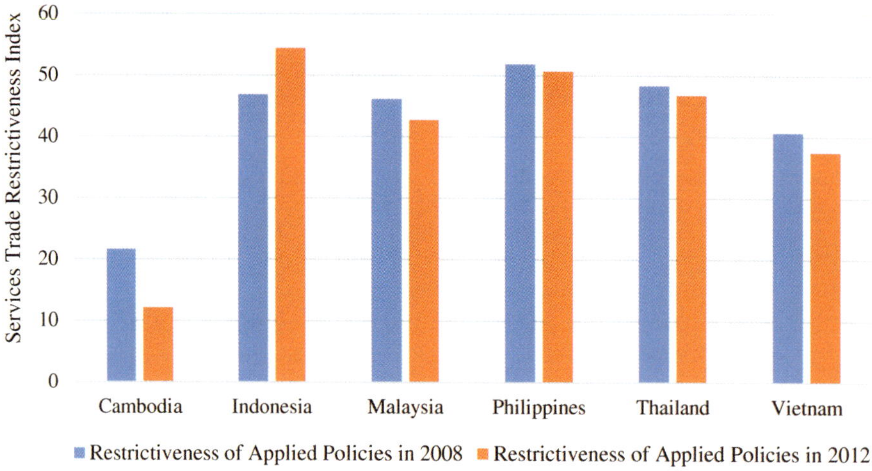

Source: ASEAN Secretariat and World Bank (2015, 64).

The World Bank's Services Trade Restrictiveness Index (STRI) allows us to assess the progress of services liberalisation in ASEAN. Constructed based on inputs from policy surveys, the STRI can be used to measure the restrictiveness of a particular set of services trade policy or liberalisation commitments. The higher the index value, the less liberal a policy is. To date, the World Bank has conducted two rounds of policy surveys in individual ASEAN member states in 2008 and 2012. The 2008 round covered Cambodia, Indonesia, Malaysia, Philippines, Thailand and Vietnam, while the 2012 round included all ASEAN countries except for Brunei.

Overall the STRI data indicates that modest progress has been made in the liberalisation of services in the regional grouping. As shown Figure 1.4, the respective STRIs reflecting services trade policies in Cambodia, Malaysia, Philippines, Thailand and Vietnam dropped in values between 2008 and 2012. This means that on average, policies applied in these countries have become more liberal over the period. Cambodia made the most significant progress while the other countries only improved marginally. In contrast, Indonesia's index increased by more than seven points due to restrictive policies in three service modes, including in the retail sector.[60]

Despite the signs of progress, the data also shows that ASEAN members are still some way from achieving the liberalisation targets set out in the AEC blueprint. The World Bank also constructed an STRI reflecting how liberal the AEC goals are, enabling direct comparison with ASEAN members' actual policies. As Figure 1.5 demonstrates, the overall STRIs for applied policies in most ASEAN members in 2012 were higher than the index value for the goals in the blueprint. Thus, on average, services trade policies applied in the region are generally more restrictive than the AEC Blueprint targets. Only Cambodia and Singapore seem to apply policies which are more liberal than the stipulated liberalisation commitments.

[60]See ASEAN Secretariat and World Bank (2015, 64).

Figure 1.5: Restrictiveness of Services Trade Policies in Selected ASEAN Economies in 2012 and Restrictiveness of the AEC Blueprint Services Liberalisation Targets

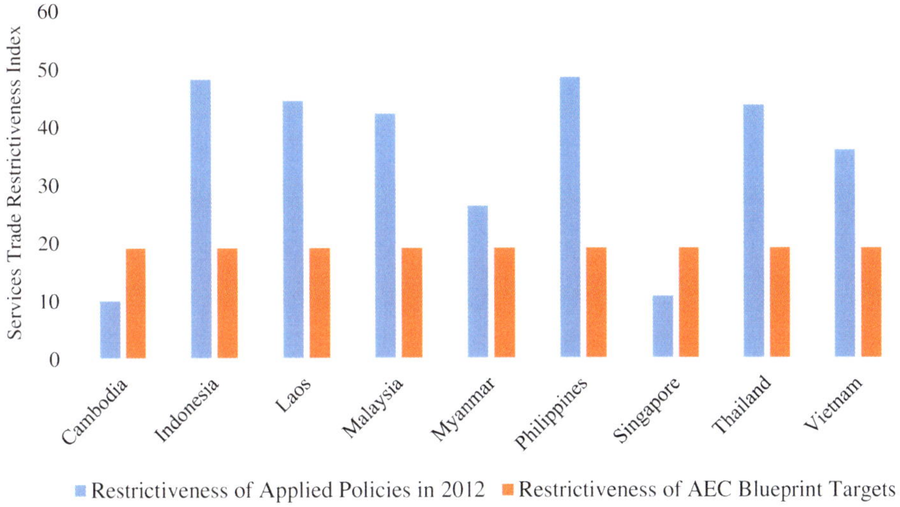

Source: ASEAN Secretariat and World Bank (2015, 76).

Skilled Labour Mobility

In tandem with liberalising services trade, ASEAN members have also committed to facilitate cross-border movement of skilled labour. However, the aspiration of "free flow of skilled labour" as stated in the blueprint does not mean unrestricted skilled labour migration. Instead, the blueprint only seeks to allow for "managed mobility" and to "facilitate the movement of natural persons engaged in trade in goods, services and investments." On balance, it is likely that we will only observe constrained cross-border movement of skilled professionals in the region.

Modest progress has been made with the signing of the ASEAN Movement of Natural Persons (MNP) Agreement in 2012 to facilitate temporary cross-border movements and stays of business visitors, intra-corporate transferees and contractual service suppliers in the services sectors. The MNP agreement is hailed as the first of its kind in the region and a key step toward enhanced intra-regional skilled labour mobility. In particular, it enables ASEAN business visitors to enjoy similar mobility opportunities as non-ASEAN citizens of countries which have signed provisions on movement of natural persons with ASEAN members.[61] However, the number of commitments pledged in the MNP agreement varies across ASEAN countries and, perhaps most strikingly, non-services sectors are excluded.[62]

As noted earlier, the main policy instruments for improving skilled labour mobility are the MRAs. In total, ASEAN members have concluded MRAs for eight professions, namely engineers, nurses, architects, surveyors, medical practitioners, dental practitioners, accountants and

[61] See ASEAN Secretariat and World Bank (2013, 112).

[62] See Fukunaga and Ishido (2015).

tourism professionals.[63] These MRAs differ significantly in their approaches. In particular, the MRAs on accountancy and surveying are mere framework MRAs aimed at facilitating further negotiations, while the MRAs on nursing, medical and dental practitioners explicitly focus on the exchange of information and best practices. Only the MRAs on engineering and architectural services provide for registration of recognised professionals at the regional level. These differences can be attributed to the political economy relations specific to each profession. Specifically, the national position of each ASEAN country in negotiating MRAs varies across professional services depending on the stances of domestic interests in each sector and the extent to which these groups are able to influence the negotiations of the MRAs.[64] This leads to the various approaches adopted in the different MRAs as observed.

The impacts of the MRAs on skilled labour mobility are likely to be limited. First, the MRAs which have been concluded only affect a fraction of the workforce in ASEAN. According to a report jointly published by the International Labour Organisation and the ADB in 2013, seven of the professions covered under the MRAs (tourism professionals excluded) jointly account for only between 0.3% and 1.4% of total employment in any country among Cambodia, Indonesia, Laos, Philippines, Thailand and Vietnam.[65] More importantly, professional services in general are still subject to domestic regulations, many of which may still impede skilled labour mobility despite the conclusion of the MRAs. Notable restrictive regulations include procedures for work visas and employment passes, numerical caps on foreign professionals, legal provisions to reserve jobs for nationals, language proficiency requirements and licensing regulations of professional bodies.[66] For instance, in order for foreign nurses to practice in Thailand, they have to pass the licensure exam which is conducted in Thai language.[67] Last but not least, differences in language, culture and religion as well as domestic social sensitivities to the influx of foreign skilled labour may constitute practical barriers beyond any legal provisions a country can impose. Thus, unless individual member state takes the initiative to ease restrictive professional regulations and works to build social support for liberalising skilled labour mobility, it is difficult to imagine how "free flow of skilled labour" will be achieved.

The ASEAN Single Aviation Market (ASAM)

The establishment of the ASAM, also known as the Open Skies policy, is a key commitment under the AEC infrastructure development agenda which has received a lot of attention in recent years. Slated to be accomplished by 2015, the ratification of the legal protocols to establish ASAM has not yet been completed. In particular, Indonesia has been holding out on key provisions to liberalise its domestic air travel market.

The vision of Open Skies in ASEAN was articulated in the Action Plan for ASEAN Air Transport Integration and Liberalisation and the accompanying Roadmap for Integration of Air

[63] See ASEAN Secretariat (2015b, 33–34).

[64] See Jones (2015).

[65] See International Labour Organisation and Asian Development Bank (2014, 91).

[66] See Chia (2013).

[67] See Basu (2016, 56).

Travel Sector (RIATS), which were both adopted in 2004. Together, these two documents lay down the policy objectives and set 2015 as the target date for achieving an effective "open skies" regime among ASEAN countries. The RIATS also stipulates that implementation is to take place based on the ASEAN-minus-x formula.[68]

Subsequently, commitments and strategic actions stipulated in RIATS were formalised into three legal agreements for ASEAN members' acceptance, namely the Multilateral Agreement on Air Services (MAAS), the Multilateral Agreement for Full Liberalisation of Passenger Air Services (MAFLPAS) and the Multilateral Agreement for Full Liberalisation of Air Freight Services (MAFLAFS). Each of the three agreements has its own set of Implementing Protocols, which set out specific commitments ASEAN members have to undertake. The stipulated provisions will only have binding effects on member states who sign and ratify the relevant protocols. In other words, the ASAM can only be considered fully established when ten ASEAN countries have signed and ratified all legal instruments, which include the three agreements as well as their Implementing Protocols.

While the majority of ASEAN members have accepted the full scope of the "Open Skies" liberalisation commitments, the formation of the ASAM is currently significantly hampered by the lack of full participation from Indonesia. Home to more than 200 million people, Indonesia has the largest air travel market in the region. Yet, it has not ratified MAFLPAS Protocols 1 and 2, which provide for unlimited third, fourth and fifth freedom traffic rights for air passenger services between all ASEAN cities. [69] In essence, this means that Indonesia has not agreed to liberalise international air passenger services in all of its cities. Instead, Indonesia has only committed to opening up Jakarta to regional carriers as it ratified MAAS Protocols 5 and 6, which provide for unlimited third, fourth and fifth freedom traffic rights between ASEAN capital cities, in 2014.[70]

Indonesia's position can be attributed to active lobbying by its leading airlines to protect their international operations against regional rivals. Their concern lies with strong competitors in other ASEAN members, particularly Singapore, Malaysia and Thailand. Furthermore, these carriers perceive a systematic imbalance for exchanging traffic flights under the current framework. In particular, Indonesia, as a huge archipelago, has potentially hundreds of points to offer international aviation while other countries have fewer. As a result, through their industry group, the Indonesian National Air Carriers Association, Indonesian airlines have opposed efforts to liberalise the ASEAN air travel market.[71]

[68]See ASEAN Secretariat (2015b, 65). The full text of RIATS is available at http://www.asean.org/?static_post=appendix-i-roadmap-for-integration-of-air-travel-sector.

[69]In aviation nomenclature, the third freedom refers to the right of an airline to fly freely from its home country to another country while carrying passengers, cargo and baggage for profit. The fourth refers to the reverse journey with the same rights. In both directions, the airline can stop over at a third country to pick up new passengers before resuming its services. This constitutes the right to carry revenue traffic between foreign countries as part of services connecting the airline's own country, or the fifth freedom. See Tan (2013, 2).

[70]As of March 2016, Indonesia has ratified MAAS, MAFLAFS and all the related Implementing Protocols for these two agreements. Indonesia's Instruments of Ratification are deposited with the ASEAN Secretariat and can be found in the list of legal instruments available at http://agreement.asean.org/.

[71]See Tan (2013).

The current struggle to establish the ASAM demonstrates a salient problem with the ASEAN-minus-x formula. Applying the formula without any caveats on future commitments of the excluded members risks prolonged delay or even exclusion of some members from an initiative altogether. This may be detrimental if, as in the case of ASAM, participation of the excluded members is essential for meaningful integration to be achieved. In this case, full participation by Indonesia as the largest ASEAN member state is vital to the meaningful formation of the ASAM.

However, pundits and policy-makers are increasingly optimistic that Indonesia will ratify the two outstanding Protocols. For instance, aviation law professor Alan Tan from the NUS, who has been following the developments closely, said that he looks forward to "Indonesia accepting this final piece of the puzzle soon."[72] Meanwhile, Singapore's Senior Minister of State for Transport, Ms Josephine Teo, stated that she is encouraged by the tone and stance that Indonesian President Joko Widodo (popularly known as President Jokowi) has adopted on the issue.[73] Aside from President Jokowi's positive attitude, there are other factors which may provide the impetus for further progress of the ASAM initiative. First and foremost is the growing confidence of Indonesian carriers such as Lion Air and Garuda. As they expand their services and increase their competitiveness, they may eventually see less of a need to resist further liberalisation of air passenger services in ASEAN, especially if they encounter limitations to their expansion in ASEAN. Secondly, provincial governments, tourism authorities and the business community in Indonesia have called for greater direct access into regional cities in Indonesia. The pressure created from these domestic groups may help create the momentum to expedite Indonesia's full participation in the ASAM.[74]

1.4.3 Can the ASEAN Economic Community be Completed by 2016?

As seen in the foregoing discussion, while the AEC was officially launched on 31st December 2015, several key commitments under the AEC blueprint have not been fulfilled and strategic goals remain unaccomplished. One initiative which may see real progress in 2016 is the establishment of the ASAM. Growing confidence of Indonesian airlines, domestic pressure for greater direct access to cities and the positive attitude of policy-makers may finally bring Indonesia, the largest air travel market in ASEAN, on board, paving the way for greater intra-regional connectivity.

Nonetheless, it is unlikely for all outstanding deliverables to be completed by the end of 2016. Firstly, elimination of all NTBs by the end of 2016 seems unlikely at best. As noted earlier, very limited progress has been made on the NTMs database and, more importantly, the current "notify-negotiate-eliminate" approach suffers from serious shortcomings. In particular, governments in ASEAN member states may have the incentive not to fully disclose their NTMs to facilitate NTBs elimination. Even if there is no such incentive misalignment, some member states may simply lack the institutional capacity to organise effective coordinating mechanisms to collect information about all NTMs which their various government agencies employ.

---

[72] See Kaur (2016a).

[73] See Kaur (2016b).

[74] See Tan (2013).

Secondly, achieving the target of 70% of foreign ownership in all services sectors by the end of 2016 also seems unrealistic. The lack of progress even in priority integration sectors such as air transport casts doubts on ASEAN countries' commitments to this particular agenda. Finally, full operationalisation of the ASW is also likely to remain outstanding. The ongoing full-fledged pilot of the ASW only involves a handful of ASEAN countries and the legal framework for its eventual implementation has yet been ratified by all members. Most importantly, the less developed members in the bloc, namely Cambodia, Laos and Myanmar, have not even operationalised their NSWs.

The AEC blueprint aside, the bigger picture of ASEAN economic integration has to be thoroughly assessed. If all AEC measures are accomplished, can one claim that the aspiration for an "economic community" is fully realised? The answer would be unlikely. For a start, full implementation of the blueprint may be undermined by various behind-the-border issues, such that completing the AEC measures does not necessarily result in the desired integration outcomes. This can be seen most clearly in the skilled labour mobility agenda. While ASEAN members have provided for mutual recognition of professional qualifications through the signing of MRAs, this has not fulfilled the desired objective of "free flow of skilled labour" as stipulated in the AEC blueprint since restrictive domestic regulations pertaining to foreign labour are still prevalent. Furthermore, cross-border movement of skilled labour may also be impeded by the diversity in languages and cultures in the region as well as social sensitivities to the influx of foreign labour in certain member states.

Meanwhile, the AEC blueprint itself leaves much to be desired. Its scope in services trade liberalisation is particularly lacking. Meeting the targets will thus only result in limited liberalisation of Mode 3 restrictions[75] as the blueprint does not mandate ASEAN members to allow for 100% foreign equity ownership. This is modest as compared to the TPP, which provides for full foreign ownership. Additionally, current initiatives to improve skilled labour mobility only affects a small fraction of total workers in ASEAN. At the same time, many flexibilities allowed under the blueprint, notably the ASEAN-minus-x formula, are vaguely worded and thus leave much room for discretion. As the struggle to establish the ASAM demonstrates, adopting the ASEAN-minus-x formula without any caveats on future commitments of the excluded members, which is the current practice, risks excluding the most important countries for an initiative to work. In this light, a more far-reaching integration agenda is required if a true economic community is to be achieved in ASEAN.

More fundamentally, the formation of the AEC and economic integration in ASEAN in general have largely been a top-down state-led initiative. Awareness of the AEC and the utilisation rates of the CEPT tariffs among the business community in ASEAN are particularly low. As noted by many pundits, since the first milestone in 1976, nine years after ASEAN's inception, economic cooperation in ASEAN has become a form of diplomacy carried out in foreign ministries in consultation with commerce or trade ministries.[76] The impetus for regional integration

[75]These refer to restrictions on commercial presence of a service provider in a foreign country.

[76]See Basu Das (2016, 64).

has thus far come from governments of the ASEAN member states and sometimes from multinational corporations (MNCs)[77] rather than indigenous businesses in the region.

For the above reasons, the AEC cannot be considered fully realised even if all outstanding measures are completed by the end of 2016, which is unlikely. The official launch of the AEC in December 2015 should therefore be seen as a stepping stone in the regional trajectory toward deeper economic integration and equitable and inclusive economic development. The realisation of an economic community will not end, but remain a work-in-progress, even when 100% of the measures have been fulfilled. In fact, as recognised by the ASEAN Secretariat, policy-makers and scholars in the region, community building in ASEAN is an ongoing and dynamic journey in which 2015 is a significant milestone.[78]

Furthermore, community building in ASEAN is not solely about the AEC. ASEAN is only a wholesome community if the people in individual ASEAN countries share a common ASEAN identity, and they feel that their and their nation's interests are being best served by being in ASEAN. To progress toward this end, it is imperative that the AEC advance in conjunction with the ASEAN Socio-Cultural Community and the ASEAN Political-Security Community. Without a shared identity and community spirit, opportunities created by an integrated region cannot be fully capitalised. Meanwhile, ASEAN can only flourish economically in an environment of peace and stability, in light of the ongoing geopolitical tussles between traditional and emerging major powers, especially the US and China, and the imminent threat of terrorism, which has amplified since the bomb attack in Jakarta on 14th January 2016.

1.5 A POST-2015 VISION FOR ASEAN

As evident in the foregoing assessment, while the AEC was formally launched in 31st December 2015, much remains to be done for ASEAN to become a truly globally integrated, equitable and competitive "single market and production base." On the one hand, several key deliverables under the AEC agenda have not been completed by the 2015 deadline, namely (i) NTBs to trade, (ii) trade facilitation through the establishment of the ASW, (iii) liberalisation of trade in services, (iv) free movement of skilled labour, and (v) the formation of the ASAM, or the Open Skies policy. On the other hand, as recognised by the ASEAN Secretariat in its own assessment of the progress of the AEC published in November 2015, regional community building is an "ongoing and dynamic process" in which the establishment of the AEC in 2015 is a milestone rather than the end.[79]

In fact, ASEAN leaders have started to develop the post-2015 vision for the AEC, presently termed "AEC 2025," which builds upon what has been achieved by the 2015 milestone. In particular, AEC 2025 will focus on: further reduction of NTBs, trade facilitation through the full implementation the ASW, liberalisation of the services sector and deeper financial integration.[80]

---

[77] See Postigo (2013, 12).

[78] See, for instance, ASEAN Secretariat (2015a), Lim (2015) and Ong (2014).

[79] See ASEAN Secretariat (2015a, 30).

[80] See ASEAN Secretariat (2015a, 30–32).

Of central importance in the AEC 2025 vision are cultivating a collective ASEAN identity and preserving ASEAN centrality in the ever-evolving global and regional trade architecture.

The issue of ASEAN centrality has become more salient than ever with the proliferation of bilateral FTAs since 2000 when little progress had been made with respect to multilateral liberalisation, particularly with the Doha Round of the WTO on the verge of faltering. By 2010, Singapore had concluded and implemented 18 bilateral FTAs, followed by Thailand with 11, Brunei, Malaysia and Laos with 8, Indonesia and Vietnam with 7 and Cambodia with 6.[81] While ASEAN members' active participation in bilateral FTAs, especially Singapore and Thailand, did not pose much threat to the viability of ASEAN, it did call into question whether ASEAN was still relevant, and to what extent membership in the regional bloc aided economic development in individual member economies.

The initiation of FTA negotiations by ASEAN countries is typically driven by three principal factors. First, there has been a growing consensus among policy-makers that FTAs, if designed with a wider scope than stipulated under the WTO framework, can support trade and FDI inflows through reduction of cross-border impediments and harmonisation of rules, standards and procedures. To that extent, FTAs are widely regarded as crucial in enhancing countries' presence in global and regional value chains and production networks. Second, in the face of the virtually stalled WTO process, the successful launch of the EU and the North American Free Trade Agreement (NAFTA) and the proliferation of bilateral FTAs outside the region, ASEAN members increasingly found it hard to gain access to sizeable developed markets without jumping on the bandwagon of signing bilateral FTAs with WTO-Plus commitments. As such, ASEAN leaders pursue an FTA policy in order to avoid trade diversion caused by formation of trade blocs elsewhere in the world. Furthermore, forging closer economic alliances with major powers would also enhance a country's political and security standing. Third, some governments view FTAs as an important catalyst for domestic economic reforms or restructuring.

Recognising the need for both deeper economic integration with extra-regional powers and positioning itself as a "hub" in such efforts, ASEAN as a group has been pursuing FTA negotiations with its major trading partners, although the pace and scope have been lagging most bilateral FTAs signed by the individual member states. Starting with China in 2002, it was not until 2006 that ASEAN signed the second FTA with South Korea, which was followed by a Comprehensive Economic Partnership (CEP) with Japan in 2008[82] and two other FTAs with Australia and New Zealand and India both in 2009. These ASEAN+1 FTAs vary in both negotiation approach and coverage. For instance, the ASEAN-Australia and New Zealand FTA and ASEAN-Japan CEP are implemented on a single undertaking basis, while the remaining

---

[81] See Hill and Menon (2010, 20).

[82] The Japanese government promotes economic partnership agreements which go beyond typical components of FTAs to include elements such as: development of logistics system and infrastructure, simplification of customs clearance, movement of skilled workers and provision of training programmes, standardisation of investment rules and regulations, and cooperation in human resource development, environmental and energy conservation. See Kawai and Wignaraja (2008, 4).

three FTAs follow a sequential approach.[83] In terms of coverage, vast differences typically prevail in ROOs,[84] the extent of tariff elimination and reduction of NTBs, timeframe for liberalisation as well as harmonisation of policies and rules such as intellectual property. Tables 1.7 and 1.8, for instance, show tariff elimination coverage (as a percentage of total tariff lines) by ASEAN members and liberalisation timeframes for ASEAN-6, CLMV and partner countries, under ASEAN+1 FTAs.

Table 1.7: Tariff Elimination Coverage under ASEAN+1 Free Trade Agreements (Percentage of Total Tariff Lines)

	ASEAN-China	ASEAN-South Korea	ASEAN-Japan	ASEAN-Australia and New Zealand	ASEAN-India
ASEAN					
Brunei	98.3	99.2	97.7	99.2	85.3
Cambodia	89.9	97.1	85.7	89.1	88.4
Indonesia	92.3	91.2	91.2	93.7	48.7
Laos	97.6	90.0	86.9	91.9	80.1
Malaysia	93.4	95.5	94.1	97.4	79.8
Myanmar	94.5	92.2	85.2	88.1	76.6
Philippines	93.0	99.0	97.4	95.1	80.9
Singapore	100	100	100	100	100
Thailand	93.5	95.6	96.8	98.9	78.1
Vietnam	NA	89.4	94.4	94.8	79.5
Partner countries					
China	94.1				
South Korea		90.5			
Japan			91.9		
Australia				100	
New Zealand				100	
India					78.8
Average	94.7	94.5	92.8	95.7	79.6

Note: Data on Vietnam under the ASEAN-China FTA are not available.
Source: Fukunaga and Isono (2013, 8).

[83] Single undertaking means negotiations proceed simultaneously, and virtually every item of the negotiation is part of a whole and indivisible package and cannot be agreed separately. In a sequential negotiation process, members go through several rounds with subsets of countries participating at different stages.

[84] Rules of origin are used to determine whether a product originates from the members of a FTA and is thus eligible for preferential benefits.

Table 1.8: Liberalisation Timeframe for ASEAN+1 Free Trade Agreements

	ASEAN-6	CLMV	Partner(s)
ASEAN-China	2007	2010	2007
ASEAN-South Korea	2010	2015 (flexibility allowed)	2008
ASEAN-Japan	2012	2018	2010
ASEAN-Australia and New Zealand	2017	2017	2010
ASEAN-India	2011 (excluding Philippines)	2016 (including Philippines)	2011

Source: Kawai and Wignaraja (2008, 24).

Notwithstanding ASEAN's collective efforts in entering into FTAs with key trading partners, a "domino effect" had already been in operation as member economies, driven by peer pressure and fear of trade discrimination, still seek to hammer out their own trade deals with willing countries, which are typically more comprehensive and of higher standard than ASEAN+1 agreements. This "noodle bowl" of FTAs threatens to result in economic fragmentation rather than integration in ASEAN as well as the broader Asia-Pacific region. Furthermore, as provisions under these FTAs are both overlapping and inconsistent, notably with respect to ROOs, the proliferation of FTAs may increase trade costs instead of curbing them, undermining the trade creation effect associated with such agreements. In fact, despite the myriad of FTAs in force in the region, utilisation rates have been low. In a survey of 841 export-oriented manufacturing firms in Asia conducted by the Asian Development Bank (ADB) Institute, it was found that the share of firms making use of FTAs in Thailand is 25% as compared to 45% in China. Singapore's utilisation rate of FTAs is even lower at 17%. Companies cited lack of information, low margin of preference, prevalence of NTBs and exclusion lists, multiple ROOs and high administrative costs as primary reasons for not taking advantage of FTAs.[85]

In light of the administrative complexity and increased transaction costs arising from the "noodle bowl" of bilateral FTAs, several proposals for consolidating trade pacts under wider regional arrangements have emerged. One such instance was the proposed Comprehensive Economic Partnership for East Asia (CEPEA) following the formation of the EAS in December 2005. Nonetheless, this proposal failed to gain traction at the time. Meanwhile in 2006, four small, very open APEC economies, namely Brunei, Chile, New Zealand and Singapore (the P4) entered into the Trans-Pacific Strategic Economic Partnership Agreement,[86] paving the way for negotiations to establish the TPP among a larger set of APEC economies which began in 2007. The US joined the TPP negotiations in 2008 and since then, the Obama administration had placed special emphasis on the successful conclusion of the agreement as part of its "pivot to Asia" policy. The TPP is thus widely regarded as a US-led initiative.

[85] See Kawai and Wignaraja (2011).

[86] Prior to the establishment of the Trans-Pacific Strategic Economic Partnership among the P4 economies, Singapore and New Zealand had signed a "Closer Economic Partnership" agreement in 2001.

1.5.1 Trans-Pacific Partnership (TPP)

With 12 participating economies representing 37% of global GDP and 28% of world trade,[87] TPP is currently the largest trade pact since the 1994 Uruguay Round which led to the establishment of the WTO. The negotiations were concluded on 5th October 2015, and the agreement was signed on 4th February 2016. Following the release of the TPP text on 5th November 2015, the agreement is now pending ratification by the respective legislatures of the 12 member economies, which is expected to be completed within two years. South Korea, Indonesia, Thailand and the Philippines have expressed their keen interest in participating, while China is also contemplating joining the US-led trade pact at an appropriate time.[88]

The TPP comprises 30 chapters with coverage far beyond traditional FTAs, addressing multiple so-called next-generation issues and thus setting a new standard for global trade.[89] These include: reduction or elimination of tariffs and NTBs across substantially all trade in goods and services, customs and trade facilitation, sanitary and phytosanitary measures, technical barriers to trade, trade remedies, investment, government procurement, competition policy, state-owned enterprises, intellectual property, labour, environment, regulatory coherence, dispute settlement, capacity building and SME development. The TPP also has a single set of ROOs that applies to all 12 members, which stipulates that inputs from all TPP members are treated the same way if used to produce a product in any TPP member.

According to the latest CGE estimates conducted by Petri and Plummer (2016) (see Table 1.9), annual income gains generated by the TPP by 2030 are projected to reach US$429 billion for the world. The US stands to gain the most among the 12 members with US$131 billion per annum, or 0.5% of baseline GDP. Sizeable gains are also expected for Malaysia and Vietnam at US$52 billion and US$41 billion, equivalent to 7.6% and 8.1%, respectively, of baseline GDP. Annual exports are estimated to increase by US$1,025 billion, or 11.5%, over the 2030 baseline, for all TPP countries, with the US, Japan, Vietnam and Malaysia leading the list. Inward investment stocks in all TPP members are projected to expand by US$446 billion or 3.5% over the baseline scenario.

While the TPP is yet to be ratified and implemented, its impacts are expected to extend well beyond tangible welfare gains. First, the TPP may help to revive the momentum towards global integration and thus reshape the global trade architecture. Particularly, many pundits believe the TPP provides the much needed impetus to rejuvenate the presently stalled WTO negotiations. Second, the fact that only four ASEAN economies are members of the TPP may pressure the rest of the bloc to step up their integration efforts, such as the ASEAN+6 RCEP. Third, for Asian countries which are apprehensive about China's growing economic and geopolitical clout in the region,

---

[87]The 12 member economies are: Brunei, Chile, New Zealand, Singapore, Australia, Canada, Japan, Malaysia, Mexico, Peru, the US and Vietnam.

[88]See Kyodo (2015), Agence France-Presse (2015), Parameswaran (2015a), The White House (2012) and Blanchard (2015).

[89]The full text of TPP is accessible from: https://ustr.gov/trade-agreements/free-trade-agreements/trans-pacific-partnership/tpp-full-text.

Table 1.9: Real Income Effects of the Trans-Pacific Partnership (TPP)

Country	Baseline (Billions of 2015 US Dollars)		Change with TPP (Billions of 2015 US Dollars)		Percentage Change from Baseline	
	2020	2030	2020	2030	2020	2030
Australia	1,986	2,590	1	15	0.0	0.6
Brunei	24	31	0	2	1.1	5.9
Canada	2,227	2,717	8	37	0.4	1.3
Chile	329	463	0	4	0.1	0.9
Japan	4,462	4,924	39	125	0.9	2.5
Malaysia	444	675	7	52	1.6	7.6
Mexico	1,598	2,169	3	22	0.2	1.0
New Zealand	217	264	1	6	0.5	2.2
Peru	287	442	1	11	0.4	2.6
Singapore	380	485	2	19	0.5	3.9
United States	20,736	25,754	29	131	0.1	0.5
Vietnam	281	497	7	41	2.3	8.1

Source: Petri and Plummer (2016, 20).

the US's participation in the TPP is, to some extent, comforting.[90] This is also evident in other regionalism efforts in Asia as it looks to extra-regional powers. The EAS, for instance, includes Australia, Zealand, the US and Russia, instead of limiting itself geographically to only East Asia.

In the longer term, the impacts of the TPP may evolve if other economies seek to join. The TPP's "Final Provisions" chapter specifies that it is open to accession by APEC members and other economies as agreed by the current members. If this open accession clause is realised, the TPP could become the building block toward a much larger trade pact that brings together the remaining ASEAN and APEC members under one umbrella.

1.5.2 Regional Comprehensive Economic Partnership (RCEP)

RCEP was introduced by ASEAN at the 19th Summit in Bali in November 2011 with a view to untangling the "noodle bowl" of existing ASEAN+1 FTAs as well as deepening economic integration between ASEAN and its six FTA partners (China, Japan, South Korea, India, Australia and New Zealand). During the November 2012 Summit, the leaders of ASEAN+6 endorsed the guiding principles for negotiating RCEP, which are to achieve a "modern, comprehensive, high-quality and mutually beneficial"[91] partnership agreement for all member

[90] See Perlez (2015).

[91] The guiding principles for negotiating RCEP are accessible from: https://www.mfat.govt.nz/assets/_securedfiles/FTAs-in-negotiations/RCEP/Guiding-Principles-and-Objectives-for-Negotiating-the-Regional-Comprehensive-Economic-Partnership.pdf.

countries. The RCEP agreement is therefore designed to facilitate broader and deeper engagement among participating countries than the prevailing ASEAN+1 FTAs, aiming in particular to be more comprehensive but less stringent in terms of requirements and commitments than the TPP as it recognises the diverse economic circumstances of member countries. The RCEP's coverage includes trade in goods, trade in services, investment, intellectual property, competition, dispute settlement and economic and technical cooperation. The last component, economic and technical cooperation, reflects the development orientation of the RCEP, which is not part of the TPP. Open accession is also adopted, which means membership can be expanded subsequently to those that sign bilateral and multilateral FTAs with ASEAN.

According to an empirical exercise conducted by the ADB, the ASEAN+6 FTA scenario, or the RCEP, offers a total economic gain of US$260 billion to world income.[92] The ten ASEAN economies are projected to experience a US$67 million income gain, with Thailand, Vietnam and Malaysia standing to benefit the most in terms of percentage change over the 2017 baseline scenario from participating in the RCEP (see Table 1.10). Among the six partner countries, Japan

Table 1.10: Real Income Effects of the Regional Comprehensive Economic Partnership (RCEP)

Country	**Compared to 2017 baseline**	
	Value (US$ million)	**Percentage change**
ASEAN		
Cambodia	109	1.21
Indonesia	8,588	2.86
Malaysia	11,869	6.33
Philippines	3,431	2.85
Singapore	9,002	5.43
Thailand	28,346	12.84
Vietnam	5,490	7.63
Others	370	0.33
Partner countries		
Japan	77,137	1.59
South Korea	51,351	6.43
China	43,598	1.33
Australia	22,546	3.91
New Zealand	4,136	5.24
India	19,270	2.42

Source: Kawai and Wignaraja (2008, 37–38).

[92] See Kawai and Wignaraja (2008, 35).

is expected to reap the largest gain in absolute term at US$77 billion while South Korea is projected to benefit the most in terms of percentage change from 2017 baseline income at 6.43%.

Apart from potential trade and economic benefits, the RCEP, as led by ASEAN, is expected to entrench ASEAN centrality in the wider Asia-Pacific regional architecture. The notion of "centrality" places ASEAN, instead of the major powers such as the US, Japan or China, at the "hub" of Asia-Pacific, driving integration initiatives in the region as well as harnessing them toward strengthening economic development in the ten ASEAN member states. Such effort has been challenged by the rapid pace of intra- and extra-regional economic cooperation arrangements proliferating in Asia-Pacific, particularly with the signing of the US-led TPP in February 2016. Once concluded, the RCEP would demonstrate ASEAN's unity and capability in forging common positions, thus enhancing ASEAN's geopolitical position as well as bargaining power *vis-à-vis* major powers and other regional groupings.

Nonetheless, RCEP negotiations have been met with many economic and political obstacles. First, while ASEAN as a bloc has entered into FTAs with each of the six extra-regional partners, these partner countries may not have existing comprehensive trade agreements with one another. Therefore, specific provisions under the RCEP could only be decided upon the conclusion of trade negotiations among the Plus Six countries. In this regard, the China–Japan–South Korea tripartite FTA (CJK-FTA) would have a significant impact as the three countries account for a dominant share of GDP and trade in the trading bloc. However, political tensions over historical conflicts and territorial disputes over the Senkaku or Diaoyu islands may hamper the negotiation of the CJK-FTA, which in turn would impede the conclusion of the RCEP.

Second, harmonising provisions under the five ASEAN+1 FTAs is particularly challenging in light of the wide gaps in commitments under the ASEAN+India FTA as compared to the remaining four. For instance, the ASEAN+India FTA has the lowest tariff elimination coverage, reaching only 79.6%, since India only commits to 78.8%, far less than the other five partner countries which commit between 90.5% (South Korea) and 100% (Australia and New Zealand) (see Table 1.7). Furthermore, India, while being widely held as a rising economic power, has been conservative in multilateral trade negotiations as evident in both the Uruguay and the Doha Round.[93]

Third, diversity in levels of economic development also poses critical challenges to the RCEP in pursuing harmonisation and consensus. While a flexibility clause has been built into the RCEP framework, which helps to protect disparate national interests and thus resolve deadlocks, it could result in a "race to the bottom." In other words, a "lowest common denominator" outcome may eventually emerge, far from the "modern, comprehensive, high-quality and mutually beneficial FTA" as stipulated in the RCEP's guiding principles.

It has also been argued that the RCEP may be rendered irrelevant in light of the speedy progress of the TPP, which sets much higher requirements for participation and covers a wider

[93]See Srinivasan (2003) and Karmakar (2009) for details on India's hardline positions in the WTO's Uruguay and Doha Round.

range of deep integration issues,[94] or the two trade pacts will coexist but engage in rivalry.[95] This is partly due to the fact that the TPP has been touted as the "only game in town" with "21st century gold standards" as part of the US's active promotion, which seems to have cornered the members of the RCEP. In addition, the TPP currently excludes China, while RCEP includes both China and India, the two emerging economic powers, but excludes the US. China's rhetorics on the TPP up until May 2013 had been cautious as to whether the TPP was designed to exclude and disadvantage China.[96] Furthermore, as only four ASEAN members are part of the TPP, the US-led trade pact may accentuate the bifurcation within ASEAN in terms of preferential access to important extra-regional markets and thus economic benefits, which, to a certain extent, goes against the principle of ASEAN centrality.[97] The RCEP, in contrast, emphasises on economic and technical cooperation with generous consideration of the varied levels of development, which would ensure more equitable distribution of benefits to all ten ASEAN members.

Notwithstanding the differences, it has to be recognised that the TPP and the RCEP are in fact complementary in nature as critical trajectories toward deeper economic integration. Less developed countries which are ready for limited liberalisation would be more comfortable to join the RCEP, while those ready to undertake deeper liberalisation can aim to become TPP members. This is primarily owing to the different approaches that the two trade pacts espouse. In particular, while the TPP adopts a stringent "no gold standard, no deal" approach, the RCEP is more pragmatic and practical in allowing potential members to participate in accordance to their needs for market access and acceptable levels of commitment. This is certainly more appealing to the less developed ASEAN members who are currently not part of, or keen on joining, the TPP. Nevertheless, as South Korea, Indonesia, Thailand and the Philippines have expressed their interests in becoming TPP members,[98] and China is considering the option of participating in the TPP, there will be potentially significant overlapping of membership between the two trade pacts, which would enhance economic integration in the broader Asia-Pacific region. This will also give ASEAN more credence as it establishes itself at the centre of Asian regionalism and regional institution-building.

The complementarity between the TPP and the RCEP has been highlighted by various members of both agreements. When attending the US-ASEAN Ministerial Meeting in August 2012, the United States Trade Representative Ron Kirk said, "[We] think they [the TPP and RCEP] are complementary, not necessarily competitive."[99] At the 2014 Pacific Economic Cooperation Council (PECC) Conference held in Singapore, Mr Lim Hng Kiang, the Singapore's Minister

---

[94] See Basu Das (2014) and Fukunaga (2015).

[95] See Wu and Mealy (2012) and Asian Development Bank (2013).

[96] China conveyed its interest in the TPP in May 2013, much to the surprise of most observers. According to a statement by Mr Shen Danyang, a China's Ministry of Commerce spokesman, published on the Ministry's website, China will analyse "the advantages, disadvantages and the possibility of joining the TPP" — see Bloomberg News (2013) and Lewis (2013).

[97] See Fukunaga (2015).

[98] See Asian Development Bank Institute (2014, 24).

[99] See Wang (2013).

for Trade and Industry, also stressed that the TPP and the RCEP are two key "mutually reinforcing pathways" toward the establishment of the FTAAP that include all APEC economies.[100] China also endorses the same vision as during the same conference, Mr La Yifan, Vice Chairman of China's National Organising Committee for APEC 2014, remarked that "members of the Asia-Pacific community should work together to embark on the FTAAP process."[101] With the RCEP negotiations failing to be completed by end 2015 as originally planned, it is imperative that ASEAN step up its efforts to conclude the RCEP as part of its post-2015 priorities if it is to uphold ASEAN centrality in the ever-evolving global trade architecture.

1.5.3 Implications for Post-2015 ASEAN

While implementing the unfinished AEC 2015 agenda, as discussed in Section 1.4, remains the top and immediate priority for ASEAN in 2016 and possibly beyond, stepping stones for a post-2015 regional trajectory need to also be carefully and firmly laid. Chief among them is the conclusion of the RCEP, which would see to a more integrated ASEAN with major trading partners outside the region. Of crucial importance is that RCEP negotiations should be carried out in due consideration of the potential formation of the FTAAP covering APEC economies, which would further enhance ASEAN's economic cooperation with other Asia-Pacific economies including long-standing and emerging major powers.

Equally imperative is the need for accelerating and deepening domestic reforms in certain ASEAN member states, particularly Indonesia, as the traditional leader of ASEAN, and Vietnam, as the leader of the newer and less developed grouping of CLMV. For Indonesia, its leadership has been largely feeble throughout the process of establishing the AEC. The Jokowi administration, in particular, has been primarily inward-looking, focusing on national economic issues including weak GDP growth and a weakened rupiah, since taking office in 2014. With regard to Vietnam, while the country has surprised most observers by signing on to the high-standard TPP in February 2016, it continues to grapple with numerous challenges such as the ailing state-owned sector and the ineffectual reform toward a full-fledged market economy.

Toward the Free Trade Area of the Asia-Pacific (FTAAP)

The idea of establishing an Asia-Pacific free trade area was first mooted in 2004 by the APEC Business Advisory Council (ABAC) during the 12th APEC Economic Leaders' Meeting, as a way to "accelerate progress toward achievement of the Bogor Goals[102] and full global liberalisation in the WTO" and to minimise "the possible ill effects associated with the increasingly complex web" of FTAs in the APEC region.[103] Nonetheless, as the proposal calls for a legally binding agreement among the APEC members, which goes against APEC's long-standing voluntary,

---

[100] See Lim (2014).

[101] See La (2014).

[102] Adopted by the APEC leaders during a meeting in 1994 in Bogor, Indonesia, the Bogor Goals aim for free and open trade and investment in the Asia-Pacific by 2010 for industrialised economies, and by 2020 for developing economies.

[103] See APEC Business Advisory Council (2004, 12).

non-binding and open regionalism, it did not gain traction among the leaders at the time. Furthermore, the successful conclusion of an Asia–Pacific-wide FTA seemed unlikely given the lack of plurilateral FTAs among major powers, notably the US, Japan and China.

Nonetheless, a decision was made during the 14th APEC Economic Leaders Meeting in 2006 to undertake a feasibility study on FTAAP, which by then had received strong backing from the US, with the US President at the time George W. Bush noting that the idea "deserves serious consideration."[104] The first EAS held in Malaysia in December 2005 had provided an extra boost to the integration dynamics in East Asia, which in turn catalysed APEC's collective rethink about the FTAAP. There had also been regular discussions on establishing an East Asian free trade area initiated by Japan and China with a view to untangling the inefficient "noodle bowl" of FTAs.

The *raison d'être* for the establishment of the FTAAP is four-fold. First, the continued proliferation of bilateral and plurilateral FTAs in the Asia-Pacific is threatening to lead to economic disintegration, particularly between East Asia and the Western hemisphere which have been engaged in erecting discriminatory trade barriers against each other, causing economic as well as political strains for many countries in the region. APEC is the only prevailing institutional mechanism that could spearhead a consolidation of FTAs within the Asia-Pacific through the FTAAP as the modality for achieving the Bogor Goals. With the formation of the FTAAP, all existing and planned FTAs could be brought under one umbrella, thereby eliminating the "noodle bowl" of FTAs and their conflicting ROOs as well as other overlapping and contradictory provisions. It is thus the only initiative that could stop the ongoing "domino effect" of signing FTAs since if all APEC economies are included under one umbrella, it would no longer be necessary for individual economies to pursue bilateral deals in order to protect their position.[105]

Second, the implementation of the FTAAP would bring about sizeable and, more importantly, non-discriminatory, income gains to all APEC members. A CGE study by Petri, Plummer and Zhai (2011) found that potential income gains from the FTAAP far exceeded the total income gains from the TPP and the East Asian track[106] (see Table 1.11). In particular, the FTAAP is projected to bring about an additional US$862.2 billion, while the TPP and the East Asian track are expected to contribute US$104.3 billion and US$215.3 billion, respectively, to global GDP over the 2025 baseline projection. Income gains accrued to individual APEC member economies would also far outweigh those resulting from the TPP and the East Asian track, with China, Russia, Japan and the US leading the list. Amongst the ASEAN members, Vietnam is expected to benefit the most from the FTAAP, followed by Malaysia. It is also evident in the empirical findings that the FTAAP can avert trade diversion effects caused by the TPP and the East Asian track, resulting in economic gains for all APEC economies. Likewise, trade diversion brought about by the RCEP will also be eliminated once the FTAAP is concluded.

---

[104] See Bush (2006).

[105] See Scollay (2004).

[106] The East Asian track builds on ASEAN integration efforts, which include the ASEAN-10, Japan, China and Korea. However, it excludes important ASEAN's trading partners such as Australia, New Zealand and India. See Petri, Plummer and Zhai (2012, 19).

Table 1.11: Real Income Effects of the Trans-Pacific Partnership (TPP), the East Asian Track and the Free Trade Area of the Asia-Pacific (FTAPP)

APEC Country	Baseline GDP 2025	Income Gains in 2025			Percentage of Baseline GDP		
		TPP	East Asian Track	FTAAP	TPP	East Asian Track	FTAAP
Australia	1,426	2.4	0.1	7.9	0.17	0.01	0.55
Brunei	22	0.1	0.3	0.4	0.48	1.31	1.64
Canada	1,982	2.3	0.4	7.3	0.12	0.02	0.37
Chile	289	2.3	0.1	5.0	0.78	0.04	1.72
China	16,834	−15.7	84.8	318.7	−0.09	0.50	1.89
Indonesia	1,473	−1.5	8.1	17.3	−0.10	0.55	1.18
Japan	5,332	30.7	42.8	73.6	0.58	0.80	1.38
South Korea	2,063	15.1	33.5	48.2	0.73	1.63	2.34
Malaysia	422	9.4	6.5	22.2	2.24	1.53	5.25
Mexico	1,999	11.7	2.8	41.3	0.58	0.14	2.07
New Zealand	206	1.7	0.0	2.5	0.83	0.02	1.22
Peru	311	6.6	0.1	7.9	2.12	0.02	2.53
Philippines	308	−0.4	3.8	8.9	−0.12	1.25	2.90
Russia	2,790	−1.0	−1.3	152.3	−0.03	−0.05	5.46
Singapore	386	1.4	0.0	1.6	0.35	−0.01	0.42
Thailand	520	−1.4	8.3	17.6	−0.28	1.60	3.38
United States	20,337	13.9	3.2	62.9	0.07	0.02	0.31
Vietnam	235	33.5	15.2	61.4	14.27	6.49	26.14

Note: East Asian track includes ASEAN-10, Japan, China and South Korea.
Source: Petri, Plummer and Zhai (2011, 26).

Third, the pursuit of the FTAAP by APEC is likely to reinvigorate the currently stalled Doha Round. Once the FTAAP is successfully concluded, the risk of facing substantial trade discrimination would become more imminent to countries that have been maintaining a hard-line position during the Doha Round negotiations. If the Doha Round eventually collapses, the FTAAP would provide a viable alternative to restore the momentum of trade liberalisation among APEC economies, which account for more than half of world output and trade, resulting in in a larger pay-off than any realistic Doha outcome.[107]

Last but not least, the success of the FTAAP would revitalise APEC itself. Established in 1989, APEC, with the participation of major powers including the US, Japan and China, has enormous potential to become one of the most important regional groupings in the world. However, even in the face of the faltering Doha Round, APEC has not implemented any significant trade agenda other than supporting the WTO. This, to a large extent, has led to the independent pursuit of

[107] See Bergsten (2007).

bilateral and plurilateral FTAs by individual APEC members notwithstanding APEC's non-binding principle. Meanwhile, APEC's inability to achieve its own Bogor Goals by 2010 for industrialised economies has become apparent. The launch of the FTAAP will thus help to restore APEC's credibility and relevance as a vehicle for furthering multilateral liberalisation.

For the ten ASEAN member states in particular, reviving APEC through the conclusion of the FTAAP is of paramount importance given the recent China's provocations in the South China Sea and the tensions between the US and China. Whether or not China is pursuing its hegemonic ambitions, a more cohesive APEC would enhance US engagement in Asia as a "hedging strategy" against possible further aggressions by China. The potential impacts of the FTAAP thus extend beyond the economic sphere as it could serve as an effective institutional mechanism for maintaining security across the Pacific. For ASEAN to effectively revitalise APEC and drive the FTAAP negotiations, it is imperative that the RCEP as led by ASEAN is successfully established. The conclusion of RCEP, which aims to be more comprehensive and of higher standard than the existing ASEAN+1 FTAs, would deepen integration between ASEAN, Japan and China, thereby paving the way for the FTAAP process which involves all three and the US. Equally important is the fact that the success of RCEP will further ingrain ASEAN centrality in the broader Asia-Pacific region.

As the FTAAP commitments, like other FTAs, are legally binding, APEC's long-standing voluntary and non-binding tradition would render ineffective in the pursuit of a free trade area. Thus, the negotiation of the FTAAP should be a separate process, distinct from the main APEC process. In doing so, neither the binding nature of the FTAAP or the non-binding, open regionalism approach of APEC will be compromised. In fact, if the FTAAP process is independent from APEC, it will also help to circumvent the problem that some of the less developed ASEAN members, namely Cambodia, Laos and Myanmar, are currently not part of APEC. An open accession clause could be incorporated in the FTAAP agreement, allowing these countries, which are already members of the RCEP, to join the Asia-Pacific trade umbrella.

Most importantly, for the FTAAP to yield significant pay-offs to the APEC economies, it must be comprehensive and high-standard in scope, incorporating and addressing traditional as well as "next generation" trade and investment issues. A CGE analysis undertaken by Kim *et al.* (2013) surmises that the FTAAP would produce the most substantial economic gains if it assumes all the following: (i) comprehensive tariff elimination in goods trade without any excluded sectors, (ii) reduction in tariff equivalents of services by 10% and (iii) a 5% reduction in trade cost through enhanced trade facilitation. In particular, such deep integration scenario is projected to yield an additional income of US$284.8 billion for APEC as a whole, as compared to US$55.4 billion if (i) is achieved, and US$75.6 billion if (i) and (ii) are achieved.[108] The study thus concludes that APEC should pursue a high-quality and comprehensive agreement for maximising welfare gains, particularly in terms of liberalisation of trade in services and trade facilitation.

In this respect, the FTAAP should draw on the strengths of existing plurilateral trade agreements. The establishment of the AEC in December 2015 and the signing of the high-standard

[108] See Kim *et al.* (2013, 20).

TPP in February 2016 provide crucial impetus to drive the FTAAP process. Likewise, the significance of the ongoing RCEP, CJK-FTA and the China–US Bilateral Investment Treaty negotiations cannot be understated. How these initiatives unfold would have significant impacts on the outcome of the FTAAP. As reiterated by the PECC, the only non-governmental official observer of APEC, the FTAAP is "not starting from a zero base"[109] as it builds on "ongoing regional undertakings including TPP, RCEP and AEC."[110]

In fact, the TPP and the RCEP are the only two realistically viable pathways toward the formation of the FTAAP given their common objectives and extensive membership. For the TPP and the RCEP to eventually merge to form the FTAAP, the RCEP, which is being negotiated, must also be high-quality and wide-ranging in scope. In other words, the levels of ambition between the two trade pacts should not be too wide as if this is the case, the merger would be more difficult. While negotiators of the RCEP need to keep in mind and accommodate the diverse levels of economic development amongst its members, they must avoid a "race to the bottom" where the lowest common denominator rules. This means that the RCEP must be negotiated with due consideration of the FTAAP.

In this respect, as highlighted by the PECC in the 2015 Symposium on "FTAAP: Asia-Pacific Economic Cooperation by 2020 and Beyond" in Beijing, while the FTAAP should address next generation issues including intellectual property protection, competition policy and SOEs, capacity building should also be emphasised. For potential FTAAP members, particularly the less developed ones, deeper integration through liberalisation of services and investment will only prove beneficial if it is accompanied by a set of capacity building measures. In terms of the negotiation process, the FTAAP should proceed on a step-by-step, flexible, transparent and consensus-based approach, in view of the heterogeneity of the Asia-Pacific. Since the level of ambition will differ across issues and member economies, there should be ample assistance for transition.

Bridging the ambition gaps within ASEAN is thus crucial if it is to achieve a meaningful, high-standard RCEP agreement that could serve as a stepping stone toward the FTAAP while maintaining ASEAN centrality in the process. As only four ASEAN member states are currently part of the TPP, namely Brunei, Malaysia, Singapore and Vietnam, much work remains to be done to narrow the development gap so that the fruits of deep integration could be equitably shared by all members. Economic and technical cooperation should thus become one of the top priorities under the post-2015 AEC agenda for ASEAN to successfully conclude the RCEP as well as fulfil the AEC visions.

The Role of Indonesia

Indonesia is widely regarded as the *de facto* leader of ASEAN given its size, population, strategic position as well as its track record in shaping the direction of ASEAN, mediating disputes and maintaining cohesion within the regional bloc. In particular, in 1976 Indonesia put forth the

[109] See Lim (2014).

[110] See Pacific Economic Cooperation Council (2015).

Treaty of Amity and Cooperation (TAC), also referred to as Bali Concord I, which institutionalised norms of conduct between ASEAN member states as well as with their dialogue partners. Another case in point is Indonesia's role in managing the territorial dispute between Cambodia and Thailand over the Preah Vihear temple in 2011. During the 2012 ASEAN Summit chaired by Cambodia, Indonesia's "shuttle diplomacy" helped to restore ASEAN's unity on the issue of the South China Sea through the adoption of the ministerial statement on the "sixth-point principles,"[111] which stresses the full implementation of the Declaration of the Conduct of Parties in the South China Sea of 2002 and the respect of international law as ASEAN's collective position, after Cambodia refused to issue a joint communique — the first time in the grouping's 45-year history.

Indonesia's leadership has extended beyond security with notable contributions in forging regional cooperation. Under its chairmanship in 2003, Indonesia initiated and formalised the concept of the ASEAN Community with its three pillars through the Bali Concord II. Indonesia's vision of an EAS which extended beyond ASEAN+3 (Japan, South Korea and China) came into fruition when Australia, New Zealand, and India joined in 2005, followed by the US and Russia in 2011.[112] In 2012, Indonesia took regional cooperation to a greater height with the signing of the Bali Concord III Plan of Action on ASEAN Community in a Global Community of Nations 2013–2017, which outlines strategies for strengthening the ASEAN Community's three pillars and envisages a common ASEAN approach for addressing global issues. Indonesia's leadership in ASEAN has thus proved to be for the most part effective as the driving force of key regional integration initiatives, notably the ASEAN Community.

While the current administration of President Jokowi seems to be pursuing a highly ambitious foreign policy with a vision to turn Indonesia into a "global maritime fulcrum" between the Pacific and the Indian Oceans,[113] many experts have asserted that Indonesia is turning away from ASEAN and focusing on bilateral ties, particularly with major powers, which benefit its domestic-oriented policies.[114] Some even argue that President Jokowi's high-profile sinking of illegal fishing vessels in Indonesian waters, the majority of which came from Vietnam, Philippines, Thailand and Malaysia, signals that he has side-lined ASEAN from Indonesia's foreign policy priorities in pursuit of a more nationalistic approach.[115] Furthermore, President Jokowi's administration had not been playing an active role in the establishment of the AEC despite Indonesia being the proponent of the ASEAN Community. Indonesia's grappling with growth slowdown, declining commodity prices and a sharply weakened rupiah throughout 2015 has further raised doubts about the future of Indonesia's leadership in ASEAN.

---

[111] The Statement on ASEAN's Sixth-Point Principles on the South China Sea is accessible from: http://cil.nus.edu.sg/rp/pdf/2012%20ASEANs%20Six-Point%20Principles%20on%20the%20South%20China% 20Sea-pdf.pdf.

[112] See Natalegawa (2015).

[113] President Jokowi's maritime vision comprises five pillars: (i) reviving Indonesia's maritime culture; (ii) revitalising management of maritime resources; (iii) prioritising infrastructure for maritime connectivity; (iv) focusing on maritime diplomacy for resolving conflicts; and (v) strengthening maritime defence power. See Tan *et al.* (2016, forthcoming) for an in-depth analysis.

[114] See Syailendra (2015) and Yahya (2015).

[115] See Parameswaran (2015b), Hamzah (2015) and Yahya (2015).

Nonetheless, in his speech at the ASEAN Summit in November 2014 in Nay Pyi Taw, President Jokowi affirmed his commitment to contribute to ASEAN integration:

> "In this 25th ASEAN Summit, which is my first Summit to attend, I want to re-emphasize our government's commitment to promote ASEAN cooperation. For Indonesia, ASEAN is a forum for building cooperation which brings benefit for our people, for development in our countries and for peace and stability in our region."[116]

He proposed three ways that ASEAN should adopt in order to be able to reap concrete and tangible benefits from its integration efforts, namely: accelerating infrastructure and connectivity development through the speedy implementation of the Master Plan on ASEAN Connectivity (MPAC); increased intra-regional trade; and increased cooperation in investment, industry and manufacturing among ASEAN member states. President Jokowi is also reported to place special emphasis on the post-2015 ASEAN agenda, aiming for it to focus on: upgrading ASEAN's institutional capacity, enhancing its credibility and strengthening its unity.[117]

Furthermore, President Jokowi seems to be making a more decisive outward-looking move toward deep integration with the global economy as he announced Indonesia's intention to join the TPP during his visit to the White House in October 2015.[118] His presence at the US-hosted US-ASEAN Summit in Sunnylands, California, on 15th and 16th February 2016, where he led a discussion on terrorism and peace-keeping in ASEAN, further reaffirms his regional outreach following his largely inward-looking first year in office.[119] Internally, President Jokowi's policies have centred upon improving Indonesia's investment climate through a number of deregulation packages with a view to attracting FDI in addition to accelerating infrastructure development.[120]

It can thus be surmised that Indonesia under the Jokowi administration is steadily reasserting itself in the regional and global stage, particularly in relation to initiatives that are vital to the future of ASEAN. While much remains to be seen as to how Indonesia will follow through its plan to join the TPP, Indonesia's intention seems to signal its commitment to high-quality, wide-ranging trade agreements, which would certainly benefit the ongoing RCEP negotiations. One can therefore expect that Indonesia will play a more prominent role in driving ASEAN's integration efforts in the economic sphere, especially with regard to the RCEP, as well as in the formulation and implementation of the post-2015 ASEAN agenda.

To Indonesia, continued leadership in ASEAN integration is paramount to its vision to become the "global maritime axis." As seen in the foregoing discussion, Indonesia's leadership in ASEAN has been earned and elevated through many bold visions and initiatives mooted by its various administrations before President Jokowi. For the Jokowi administration to take Indonesia to a greater height globally, its leadership within ASEAN would be the crucial starting

[116] See Widodo (2014).
[117] See Hermawan (2015).
[118] See Davis (2015).
[119] See Amindoni (2016).
[120] See Asian Development Bank (2015) and Ismail (2016).

point. As ASEAN further entrenches its linkage with Asia-Pacific through the TPP, the RCEP and potentially the FTAAP, Indonesia's relevance in the global stage will primarily depend on its ability to turn ASEAN into a meaningful and credible Economic Community beyond the 2015 milestone, which would include, among others, the successful conclusion of the RCEP and the FTAAP.

Furthermore, in light of the ongoing political instability in Thailand, the high-profile corruption scandal involving Prime Minister Najib Razak in Malaysia and the upcoming presidential election in 2016 in Philippines, the vacuum of leadership in ASEAN is for Indonesia to fill. With President Jokowi's renewed focus on trade diplomacy, which is also pivotal to Indonesia's growth trajectory, Indonesia should rise up to the challenge, leading ASEAN toward the complete formation of the AEC and beyond. As the only ASEAN member in the G20 grouping, Indonesia has substantial leverage to advance integration initiatives in ASEAN as well as the Asia-Pacific, which it should make full use of.

Domestic Reforms

While regional economic integration could bring about both tangible economic gains and intangible political leverage, especially in today's increasingly globalised world of global value chains, it is no magic bullet to economic development if a country is not earnest in formulating and implementing its domestic reforms for enhanced competitiveness. Chapter 2 of this book provides a roadmap to such endeavours by not only ranking the ten ASEAN member states in terms of competitiveness but also offering in-depth analyses on strengths and weaknesses for each economy, drawing on an extensive database of 121 indicators and a robust, comprehensive methodology. Our research publications on Indonesia's regional and provincial competitiveness analysis are particularly useful for bridging the development gaps at the sub-national level.[121]

Indonesia as the leader of ASEAN-6

For Indonesia as the largest economy and the *de facto* leader of ASEAN, its reform progress would have significant impacts on ASEAN integration as well as its future leadership of the regional grouping. As noted, Indonesia was grappling with sluggish growth, low demand for commodities and a weakened rupiah throughout 2015. Plummeting global commodity prices created a strain on Indonesia's revenue as palm oil, metal, and fossil fuels accounted for half of the country's exports. The rupiah further weakened given the strengthening of the US dollar as the prospect of a rate hike by US Federal Reserve became imminent. Indonesia's reported quarterly GDP growth rate in 2015 hovered between 4.5 and 5%, much lower than President Jokowi's 7% target.[122] While measures have been put in place to improve the country's investment climate and attract FDI, much remains to be done to reinvigorate the currently lacklustre Indonesian economy.

[121] See Tan *et al.* (2013b, 2015d, 2015e) for ACI's publications on Indonesia.

[122] See Tan and Merdikawati (2016, forthcoming, Chapter 2).

Against this backdrop, on eighth September 2015, President Jokowi unveiled a new economic reform package to bolster the flagging economy. He announced measures to overhaul 89 regulations impeding investment including streamlining licensing and land acquisition for infrastructure projects and easing rules for foreigners to open foreign currency bank accounts in Indonesia.[123] While the one-stop investment service has been implemented at the national level, further synchronisation of local licences under the jurisdiction of provinces, districts, and cities will be put in place. In terms of infrastructure development, a mega power plant project is in the pipeline to produce an additional 35,000 megawatt of electricity by 2019 in more than 200 places across Indonesia.[124] Emphasis is placed on using renewable energy such as geothermal and wind power for Indonesia to lower its reliance on fossil fuel.

Nevertheless, reform has to be deepened, particularly in areas of manufacturing and services sectors as well as infrastructure connectivity, if the country is to achieve and sustain its growth target of 7%. Compared to many of its ASEAN peers, Indonesia largely remains outside Asia's production networks as its share of manufacturing in total exports and GDP is by far the smallest in the regional bloc.[125] While Indonesia is already part of some of the basic manufacturing value chains, Indonesia-based producers of intermediate goods are yet to emerge as reliable suppliers for multinational corporations. Attracting FDI and strengthening local firms in upstream and complementary industries are equally important as part of the country's economic reform.

Additionally, Indonesia's infrastructure spending as a percentage of GDP has not returned to its pre-AFC levels.[126] Rampant infrastructural bottlenecks continue to hamper FDI as well as domestic investment, creating substantial economic development divide at the provincial and regional level as many islands remain isolated from the nation-wide economic production and distribution networks. It is estimated that logistics cost accounts for 18–22% of firms' production cost and up to 24% of Indonesia's GDP. While streamlining bureaucracy is necessary to drive down cost of doing business, more measures are required for improving transport, energy and industry infrastructure.

As the largest economy in ASEAN and the natural leader of the original ASEAN-6, Indonesia plays a pivotal role in the completion of the AEC and the post-2015 ASEAN agenda. In particular, a well-connected archipelagic Indonesia, both in terms of land and sea transport, is vital to regional connectivity as envisaged in the MPAC as well as by ACI through the Asia Economic Connectivity Vision 2030 (AECV2030).[127] Furthermore, the progress and outcome of the RCEP are likely to be greatly affected by Indonesia's trade talks with current RCEP members and potential FTAAP members, namely South Korea, Australia, India, Chile, which were resumed in 2015.[128] As three of the original ASEAN-6 are already committed to high-standard trade agreements as part of the TPP, namely Brunei, Malaysia and Singapore, the prospect of ASEAN

---

[123] See Chilkoti and Hidayat (2015).

[124] See Tan *et al.* (2016, forthcoming, Chapter 2).

[125] See Armstrong (2014).

[126] See Tan and Merdikawati (2016, forthcoming, Chapter 1).

[127] See Tan *et al.* (2015a, Chapter 1) for details on the AECV2030.

[128] See Amianti and Amirio (2016).

integration would now depend on Indonesia's commitment to and implementation of domestic reform and trade negotiations.

Not less paramount is the pressing need for Indonesia to stamp out terrorism in view of the recent terrorist attack in Jakarta on 14th January 2016. This was the first attack in Indonesia and ASEAN carried out by the Islamic State, a well-organised jihadist group that has seized large swathes of territory in Syria and Iraq. Indeed, the rise of the Islamic State has posed a formidable threat to security not only in the West but also in ASEAN where Islam is one of the major religions with Indonesia being the largest Islamic country in the world. The attack in Jakarta has sent shock waves across ASEAN as the risk of terrorism looms larger than ever.

In the aftermath of the attack, the Jokowi administration has moved rather swiftly to revise Indonesia's terrorism laws, which broaden the definition of terrorism, allow for longer detention of terrorist suspects for up to 120 days, and enable Indonesian authorities to target anyone who recruits members for, or cooperates with a militant group.[129] Nonetheless, more has to be done to revamp Indonesia's intelligence system to effectively detect potential attacks. Measures for intra-regional cooperation in anti-terrorism intelligence should be put in place as part of the ASEAN Political-Security Community; otherwise, the ASEAN economy as a whole will risk being derailed by terrorism threats. Security collaboration with global military powers should be explored and followed through. For instance, while the issues of terrorism and violent extremism were discussed in the US-ASEAN Summit in Sunnylands on 15th and 16th February 2016,[130] concrete plans for US-ASEAN cooperation are yet to be carved out.

Vietnam as the front-runner of CLMV

Vietnam as the front-runner transitional economy in ASEAN holds great significance both in the area of narrowing the development gap within the regional bloc and as the driver of economic integration within the CLMV grouping. In October 2015, Vietnam impressed observers by participating in the high-standard TPP together with three of the more developed ASEAN peers (Brunei, Malaysia and Singapore), through which the country commits to a set of reform measures covering not only trade and investment but also labour, environment and notably SOEs. While the TPP is expected to provide the critical impetus to Vietnam's current lacklustre economic performance, it is not the "holy grail" for the country's many structural weaknesses, as argued by ACI in an editorial opinion for the Strait Times on 25th February 2016.[131]

In particular, whether the TPP will be make or break for Vietnam depends on the country's ability to resolve its most exigent economic malaises before the waves of global competition sweep in. Notwithstanding three decades of *Doi Moi* (renovation) reform toward a market economy, market distortions in favour of SOEs remain widespread, which perpetuates unfair competition between the private and the state-owned sector. While the promulgation of the third

[129] See Da Costa and Kapoor (2016).

[130] See The White House (2016).

[131] See Tan *et al.* (2016). See also Appendix 1 for the related policy brief.

Enterprise Law and the launch of a new round of SOE restructuring in 2005 raised new hopes for private sector development in Vietnam, the progress has been notably slow with the state-owned sector still accounting for a significant 40% of annual total investment. While the pace of the SOE reform has to be quickened, more fundamental is the elimination of monopoly power and special privileges SOEs have been enjoying at the expense of private enterprises. To that extent, Vietnam's SOE reform has to be deepened, which essentially entails relinquishing the state's direct involvement in the economy, levelling the playing field and moving towards a rule-based, transparent and competitive economic environment.

Equally imperative is the need to build strong and transparent corporate governance, which is another long-standing ailment plaguing the Vietnamese economy. Without competitive enterprises, it is hard to envisage how Vietnam could benefit from the TPP, particularly when joining the trade bloc means domestic firms will have to compete with firms from the other 11 members. Nonetheless, corporate governance in Vietnam remains problematic notwithstanding the rapid development of the legal and regulatory framework for investor protection. For instance, new laws on independent audit was enacted in 2011, and in 2013 the State Securities Commission of Vietnam revised Corporate Governance Regulations and a Model Charter that detail good practice for listed companies. Nonetheless, weaknesses remain in terms of poor disclosure, outdated accounting standards, lack of internal control mechanisms and limited accountability of firm management. In fact, current practices in many public enterprises fall far short of what is required in the revised Corporate Governance Regulations, which also highlights the government's difficulties in enforcing the rule of law.

Furthermore, upgrading labour skills is paramount for Vietnam to enhance competitiveness in the face of rising competition. After 30 years of reform, firms still find it hard to find skilled workers to fill the jobs, particularly managers, executives and technicians. Foreign corporations have also been complaining about the lack of reliable local suppliers, which is a testament to the wide capability gap between the FDI and the local sector. Without concrete policies and programmes for improving labour quality, Vietnam will certainly risk being dominated by foreign firms. Specifically, technical education and training must be revamped in alignment with industry needs. Measures to support and upgrade best-performing local firms, as well as pair them up with multinational corporations, should be put in place. Tax incentives could be introduced to foreign firms engaging in technology transfer with local suppliers.

As with other transitional economies in ASEAN, the central thrust of Vietnam's *Doi Moi* has been resource mobilisation from agriculture to manufacturing and services on the back of a large young, literate and low-cost workforce. Nonetheless, such model is not sustainable as competing based on cheap labour is simply a race to the bottom, especially when Vietnam's youth population is shrinking. Vietnam's experience with reforms and liberalisation through the TPP would therefore have significant implications to the rest of the CLMV grouping as Cambodia, Laos and Myanmar embark on deeper integration within the AEC and subsequently as part of the RCEP and potentially FTAAP. Most importantly, the case of Vietnam highlights the critical need for far-reaching domestic reforms in concurrence with efforts to integrate regionally and globally. Without sound economic fundamentals in place, the benefits of economic integration would be at best elusive.

1.6 CONCLUDING REMARKS

After nearly five decades of remarkable growth, bolstered by wide-ranging regional cooperation initiatives, ASEAN has taken regional economic integration to new heights with the official establishment of the AEC on 31st December 2015, paving the way for potential significant transformation that could propel ASEAN forward as the fourth largest economy by 2030 and place the grouping at the centre of the global economic architecture. While the AEC aspires to become "a single market and production base" by the 2015 milestone, only about 80% of the stipulated measures under the blueprint were completed by then. This chapter seeks to shed light on the challenges in completing the final mile of the AEC journey and answer the question as to whether the full formation of the AEC would be feasible by the end of 2016.

Our extensive analysis as detailed in Section 1.4 has shown that, while progress may be made in certain stipulated requirements, it is unlikely for all outstanding deliverables to be fully implemented by the end of 2016. Thorny issues, where implementation has been most problematic include: (i) NTBs to trade, (ii) trade facilitation through the establishment of the ASW, (iii) trade in services, (iv) free movement of skilled labour, and (v) the launch of the ASAM, or the Open Skies policy. In the year 2016, the initiative which may see substantial progress is likely to be the ASAM as domestic pressure has mounted for Indonesia to ratify the Open Skies agreement. Growing confidence amongst Indonesian airlines may finally lead Indonesia to commit to the Open Skies policy. In fact, many pundits have expressed optimism that Indonesia will ratify the ASAM agreements soon.

Nonetheless, challenges abound in the implementation of several other key targets specified in the AEC blueprint. First, with regard to NTBs the current "notify-negotiate-eliminate" approach is plagued with serious shortcomings in terms of incentive misalignments as well as institutional weaknesses in some of the ASEAN member states. In particular, there is an apparent conflict of interest as many ASEAN member states have the incentive to hoard information to avoid complete elimination of NTMs. Government agencies also tend to under-report their regulatory measures for fears of being admonished in case full disclosure causes a drop in the country's rank in international competitiveness or ease of doing business rankings. Furthermore, limited institutional capacity may hamper the member states' effort to compile information on their NTMs, which usually fall under the purview of various departments and agencies.

Second, achieving the 70% target of foreign ownership in all services sectors by 2016 also seems unrealistic in view of the lack of progress even in priority integration sectors such as air transport. While the limit on foreign ownership for air transport businesses should have been increased to 51% by 2008 and 70% by 2012, the maximum foreign ownership allowed in ASEAN member states, except for Singapore, is only 49%. Furthermore, according to policy surveys conducted by the World Bank, overall policies on trade in services adopted in the region still remain more restrictive than what the AEC blueprint requires, with the exception of Singapore and, to a lesser extent, Cambodia.

Third, full operationalisation of the ASW is also likely to remain outstanding as the ongoing full-fledged pilot of the ASEAN only involves Indonesia, Malaysia, Thailand, Singapore and Vietnam, while Brunei and Philippines have opted to join by the second quarter of 2016.

Additionally, the legal framework for the eventual implementation of the ASW has not been ratified by all members. The NSWs have not even been fully established in Cambodia, Laos and Myanmar. Existing technical as well as legal gaps pertaining to e-banking, e-commerce, digital documents and so on make it impractical for all ASEAN member states to operationalise the ASW by the end of 2016.

Fourth, in the area of skilled labour mobility, notwithstanding the conclusion of the various MRAs for eight priority professions, movements of skilled professionals are defined in different ways in different MRAs. Furthermore, restrictive domestic regulations remain, particularly in issuance of work visas, quotas on foreign labour, licensing requirements, impeding the cross-border movement of professionals in the region. Unless individual member governments ease the restrictive regulations with respect to foreign labour, it is hard to envisage how "free flow of skilled labour," even just within the eight stipulated professions, can be achieved.

Furthermore, the AEC blueprint itself leaves much to be desired. For instance, its scope in services trade liberalisation is particularly limited as only 70% of foreign ownership is targeted, as compared to high-standard FTAs such as the TPP which aims for 100%. Additionally, measures for facilitating skilled labour mobility only affect a small fraction of total workers in ASEAN. At the same time, many flexibilities allowed under the blueprint, such as the ASEAN-minus-x formula, are vaguely worded and thus leave ample room for discretion, as exemplified by the case of Indonesia.

Equally important is the fact that establishment of the AEC has largely been a state-led top-down project with limited inputs and awareness amongst the private sector in the region. A number of surveys have shown that the business community in ASEAN has not been taking advantage of liberalisation measures under the AEC. More needs to be done to enhance awareness and consciousness of the AEC amongst its people, as noted by Mr Ong Keng Yong, former ASEAN Secretary-General in a speech in 2014.[132]

The formal launch of the AEC in December 2015 should thus be seen as a stepping stone, rather than the end, in ASEAN's trajectory toward deeper economic integration. Given the multiple flexibilities allowed both in the design of the AEC blueprint and in its implementation, the realisation of a "single market and production base" will not end even as 100% of the requirements have been fulfilled. In fact, as recognised by the ASEAN Secretariat, policy-makers and scholars in the region, community building in ASEAN is an ongoing and dynamic journey in which 2015 is a significant milestone.[133]

The AEC will only be truly formed when each ASEAN member state and their people feel that they are sharing the economic dividends of being in the community, and take a greater ownership and stake in shaping its direction. Without a shared sense of ASEAN identity and community spirit, opportunities arising from an integrated regional grouping will not be fully capitalised. Therefore, the AEC must advance in conjunction with the ASEAN Political-Security Community and the ASEAN Socio-Cultural Community. ASEAN can only flourish economically in an

[132]See Ong (2014).

[133]See, for instance, ASEAN Secretariat (2015a), Lim (2015), Ong (2014).

environment of peace and stability, in light of the ongoing geopolitical tussles between traditional and emerging major powers, especially the US and China, and the imminent threat of terrorism, which has amplified since the bomb attack in Jakarta on 14th January 2016.

Beyond the 2015 milestone, enormous work awaits the 10 ASEAN member states to realise the common dream of inclusive and sustainable prosperity. For one, narrowing the development divide remains a work in progress, which requires not only capacity building and technical assistance from the more mature economies in the bloc but also, and even more importantly, far-reaching domestic reforms, particularly in the less developed CLMV grouping. The experience of Vietnam as analysed in the section on Domestic Reforms, shows that global integration is not the "holy grail" for domestic structural weaknesses. Without decisive wide-ranging domestic reforms and sound economic fundamentals in place, the benefits of economic integration would be elusive at best.

On the other hand, economic convergence could be expedited if ASEAN steps up to the challenge and takes the lead in trade and investment initiatives in the broader Asia-Pacific region. Of particular emphasis are the RCEP and the FTAAP, which have both grained traction as critical vehicles for enhancing economic growth in Asia-Pacific in light of the currently stalled Doha Round. In other words, domestic reforms and global integration must go hand in hand as a two-pronged strategy in the pursuit of inclusive, balanced and sustainable growth.

In this light, it is imperative that ASEAN take regional integration to the next level by driving the formation of the RCEP, which failed to be concluded by the 2015 deadline. As discussed in-depth in Section 1.5.2, for the RCEP to achieve meaningful and impactful integration outcomes, it has to be comprehensive and of high-quality with a broader and deeper set of commitments than the prevailing ASEAN+1 FTAs, while recognising the need for narrowing the income gaps amongst its members through provisions for economic and technical cooperation.

While the RCEP may present an alternative path of integration for less developed countries which are only ready for limited liberalisation, the ambition gaps between the RCEP and the TPP should not be too wide. As demonstrated in the section, Towards the FTAAP, the TPP and the RCEP are the two realistically viable pathways toward the formation of a larger, more inclusive and beneficial FTAAP, given their common objectives and extensive membership. Empirical studies have shown that the FTAAP would yield the most significant pay-offs to the APEC economies if it is comprehensive and high-standard in scope, incorporating and addressing traditional as well as "next generation" trade and investment issues. Therefore, while the diverse levels of economic development should be accommodated under the RCEP, a "race to the bottom" where the lowest common denominator rules must be avoided. This means that the RCEP must be negotiated with due consideration of the FTAAP as the ultimate goal.

Insofar as the RCEP and the post-2015 ASEAN agenda are concerned, Indonesia's leadership would be pivotal. The ongoing political instability in Thailand, the high-profile corruption scandal involving Prime Minister Najib Razak in Malaysia and the upcoming presidential election in 2016 in Philippines have created a vacuum of leadership in ASEAN for Indonesia to fill. With President Jokowi's renewed focus on trade diplomacy as evident in his intention of joining the TPP, Indonesia should rise up to the challenge, leading ASEAN toward the conclusion of the RCEP as well as the fulfilment of the post-2015 agenda. After all, for Jokowi's maritime vision to be realised, its leadership in ASEAN would be the crucial starting point as the regional grouping is seeking to deepen economic ties with major powers in Asia-Pacific and beyond. Furthermore, Indonesia, as the only ASEAN member in the G20 grouping, has substantial leverage to advance integration initiatives in ASEAN as well as the Asia-Pacific, which it should make full use of.

Central in ASEAN's trajectory toward an inclusive, sustainable and globally integrated economic community is the need for the regional bloc to maintain centrality and unity amidst the ongoing geopolitical uncertainties. In particular, the "Sino-American rivalry for influence" in the region is threatening to turn ASEAN "into a house divided against itself" if member states begin to choose sides between the two powers.[134] A case in point is the failure of ASEAN's foreign ministers to produce a joint communiqué at the ASEAN Summit chaired by Cambodia in July 2012 — the first time in the bloc's 45-year history — due to Cambodia's opposition to the inclusion of references to the South China Sea issue, which is widely believed to be a reflection of China's expanding influence in ASEAN's agenda. Such potential threats of disintegration must be artfully managed and mitigated; otherwise, community building efforts will be rendered fruitless.

It is thus paramount that ASEAN must speak with one voice in articulating its collective interests in the multiple forums and initiatives led by the bloc itself as well as by major powers, for instance the China-led Belt & Road. Furthermore, with the establishment of the Asia Infrastructure Investment Bank (AIIB) in December 2015, the ten ASEAN economies must leverage on their positions as founding members of the Bank to shape its governance structure and operational guidelines and channel this new infrastructural funding mechanism toward fostering intra-ASEAN connectivity. In brief, balancing power rivalry while driving global cooperation efforts toward bolstering economic development in the region is the way forward for ASEAN in its post-2015 journey.

REFERENCES

Acharya, Amitav. 2001. *Constructing A Security Community in Southeast Asia: ASEAN and the Problem of Regional Order*. London: Routledge.

Agence France-Presse. 2015. "Indonesia will join trans-pacific partnership, Jokowi tells Obama." *The Guardian*, October 27. http://www.theguardian.com/world/2015/oct/27/ indonesia-will-join-trans-pacific-partnership-jokowi-tells-obama.

Amianti, Grade D. and Dylan Amirio. 2015. "RI to resume free-trade talks." *The Jakarta Post*, March 19. www.thejakartapost.com/news/2015/03/19/ri-resume-free-trade-talks.html.

Amindoni, Ayomi. 2016. "Indonesia to highlight partnership for peace, prosperity at ASEAN-US summit." *The Jakarta Post*, February 14. http://www.thejakartapost.com/news/2016/02/14/indonesia-highlight-partnership-peace-prosperity-asean-us-summit.html.

APEC Business Advisory Council. 2004. *Bridging the Pacific: Coping with the Challenges of Globalization*. Report to APEC Economic Leaders. Santiago: Asia-Pacific Economic Cooperation.

[134]See Mahbubani (2015, 140).

Armstrong, Shiro. 2014. "Can Jokowi transform Indonesia's economy?" *East Asia Forum*, August 17. http://www.eastasiaforum.org/2014/08/17/can-jokowi-transform-indonesias-economy/.

ASEAN Secretariat. 1997. "ASEAN vision 2020." http://www.asean.org/?static_post=asean-vision-2020.

ASEAN Secretariat. 2007. "The ASEAN charter." http://www.asean.org/storage/images/ ASEAN_RTK_2014/ASEAN_Charter.pdf.

ASEAN Secretariat. 2008. *ASEAN Economic Community Blueprint*. Jakarta: ASEAN Secretariat.

ASEAN Secretariat and World Bank. 2013. *ASEAN Integration Monitoring Report*. Washington DC: ASEAN Secretariat and World Bank.

ASEAN Secretariat. 2015a. *A Blueprint for Growth — ASEAN Economic Community 2015: Progress and Key Achievements*. Jakarta: ASEAN Secretariat.

ASEAN Secretariat. 2015b. *ASEAN Integration Report 2015*. Jakarta: ASEAN Secretariat.

ASEAN Secretariat and World Bank. 2015. *ASEAN Services Integration Report*. Washington DC: ASEAN Secretariat and World Bank.

Asian Development Bank. 2013. *Asian Economic Integration Monitor — October 2013*. Manila: Asian Development Bank.

Asian Development Bank. 2015. *Asian Development Outlook 2015 Update: Enabling Women, Energizing Asia*. Manila: Asian Development Bank.

Asian Development Bank Institute. 2014. *ASEAN 2030: Toward a Borderless Economic Community*. Tokyo: Asian Development Bank Institute.

Balassa, Bela. 1961. *The Theory of Economic Integration*. Homewood, Illinois: Richard D. Irwin Inc.

Basu Das, Sanchita. 2014. "Asia's regional comprehensive economic partnership." *East Asia Forum*, August 27. http://www.eastasiaforum.org/2012/08/27/asias-regional-comprehensive-economic-partnership/.

Basu Das, Sanchita. 2015. "The ASEAN Economic Community: An Economic and Strategic Project." In *ISEAS Perspective No*. 4–2015, Singapore: Institute of Southeast Asian Studies.

Basu Das, Sanchita. 2016. *The ASEAN Economic Community and Beyond: Myths and Realities*. Singapore: ISEAS Publishing.

Bergsten, C. Fred. 2007. "A Free Trade Area of the Asia-Pacific in the Wake of the Faltering Doha Round: Trade Policy Alternatives for APEC." In *An APEC Trade Agenda? The Political Economy of a Free Trade Area of the Asia-Pacific*, edited by Charles E. Morrison and Eduardo Pedrosa. Singapore: Institute of Southeast Asian Studies.

Bhagwati, Jagdish. 1968. "Trade Liberalization Among LCDs, Trade Theory and GATT Rules." In *Value, Capital and Growth: Essays in Honour of Sir John Hicks*, edited by J. N. Wolfe, 21–43. Edinburgh: University of Edinburgh Press.

Blanchard, Ben. 2015. "China communist party paper says country should join U.S.-led trade pact." *Reuters*, October 24. http://www.reuters.com/article/us-china-trade-tpp-idUSKCN0SJ01X20151025.

Bush, George W. 2006. "Remarks by President George W. Bush at National University of Singapore." http://singapore.usembassy.gov/president_at_nus2.html.

Buszynski, Leszek. 1997. "ASEAN's New Challenges." *Pacific Affairs* 70(4): 555–577.

Cadot, Oliver, Ernawati Munadi and Lili Yan Ing. 2013. "Streamlining NTMs in ASEAN: The Way Forward". ERIA Discussion Paper 2013–24, Economic Research Institute for ASEAN and East Asia (ERIA), Jakarta.

Chia, Siow Yue. 2013. "The ASEAN Economic Community: Progress, Challenges and Prospects". ADBI Working Paper No. 440, Asian Development Bank Institute, Tokyo.

Chilkoti, Avantika and Taufan Hidayat. 2015. "Lacklustre Indonesia Tries to Boost Economy with Reforms." *Financial Times*, September 9. http://www.ft.com/cms/s /0/35e93ab8-56f6-11e5-9846-de406ccb37f2.html#axzz40xXfgto7.

Cooper, Charles A. and Benton F. Massell. 1965. "A New Look at Customs Union Theory." *The Economic Journal* 75: 742–747.

Da Costa, Agustinus Beo and Kanupriya Kapoor. 2016. "Exclusive — Indonesia plans tougher anti-terrorism laws after Jakarta attack." *Reuters*, February 16. http://uk.reuters.com/article/uk-indonesia-security-exclusive-idUKKCN0VP1KB.

Davis, Julie H. 2015. "President Joko Widodo of Indonesia joins trans-pacific partnership." *The New York Times*, October 26. http://www.nytimes.com/2015/10/27/us/politics/ president-joko-widodo-of-indonesia-joins-trans-pacific-partnership.html.

Fukunaga, Yoshifumi and Ikumo Isono. 2013. "Taking ASEAN+1 FTAs towards the RCEP: A Mapping Study." ERIA Discussion Paper 2013-02, Economic Research Institute for ASEAN and East Asia (ERIA), Jakarta.

Fukunaga, Yoshifumi. 2014. "ASEAN's Leadership in the Regional Comprehensive Economic Partnership." *Asia & the Pacific Policy Studies* 2(1): 103–115.

Fukunaga, Yoshifumi and Hikari Ishido. 2015. "Values and Limitations of the ASEAN Agreement on the Movement of Natural Persons". ERIA Discussion Paper 2015-20, Economic Research Institute for ASEAN and East Asia (ERIA), Jakarta.

Grugel, Jean B. 2004. "New Regionalism and Modes of Governance: Comparing US and EU Strategies in Latin America." *European Journal of International Relations* 10(4): 603–626.

Hamzah, B. A. 2015. "Sinking the ships: Indonesia's foreign policy under Jokowi." RSIS Commentary No. 016, S. Rajaratnam School of International Studies, Singapore. https://www.rsis.edu.sg/wp-content/uploads/2015/01/CO15016.pdf.

Heng, Swee Keat. 2015. "Singapore, a Connected and Trusted Home Base for Companies to Go Global." Opening Remarks at the 2015 Asia Competitiveness Institute Annual Conference, Singapore, November 26–27.

Hermawan, Yulius P. 2015. "China's Dual Neighbourhood Diplomacy and Indonesia's New Pragmatic Leadership: How Can ASEAN Preserve its Centrality in a New Challenging Dynamic?" *Georgetown Journal of Asian Affairs* 2(1): 31–40.

Hew, Denis. 2007. "Conclusion: Towards an ASEAN Economic Community by 2015." In *Brick by Brick: The Building of an ASEAN Economic Community*, edited by Denis Hew. Singapore: Institute of Southeast Asian Studies.

Hew, Denis, Rahul Sen, Poh Onn Lee, Maghaisvarei Sellakumaran, Sakulrat, Montreevat and Ngiam Kee Jin. 2005. "ISEAS Concept Paper on the ASEAN Economic Community." In *Roadmap to an ASEAN Economic Community*, edited by Denis Hew. Singapore: Institute of Southeast Asian Studies.

Hill, Hal and Jayant Menon. 2010. "ASEAN Economic Integration: Features, Fulfilments, Failures and the Future." ADB Working Paper Series on Regional Economic Integration No. 69, Asian Development Bank, Manila.

Hu, Albert G. 2013. "ASEAN Economic Community Business Survey". In *The ASEAN Economic Community — A Work in Progress*, edited by Sanchita Basu Das, Jayant Menon, Rodolfo Severino and Omkar Lal Shrestha. Singapore: Asian Development Bank and Institute of Southeast Asian Studies.

Intal, Ponciano Jr., Yoshifumi Fukunaga, Fukunari Kimura, Phoumin Han, Philippa Dee, Dionisius Narjoko and Sothea Oum. 2014. *ASEAN Rising: ASEAN and AEC Beyond 2015*. Jakarta: Economic Research Institute for ASEAN and East Asia (ERIA).

International Labour Organisation and Asian Development Bank. 2014. "ASEAN Economic Community 2015: Managing Integration for Better Jobs and Shared Prosperity". Bangkok: International Labour Organisation and Asian Development Bank.

Ismail, Saifulbahri. 2016. "Indonesia to Continue Easing Restrictions on Foreign Investors: President Widodo." *Channel NewsAsia*, February 18. http://www.channelnewsasia.com/ news/asiapacific/ indonesia-to-continue/2525406.html.

Johnson, Harry G. 1965. *The World Economy at the Crossroads: A Survey of Current Problems of Money, Trade and Economic Development*. Oxford: Clarendon Press.

Jones, Lee. 2015. "Explaining the Failure of the ASEAN Economic Community: The Primacy of Domestic Political Economy." *The Pacific Review*.

Karmakar, Suparna. 2009. "Rescuing the Doha Development Round: The Role of India and China in Multilateral Trade Governance." ISAS Working Paper No. 74, Institute of South Asian Studies, Singapore.

Kaur, Karamjit. 2016a. "Big jump in regional air travel poses challenge: Khaw Boon Wan". *The Straits Times*, February 16, http://www.straitstimes.com/singapore/big-jump-in-regional-air-travel-poses-challenges.

Kaur, Karamjit. 2016b. "Steps to boost Changi air traffic management". The Straits Times, February 19, http://www.straitstimes.com/singapore/steps-to-boost-changi-air-traffic-management.

Kawai, Masahiro, Peter J. Morgan, and Pradumna B. Rana, eds. 2014. *New Global Economic Architecture: The Asian Perspective*. Cheltenham: Edward Elgar.

Kawai, Masahiro and Ganeshan Wignaraja. 2008. "Regionalism as an Engine of Multilateralism: A Case for a Single East Asian FTA." ADB Working Paper Series on Regional Economic Integration No. 14, Asian Development Bank, Manila.

Kawai, Masahiro and Ganeshan Wignaraja, eds. 2011. *Asia's Free Trade Agreements: How Is Business Responding?* Tokyo: Asian Development Bank Institute.

Kemp, Murray C. and Henry V. Wan Jr. 1976. "An Elementary Proposition Concerning the Formation of Customs Unions." *Journal of International Economics* 6(1): 95–97.

Kim, Sangkyom, Innwon Park and Soonchan Park. 2013. "A Free Trade Area of the Asia-Pacific (FTAAP): Is It Desirable?" *Journal of East Asian Economic Integration* 17(1): 3–25.

Kow, Jonathan and Andrea Feldman Mowerman. 2013. "Towards a Truly Seamless Single Windows and Trade Facilitation Regime in ASEAN Beyond 2015". ERIA Discussion Paper 2013-29, Economic Research Institute for ASEAN and East Asia (ERIA), Jakarta.

Krugman, Paul. 1991. "The Move to Free Trade Zones." In *Policy Implications of Trade and Currency Zones: A Symposium*, 7–41. Kansas City: Federal Reserve Bank.

Krugman, Paul. 1993. "Regionalism Versus Multilateralism: Analytical Notes." In *New Dimensions in Regional Integration*, eds. Jaime de Melo and Arvind Panagariya, 58–79. New York: Cambridge University Press.

Kurihara, Takeshi. 2012. "Achievements of Asian Bond Markets Initiative (ABMI) in the Last Decade and Future Challenges." Presentation at the OECD-ADBI 12th Roundtable on Capital Market Reform in Asia, Tokyo, February 7. http://www.oecd.org/finance/ financial-markets/49700575.pdf.

Kyodo. 2015. "Park to Convey Korean desire to join TPP in ssummit with Abe." *The Japan Times*, October 31. http://www.japantimes.co.jp/news/2015/10/31/business/economy-business/park-convey-korean-desire-join-tpp-summit-abe/#.VsGSqubICap.

La, Yifan. 2014. Keynote Speech at the PECC Conference on "New Priorities of Regional Economic Integration — Mandate for APEC." Singapore, February 10–11.

Lawrence, Robert Z. 1996. *Regionalism, Multilateralism and Deeper Integration*. Washington DC: Brookings Institution.

Lewis, Meredith Kolsky. 2013. "The TPP and the RCEP (ASEAN+6) as Potential Paths Toward Deeper Asian Economic Integration." *Asian Journal of WTO and International Health Law and Policy* 8(351): 359–378.

Lim, Hng Kiang. 2014. Opening Speech at the PECC Conference on "New Priorities of Regional Economic Integration — Mandate for APEC." Singapore, February 10–11.

Lim, Hng Kiang. 2015. Speech at the ASEAN Conference 2015 on "Progress of the ASEAN Economic Community 2015: Opportunities, Challenges and the Road Ahead." Singapore, May 20.

Mahbubani, Kishore. 2015. "ASEAN as a Living, Breathing Modern Miracle." *CIRSD Horizons* (2): 136–148.

Mansfield, Edward D., and Helen Milner. 1999. "The New Wave of Regionalism." *International Organization* 53(3): 589–627.

Manchin, Miriam and Annette Pelkmans-Balaoing. 2007. "Rules of Origin and the Web of East Asian Free Trade Agreement". World Bank Policy Research Working Paper 4273, World Bank, Washington DC.

McKinsey Global Institute. 2014. *Southeast Asia at the crossroads: Three paths to prosperity.* http://www.mckinsey.com/insights/energy_resources_materials/three_paths_to_sustained_economic_growth_in_southeast_asia.

Menon, Jayant and Anna Cassandra Melendez. 2015. "Realizing an ASEAN Economic Community: Progress and Remaining Challenges". ADB Economics Working Paper No. 432, Asian Development Bank, Manila.

Mittelman, James H. 2000. *The Globalization Syndrome: Transformation and Resistance*. New Jersey: Princeton University Press.

Natalegawa, R.M. and Marty M. 2015. "Indonesia's Deft Diplomacy Pays Dividends." *The East Asia Forum*, June 22. http://www.eastasiaforum.org/2015/06/22/indonesias-deft-diplomacy-pays-dividends/.

Nikomborirak, Deunden and Supunnavadee Jitdumrong. 2013. "ASEAN Trade in Services". In *The ASEAN Economic Community — A Work in Progress*, edited by Sanchita Basu Das, Jayant Menon, Rodolfo Severino and Omkar Lal Shrestha. Singapore: Asian Development Bank and Institute of Southeast Asian Studies.

Okabe, Misa and Shujiro Urata. 2013. "The Impact of AFTA on Intra-AFTA Trade." ERIA Discussion Paper 2013-05, Economic Research Institute for ASEAN and East Asia (ERIA), Jakarta.

Ong, Keng Yong. 2014. Speech at the 28th Asia-Pacific Roundtable Plenary Session 6 on "ASEAN's Post-2015 Agenda: Strengthening and Deepening Community Building." Kuala Lumpur, June 2–4.

Pacific Economic Cooperation Council. 2015. *Summary of Symposium "FTAAP: Asia-Pacific Economic Cooperation by 2020 and Beyond" — Recommendations for FTAAP Collective Strategic Study*. Beijing, 15–16 October.

Parameswaran, Prashanth. 2015a. "Confirmed: Philippines wants to join TPP." *The Diplomat*, June 25. http://thediplomat.com/2015/06/confirmed-philippines-wants-to-join-tpp/.

Parameswaran, Prashanth. 2015b. "Why is Indonesia sinking 70 foreign vessels on its independence day?" *The Diplomat*, August 10. http://thediplomat.com/2015/08/why-is-indonesia-sinking-70-foreign-vessels-on-its-independence-day/.

Perlez, Jane. 2015. "U.S. Allies see trans-Pacific partnership as a check on China." *New York Times*, October 6. http://www.nytimes.com/2015/10/07/world/asia/trans-pacific-partnership-china-australia.html?_r=0.

Petri, Peter A., Michael G. Plummer and Fan Zhai. 2011. "The Trans-Pacific Partnership and Asia-Pacific Integration: A Quantitative Assessment." East-West Center Working Paper No.119, East West Center, Honolulu.

Petri, Peter A., Michael G. Plummer and Fan Zhai. 2012. "ASEAN Economic Community: A General Equilibrium Analysis." *Asian Economic Journal* 26(2): 93–118.

Petri, Peter A. and Michael G. Plummer. 2016. "The Economic Effects of the Trans-Pacific Partnership: New Estimates." Working Paper 16-2, Peterson Institute for International Economics, Washington DC.

Postigo, Antonio. 2013. "Formulation of East Asian Free Trade Agreements: Top-down, Bottom-up and Across Border". Working Paper Series 2013, No.13–147. Development Studies Institute, London School of Economics and Political Science, London.

Scollay, Robert. 2004. "Preliminary Assessment of the Proposal for a Free Trade Area of the Asia-Pacific (FTAAP)." An Issues Paper for the APEC Business Advisory Council. http://www.apec.org.au/docs/koreapapers2/sx-rs-paper.pdf.

Srinivasan, T.N. 2003. "India in the Doha Round." Unpublished Working Paper. http://www.econ.yale.edu/~srinivas/India in the Doha Round.pdf.

Sukma, Rizal. 2014. "ASEAN Beyond 2015: The Imperatives for Further Institutional Changes." ERIA Discussion Paper 2014-01, Economic Research Institute for ASEAN and East Asia (ERIA), Jakarta.

Summers, Lawrence H. 1991. "Regionalism and the World Trading System." In *Policy Implications of Trade and Currency Zones: A Symposium*, 295–301. Kansas City: Federal Reserve Bank.

Syailendra, Emirza Adi. 2015. "Consensual Leadership in ASEAN: Will It Endure Under Jokowi?" RSIS Commentary No. 005, S. Rajaratnam School of International Studies, Singapore. https://www.rsis.edu.sg/wp-content/uploads/2015/01/CO15005.pdf.

Tan, Alan Khee-Jin. 2013. "Toward a Single Aviation Market in ASEAN: Regulatory Reform and Industry Challenges". ERIA Discussion Paper 2013-22, Economic Research Institute for ASEAN and East Asia (ERIA), Jakarta.

Tan, Khee Giap, Linda Low, Kong Yam Tan and Amanda Lijuan Lim. 2013a. *Annual Analysis of Competitiveness, Development Strategies and Public Policies on ASEAN-10: 2000–2010*. Singapore: Pearson Education.

Tan, Khee Giap, Mulya Amri, Linda Low and Kong Yam Tan. 2013b. *Competitiveness Analysis and Development Strategies for 33 Indonesian Provinces*. Singapore: World Scientific.

Tan, Khee Giap, Sangiita Wei Cher Yoong, Sasidaran Gopalan and Le Phuong Anh Nguyen. 2015a. *2014 Annual Competitiveness Ranking and Simulation Study for ASEAN-10 and Development Strategies to Enhance Asia Economic Connectivity*. Singapore: World Scientific.

Tan, Khee Giap, Randong Yuan and Sangiita Wei Cher Yoong. 2015b. *2014 Annual Provincial and Regional Competitiveness Analysis for Greater China Economies: Development Strategies under the New Normal*. Singapore: World Scientific.

Tan, Khee Giap, Linda Low, Kartik Rao and Kong Yam Tan. 2015c. *2014 Regional Competitiveness Analysis and a Master Plan on Regional Development Strategies for India*. Singapore: World Scientific.

Tan, Khee Giap, Nurina Merdikawati, Mulya Amri and Kong Yam Tan. 2015d. *2014 Provincial and Inaugural Regional Competitiveness Analysis: Safeguarding Indonesia's Growth Momentum*. Singapore: World Scientific.

Tan, Khee Giap, Nurina Merdikawati, Mulya Amri and Blake Harley Berger. 2015e. *2014 Annual Competitiveness Analysis and Development Strategies for Indonesian Provinces*. Singapore: World Scientific.

Tan, Khee Giap, Nurina Merdikawati, Mulya Amri and Kong Yam Tan. 2016 (forthcoming). *2015 Annual Provincial and Regional Competitiveness Analysis in Indonesia: Optimising Layers of Government and Agricultural Productivity*. Singapore: World Scientific.

Tan, Khee Giap, Le Phuong Anh Nguyen and Trieu Duong Luu Nguyen. 2016. "Make or Break for Vietnam in the TPP Era?" *The Straits Times*, February 25. www.straitstimes.com/opinion/make-or-break-for-vietnam-in-the-tpp-era.

Tham, Siew Yean and Sanchita Basu Das. 2015. "The ASEAN Economic Community and Conflicting Domestic Interests: An Overview." *Journal of Southeast Asian Economies* 32(2): 189–201.

The White House. 2012. "Remarks by President Obama and Prime Minister Shinawatra in a joint press conference." https://www.whitehouse.gov/the-press-office/2012/11/18/ remarks-president-obama-and-prime-minister-shinawatra-joint-press-confer.

The White House. 2016. "Joint statement of the U.S.-ASEAN special leaders' summit: Sunnylands declaration." https://www.whitehouse.gov/the-press-office/2016/02/16/ joint-statement-us-asean-special-leaders-summit-sunnylands-declaration.

Vo, Tri Thanh. 2015. "Managing Domestic Consensus Building for ASEAN Community Building in Vietnam." *Journal of Southeast Asian Economies* 32(2): 275–288.

Viner, Jacob. 1950. *The Customs Union Issue*. New York: Carnegie Endowment for International Peace.

Wang, Yuzhu. 2013. "The RCEP Initiative and ASEAN 'Centrality.'" *China Institute of International Studies*, December 6. http://www.ciis.org.cn/english/2013-12/06/content_6518129.htm.

Widodo, Joko. 2014. "Speech at the 25th ASEAN Summit." Nay Pyi Taw, Myanmar, November 12.

Witular, Rendi A. 2014. "Presenting Maritime Doctrine." *The Jakarta Post*, November 14. http://www.thejakartapost.com/news/2014/11/14/presenting-maritime-doctrine.html.

Wu, Daniel and Marc Mealy. 2012. "Explaining 'Competing' Visions: ASEAN-RCEP, TPP, FTAAP." *Pacific Forum CSIS*, June 26. http://csis.org/publication/pacnet-40-explaining-competing-visions-asean-rcep-tpp-ftaap.

Yahya, Tantowi. 2015. "Has Jokowi put ASEAN on the sidelines?" *The Jakarta Post*, June 1. http://www.thejakartapost.com/news/2015/06/01/has-jokowi-put-asean-sidelines.html.

Discussant Notes 1.1

Mr Nguyen Cam Tu
Deputy Minister,
Ministry of Industry and Trade, Vietnam

First and foremost, I would like to thank the Asia Competitiveness Institute at the Lee Kuan Yew School of Public Policy for honouring me with this opportunity to address the audience of leading scholars and business representatives, on Vietnam's challenges to maintain and enhance national competitiveness in the dynamic regional and global context.

I would also like to take this opportunity to express my high appreciation of the active and valuable contribution made by the Asia Competitiveness Institute in building the intellectual leadership and network for understanding and developing competitiveness in ASEAN. I thank you all for your warm welcome.

The essential issue for any country is to maintain and enhance national competitiveness. It is obvious nowadays that this has to be done in the face of increasing international trade and investment with other countries, whether in the multilateral, regional or bilateral frameworks. The efforts and determination of the country and its industries to compete globally will shape the position of the country.

The experience of Vietnam in our pursuit of prosperity is remarkable. Since the launch of reforms nearly 30 years ago, Vietnam has successfully transformed from being a closed and centrally controlled economy into a market economy and from a poor, less-developed, and war-torn country into a middle-income and developing nation. In 25 years from 1986 to 2010, Vietnam's GDP growth rate averaged 7% per year. Even in the testing time of the global financial crisis and economic recession, Vietnam still managed to sustain the GDP at the rate of approximately 5.7% during 2011–2014. In 2015, our GDP is estimated to grow at 6–6.2%. The macroeconomic and socio-political stability have been maintained and social security has been ensured. Vietnam continues to improve the rule of law, strongly promote people's right to democracy and upgrade the market economy's institutions.

However, "at a crossroads" is the correct reflection of Vietnam's situation at the moment. Vietnam is approaching an important transition point, from economic growth based on comparative advantages, especially in terms of low cost labour force, to growth based on upgrading competitiveness. To be specific, previous integration with the global economy helped Vietnam access foreign markets while also encouraging foreign direct investment (FDI) into Vietnam. This process has driven the structural change from agriculture to manufacturing for exports, boosted productivity and brought about the current level of development for Vietnam. However, value-added within the exporting sectors remains low and there is the threat of eroding cost competitiveness because countries with large pools of low cost labour in the region are also growing. As this model has limited potential for realising Vietnam's five-year plan's goal of becoming a fundamentally industrialised and modern country by 2020, Vietnam needs to implement strategies to upgrade the existing advantages and continue to render reforms on conditions driving productivity.

Our strategies comprise of mainly two familiar elements, which are: enhanced international integration and accelerated domestic reforms.

International integration

Recognising that globalisation is imperative, Vietnam has actively integrated into the region and the world. In this course, economic integration is the focus and key driver. Economic integration delivers access to foreign markets and facilitates FDI inflows that are needed for our structural change. Our economic integration illustrates the experience of a gradual process, comprising seven key steps: (i) unilateral reforms to adapt to international rules and institutions; (ii) restoration of relation to international financial organisations, including the World Bank and International Monetary Fund; (iii) participation in international economic forums, for example APEC, then ASEM of which Vietnam was a founding member; (iv) entering into bilateral trade agreements, notably the bilateral trade agreement with the United States of America on prevailing standards and terms at that time; (v) regional economic integration, especially AFTA of ASEAN, now evolved into AEC; (vi) multilateral integration under the WTO framework; and (vii) entering into free trade agreements.

For free trade agreements in particular, Vietnam has experienced three phases: (i) signed six FTAs including AFTA as a member to ASEAN; (ii) signed two bilateral "traditional" FTAs with Japan and Chile; and (iii) negotiating and signing so-called "new generation" high-standard FTAs. In this connection, it is important to note that Vietnam is entering into the new phase of economic integration, with deeper and broader commitments, especially including those within the ASEAN Economic Community, the EU-Vietnam FTA which was substantially concluded earlier this month and the two ongoing negotiations — the Trans-Pacific Partnership (TPP) and the Regional Comprehensive Economic Partnership (RCEP). They are platforms which will open up immense opportunities for Vietnam to engage more effectively in the global production and value chains. They also put greater pressure on domestic institutions to reform for economic development.

Domestic reforms

Vietnam's efforts in this regard are extensive and wide-ranging. First, Vietnam is redoubling efforts in improving the business environment and reforming administrative procedures. Second, restructuring is being conducted across sectors of the economy, including industry, agriculture, service, investment, banking system and financial institutions, business system and state-owned enterprises. Third, the government continues to create the most favourable conditions for the development of private businesses as well as attracting quality FDI. Fourth, Vietnam will supplement policies to further encourage investment in socio-economic infrastructure development. Fifth, the government also continues to allocate sufficient resources while following the market rules in investment to spur the development of healthcare, education, especially high-quality human resource training to meet the requirements of the labour market and businesses' real demands. We believe that these efforts of domestic reforms will unleash the potential for improving productivity for prosperity in the medium and long term.

Given the time constraints, I would like to draw your attention to three inter-related areas in domestic reforms, namely industry policy, supporting industries and bridging the gap between local and FDI firms.

Industrial policy

A new Strategy for Development of Industry to 2025 with vision to 2035 has been endorsed by the Prime Minister. The key objective of this Strategy is to increase local contents of industrial products, shift from processing and assembling to manufacturing and producing. There are four groups of priority industrial sectors mentioned in this Strategy, including: (i) agro-processing, (ii) apparel and footwear, (iii) electronics and information technology, (iv) auto parts, petro-chemistry, metallurgy, manufacturing engineering, energy. Amongst those four groups of priority industrial sectors, two are related to supporting industries.

Therefore, building up the supporting industries is crucial. The Master Plan for Supporting Industries by 2020, with a vision to 2030 was issued in 2014 by the Minister of Industry and Trade. The Master Plan focuses on development in three broad categories of products including: (i) engineering, electrical or electronic parts and components; (ii) inputs for the apparel and footwear industry; (iii) specialised equipment, software and services for high-tech industries. The Master Plan provides for two main groups of measures including financial and technical assistance measures. As a result, supporting industries are entitled to specific favourable treatments relating to: (i) taxation, duties and land rent, access to credit; (ii) access to utilities, training, technology transfer, information, research and development, infrastructure development, etc.

Bridging the gap between local and foreign invested enterprises

Development of supporting industries is also a solution to the issue of the widening gap between local and foreign invested enterprises in Vietnam. The gap is obvious in a number of dimensions. For example, both value and ratio of contribution by FDI enterprises to total industry output are increasing while those of local firms are shrinking; the value and ratio of exported goods by FDI enterprises in total exports of Vietnam has grown rapidly while the value of exported goods by local firms has increased little and its ratio in total exports of Vietnam has decreased. The cause is that the majority of inputs for production of both domestic and exported goods are imported.

Development of supporting industries is meant to increase local contents or value-added of industrial products made in Vietnam. It is a win–win solution to both local and FDI enterprises. To be more specific, the government and local business community fully recognise that FDI enterprises investing and operating in Vietnam are key customers for supporting industries. Therefore, supporting industries to be built as clusters around FDI enterprises, while benefiting the FDI enterprises seeking local inputs to cut costs in logistics and transportation, could serve as the platform for local firms, including small- and medium-sized enterprises (SMEs), to participate in the global value chains. In this direction, policy priorities focus on creating an environment in which meaningful linkages and positive spillovers can emerge. Support is designed to enable supporting industries, local firms, especially SMEs, to effectively compete with imports to become suppliers to FDI enterprises.

Vietnam is determined to upgrade its national competitiveness. Enhanced international economic integration creates an environment that supports ongoing domestic reforms. In turn, domestic reforms with a priority on industrial policy, especially policies for developing supporting industries, if implemented successfully, will put Vietnam in a better position to accelerate its economic integration.

In light of this, the Seminar today and the Asia Competitiveness Institute play an important role in updating ourselves of where we are in terms of competitiveness. That would be very helpful for the purpose of assessing and identifying the areas that require more attention and efforts to put into policy-making. I, therefore, look forward to learning from subsequent presentations of experts in this field.

On this note, I thank you for your attention. May I wish you all be of good health, happiness and success.

Discussant Notes 1.2

Dr Nguyen Dinh Cung

President,
Central Institute for Economic Management (CIEM), Vietnam

There has been 30 years of *Doi Moi* in Vietnam, which is a process of transformation from the centrally-planned to market-oriented economy. The process has made differences, and all Vietnamese, especially those of my generation, have been personally experiencing that development; our living standards has been enormously improved.

On average, the rate of GDP growth for the period of 1990–2015 is about 7%. As a result, per capita income increased from more than US$100 in 1990 to about US$2000 (in current prices) in 2015. Poverty rate has also decreased significantly, from 58% in 1990 to 9% in 2014. In terms of GDP growth, Vietnam is the second best performer in the world after China; and what Vietnam has achieved in last 30 years has been recognised not only domestically but also internationally.

However, if we look more deeply into Vietnam's economic performance in last three decades, there are two different periods. The first one is between 1991 and 2007, and the second one is from 2008 up to now. The average GDP growth rate for the first period is 7.6%, and for the second period is only 5.8%, 1.8% point lower than that of the first period.

From 2008 to 2012 Vietnam's economy suffered from severe macroeconomic instability; and the economic performance slowed down significantly. Since 2013, the macroeconomic stability has recovered; and the rates of GDP growth have gradually improved. The expected rate of this year is 6.5%. However, there are some critical problems.

In terms of macroeconomic fundamentals, there are non-performing loans, large and chronic budget deficit, high rate of public debt, etc. All of them are likely to make macroeconomic stability very vulnerable.

In terms of microeconomic foundations, there is low efficiency of investment and low productivity of labour. That situation is a result of resource misallocation. Investment has flown increasingly into inefficient sectors such as financial, real estates, retail and wholesale. Investors tend to be speculators and rent-seekers rather than value-added makers. Misallocation and low efficiency of social resources will finally lead to macroeconomic imbalances and instability.

There is obviously something wrong with the incentive system that has caused markets, especially factor ones, to be distorted. Therefore, they have failed to play a decisive role in allocating resources. Why does this happen? There may be several different perspectives. But, in my opinion, the reasons are the followings:

There are inconsistent domestic market reforms. As a result, the process of transformation to full-fledged and modern market economy in Vietnam has not been completed. The current state of the market economy in Vietnam has been a product of reducing only the scope of state rather than of changing the role, functions and ways of public governance. Large size of the state sector and ineffectiveness of public governance create widespread market distortions; and markets are unequally competitive. Government interventions do not resolve market failures, but create both market and state failures.

There is an imbalance between international integration and domestic institutional reforms. The reality of the last 30 years has shown that the success of Vietnam's economy has arisen from the balance of domestic institutional reforms and international integration; they are similar to two wings of a bird. Since 2007 Vietnam has become a member of the WTO, but domestically there have not been any decisive reforms to upgrade market economy institutions. In contrary, we have seen more and more state interventions in the economy. Therefore, an inappropriate institution becomes the key bottleneck in transforming Vietnam into a full-fledged and modern market economy.

In a recent CIEM policy discussion paper, I introduced and analysed key gaps between the current market institutions and requirements of reforms for a more decisive and consistent transformation to a real market economy. My conclusion is that more strong and consistent market institutional reforms for development are imperative for Vietnam and some have called reforms of this kind as "Version 2."

My conclusion is based on two forecast scenarios for Vietnam to 2035:

(i) If GDP per capita growth was only 5% (similar to the current one), then up to 2035 Vietnam's GDP per capita would be about 83% of China and about 75% of Thailand in 2015.
(ii) If GDP per capita growth was 7% annually, then to 2035 Vietnam's GDP per capita would be 98% of Malaysia and 183% of China in 2015.

Which scenario will be realised for Vietnam depends to a large extent on the reform "Version 2", which is much more difficult than the reform Version 1 (*Doi Moi*) as seen since 1986.

Chapter 2

2015 Update on Annual Competitiveness Analysis for ASEAN-10

2.1 INTRODUCTION

The Association of Southeast Asian Nations (ASEAN) was established in 1967 in Bangkok, Thailand, with the signing of the ASEAN Declaration by the founding members, namely Indonesia, Malaysia, Philippines, Singapore and Thailand. Brunei Darussalam then joined in 1984, Vietnam in 1995, Laos and Myanmar in 1997, and Cambodia in 1999, making up the 10 member states of ASEAN. One year away from its 50th anniversary, ASEAN-10 is already the seventh largest economy in the world[1] and remains a rising star despite the 2008–2009 global financial tsunami whose repercussions are still felt and being dealt with around the world.

On the flip side, the development gap between the ASEAN-10 economies remains stark (see Table 2.1). The regional bloc comprises two high-income countries (Brunei and Singapore), one upper-middle-income country (Malaysia), five lower-middle-income countries (Indonesia, Laos, Philippines, Thailand and Vietnam) and two low-income countries (Cambodia and Myanmar).[2] Marked heterogeneity is also observed in human development with Singapore and Brunei in the leading group and Myanmar trailing in the low human development category.

While there is evidence that the process of convergence is already under way, much needs to be done before the development divide, particularly between the newer members of the group, namely Cambodia, Laos, Vietnam and Myanmar (CLMV) and the older, more mature six, can be substantially narrowed.

It is with this objective that the Asia Competitiveness Institute (ACI) at the Lee Kuan Yew School of Public Policy (LKYSPP), National University of Singapore (NUS), has crafted a comprehensive methodology to assess annually, competitiveness at the national level for the ASEAN-10. Now in its third edition,[3] ACI's competitiveness analysis enables policy-makers in the region to evaluate their own competitiveness status as well as identify the gaps that need to be addressed to achieve convergence. This chapter provides an update of the annual competitiveness ranking and simulation analysis with data from 2000 to 2012.

With this latest competitiveness ranking and simulation results, ACI also seeks to contribute to realising the ASEAN Economic Community (AEC) which aims to facilitate greater regional economic integration through constructing a competitive and sustainable "single market and

[1] See McKinsey Global Institute (2014).

[2] According to World Bank's classification.

[3] See Tan *et al.* (2013) and Tan *et al.* (2015) for the first and second edition of ACI's competitiveness analysis for ASEAN-10.

Table 2.1: ASEAN-10 in a Snapshot — 2013

Country	Population (million)	Gross National Income (GNI) Per Capita (2011 PPP$)	Human Development Index
Brunei	0.4	70,883	0.852
Cambodia	15.1	2,805	0.584
Indonesia	249.9	8,970	0.684
Laos	6.8	4,351	0.569
Malaysia	29.7	21,824	0.773
Myanmar	53.3	3,998	0.524
Philippines	98.4	6,381	0.660
Singapore	5.4	72,371	0.901
Thailand	67.0	13,364	0.722
Vietnam	89.7	4,892	0.638

Source: World Bank and United Nations Development Programme.

production base." While the AEC is scheduled to be launched at the end of 2015, the consensus is that this remains a work in progress beyond the 2015 milestone.[4] While regional efforts like the AEC can play a part in narrowing the development gap, solutions must also come from within individual economies. ACI's annual competitiveness analysis of the ASEAN-10 thus offers a useful roadmap for policy-makers in respective member states.

Following this Introduction, Section 2.2 details the methodology and data used to derive the rankings as well as the simulation analysis showing how ASEAN-10 economies can improve their competitiveness in various dimensions. This makes ACI's contribution more instructive than other competitiveness indices which merely report the rankings. Section 2.3 discusses the results of the competitiveness rankings and the competitiveness progress over time. The top 20% strengths and weaknesses of individual ASEAN economies are also presented. This is followed by *what-if* policy simulation that shows changes in the standardised scores as well as the rankings for individual economies following improvement of their top 20% weaknesses. Section 2.4 concludes with policy implications.

2.2 ACI COMPETITIVENESS FRAMEWORK AND METHODOLOGY

2.2.1 Literature Review on Competitiveness

The notion of regional or *in situ* competitiveness has been explored in various ways. Initially, in the 1980's, "competitiveness" was considered as part of the private sector domain, especially in relation to a firm's performance.[5] Over time, the discourse expanded onto the domain of macroeconomics, as shown by the establishment of "Council on Competitiveness" or similar

[4]See Chapter 1 of this book for an in-depth assessment of the completion of AEC.

[5]See Porter (1980).

institutions in various advanced industrialised countries. Some scholars acknowledge that it is primarily firms that compete, not nations, regions, or places.[6] However, the performance of those firms depends on various factors that are embedded in the firm's location. These include quality of labour and infrastructure, governance, costs of doing business, as well as the performance of competing and complementary firms around the area.[7]

Two of the most highly cited cross-country competitiveness studies at the moment are the World Competitiveness Yearbook (WCY) by the IMD World Competitiveness Centre and the World Economic Forum's Global Competitiveness Report (GCR). The WCY defines competitiveness as composed of four factors: (i) economic performance, (ii) government efficiency, (iii) business efficiency, and (iv) infrastructure. Each factor is further made up of five criteria, so that there are 20 criteria in total. The WCY uses equal weights such that each criterion contributes 5% weight, and each factor contributes 25% weight to the overall competitiveness index.[8] Most of the data used are statistical indicators sourced from international, regional and national organisations. However, substantial perception data sourced from executive surveys supplement the statistical data.

The GCR breaks down its Global Competitiveness Index into three sub-indices: (i) basic requirements, (ii) efficiency enhancers, and (iii) innovation and sophistication factors.[9] Each of the three sub-indices has a different number of pillars. The basic requirements sub-index has four pillars: (i) institutions, (ii) infrastructure, (iii) macroeconomic environment and (iv) health and primary education. The efficiency enhancers sub-index is composed of six pillars: (i) higher education and training, (ii) goods market efficiency, (iii) labour market efficiency, (iv) financial market development, (v) technological readiness, and (vi) market size. Lastly, the innovation and sophistication factors sub-index consists of two pillars: (i) business sophistication and (ii) innovation. Most of the data that make up these pillars and sub-indices come from executive opinion surveys.

2.2.2 ACI Competitiveness Framework

While the notion of competitiveness has been defined in various ways, ACI's approach to competitiveness is eclectic and comprehensive as it takes into account different factors that collectively shape the ability of a nation or region to achieve substantial and inclusive economic development over a sustained period of time.

In line with this overarching thrust, ACI defines competitiveness through four environments: (i) Macroeconomic Stability (MS), (ii) Government and Institutional Setting (GIS), (iii) Financial, Businesses and Manpower Conditions (FBMC) and (iv) Quality of Life and Infrastructure Development (QLID). These four environments carry equal weightage (25%) in the computation of the Overall Competitiveness score. Following a nested approach, each environment comprises three sub-environments of the same weightage (33.3%) (see Figure 2.1). While giving different

---

[6]See Krugman (1994).

[7]See Camagni (2002), Kitson, Martin and Tyler (2004) and Bristow (2010).

[8]See IMD World Competitiveness Center (2014).

[9]See World Economic Forum (2014).

Figure 2.1: ACI Competitiveness Framework

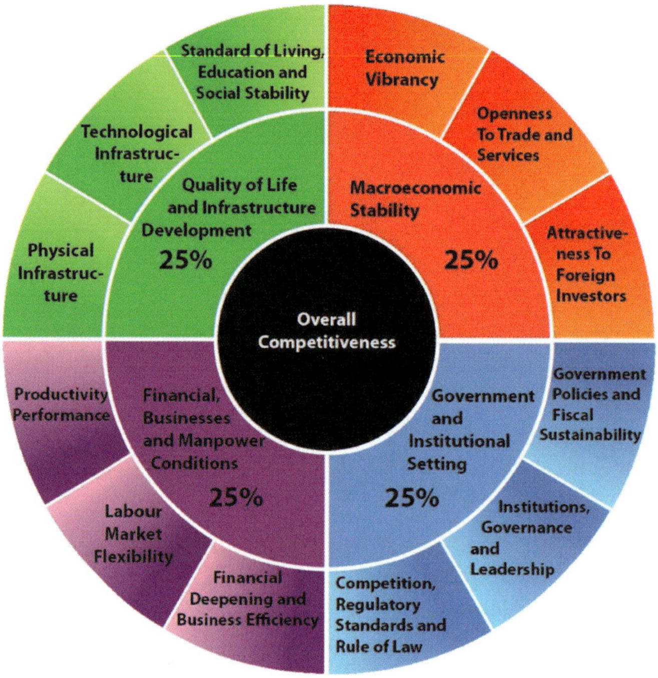

Source: Asia Competitiveness Institute.

weights to different environments and sub-environments based on their importance seems more appropriate, the method is nonetheless arbitrary and contentious. Thus, equal distribution of weight reflects a balanced view toward the different factors that make up an overall notion of competitiveness.

The ACI's framework provides a holistic, integrated and systematic way of examining national competitiveness. Two environments are specifically related to economic performance, namely Macroeconomic Stability and Financial, Businesses and Manpower Conditions. Macroeconomic Stability encompasses aggregated macroeconomic conditions which include: (i) Economic Vibrancy, (ii) Openness to Trade and Services, and (iii) Attractiveness to Foreign Investors. Financial, Businesses and Manpower Conditions, on the other hand, represents the micro-economy, particularly firms and factors that impact their performances. Financial, Businesses and Manpower Conditions thus comprises: (i) Financial Deepening and Business Efficiency, (ii) Labour Market Flexibility, and (iii) Productivity Performance. The remaining two environments are institutional and social in character. Government and Institutional Setting covers efficacy of government institutions and rule of law. This environment consists of: (i) Government Policies and Fiscal Sustainability, (ii) Institutions, Governance and Leadership and (iii) Competition, Regulatory Standards and Rule of Law. Lastly, Quality of Life and Infrastructure Development combines infrastructure, health and education services, environmental standards and social security. The three sub-environments are: (i) Physical Infrastructure, (ii) Technological Infrastructure and (iii) Standard of Living, Education and Social Stability.

There are a total of 121 indicators used in ACI's competitiveness rankings and simulations for ASEAN-10 economies in the period 2000–2012 (see Appendix 2). The extensive number of indicators ensures robust interpretation of each environment and sub-environment in the Overall Competitiveness score. Depending on data availability, some environments and sub-environments may have more indicators than others. Nonetheless, this is not an issue as they simply average out. This also allows a certain amount of flexibility in adding or removing indicators when updating the rankings in the future, so long as the overall structure of the four environments and 12 sub-environments remains unchanged.

2.2.3 Data Sources

The data for the 121 indicators used in this study were collected from various sources, of which the bulk come from official data sources such as the World Bank World Development Indicators, Asian Development Bank (ADB) and World Economic Forum Global Competitiveness Index. Other sources include Economist Intelligence Unit (EIU), Global Market Information Database (GMID), UNESCO Institute for Statistics, International Monetary Fund (IMF) World Economic Outlook Database, United Nations Conference on Trade and Development (UNCTAD), IMD World Competitiveness Yearbook, Environment Performance Index (EPI), International Telecommunication Union ICT Indicators Database, United Nations Development Programme (UNDP), Reporters Without Border and the Ciingranelli–Richards Human Rights Dataset.

2.2.4 Computation of Standardised Scores

The next important question in terms of methodology is how to aggregate the different types of data into one coherent way of analysis. Given 121 different indicators in this research, all in different value or quantity terms, such Gron Domestic Product (GDP) in millions of US dollars, or population in number, how do we resolve these different units of measurement?

The ACI uses a statistical methodology of 'standardised score.' The standardised score is a relative comparison to see how well is the performance of a certain economy in comparison to the average economy. Therefore, the unit of measurement is no longer relevant. The standardised score has no unit of measurement because it simply measures relative performance among economies, whatever the indicator is.

In statistical terms, it measures how many standard deviations away is each economy from the average economy (see Appendix 3 for technical explanation). If an economy has a standardised score of zero, it is basically an average performer in terms of that particular indicator. Having a negative score means the economy performs below average. Having a positive score, on the other hand, means above-average performance.

The further away the score is from zero, the further away is the performance of said economy from the ASEAN-10 average. If an economy has a high positive score, it means it is much above the average score. The standardised scores for each indicator are then aggregated at the sub-environment level, and then re-aggregated at the environment level, and finally aggregated again at the overall level. This allows comparison of performance of ASEAN 10-economies at different levels, from the Overall Competitiveness to specific indicators in four environments.

2.2.5 *What-if* Competitiveness Simulation Analysis

A competitiveness ranking in itself is like a beauty contest; it merely identifies who are doing well and who are facing challenges, but stops short of giving more constructive advice on improving the rankings. The ACI index goes one step further by tackling the 'so what' question. What is the policy implication of a competitiveness ranking result for a particular economy? The data used in the ACI index allows us to do an in-depth analysis of the performance of each economy according to the different indicators, sub-environments and environments. By analysing the said data we are able to identify not just the Overall Competitiveness ranking, but also specific indicators for which the economy is doing well or struggling. This allows us to come up with policy recommendations for each economy.

The ACI has *what-if* competitiveness simulation that is based on an improvement of each economy's top 20% weakest indicators[10] and a re-calculation of the standardised score based on such improvement. To do this, we first looked at each economy and sorted the different indicators for each economy from the highest to the lowest score. This allowed us to identify the 24 weakest indicators that each economy has. Why 24? Because we are looking from the bottom 20% of the 121 indicators used in the ASEAN study.

Next, we raised the scores of these 24 weakest indicators moderately, to match the average score. This means their average score is raised to zero if it was negative previously. For other economies which were already competitive, we leave their scores alone. Their scores are not discounted if they are already competitive. Once these scores are raised, we re-calculate the ranking, assuming that other economies stay constant. So, we do this simulation one-by-one, for each economy. Each economy would then have a new, improved standardised score. This allows us to answer the question: if a particular economy improves on its set of weakest indicators and assuming other economies remain constant, how would that economy's competitiveness ranking improve?

2.3 COMPETITIVENESS ANALYSIS OF ASEAN-10 ECONOMIES, 2000–2012

In this section, we present the Overall Competitiveness rankings as well as the rankings for the four environments as derived from the standardised scores for each economy over the period 2000–2012. The scores for individual sub-environments are visualised in the median and maximum web analysis. Findings on the *what-if* competitiveness simulation are also discussed.

2.3.1 Overall Competitiveness

Table 2.2 and Figure 2.2 show the Overall Competitiveness rankings along with each economy's scores for the period 2000–2012. As of the year 2012, Singapore remained at the top for the 13th consecutive year with the standardised score of 2.0806. While Malaysia's scores have improved

[10]Counting from the bottom of the given list of 121 indications, the bottom 20% will emerge as the weakest, while the top 20% will emerage as the strongest.

Table 2.2: Overall Competitiveness Rankings and Standardised Scores, 2000–2012

Overall										
Rank	**1**		**2**		**3**		**4**		**5**	
	Economy	**Score**	**Economy**	**Score**	**Economy**	**Score**	**Economy**	**Score**	**Economy**	**Score**
2000	Singapore	2.1239	Malaysia	0.9908	Thailand	0.5458	Brunei	0.3689	Philippines	–0.1195
2001	Singapore	2.1000	Malaysia	0.9536	Thailand	0.6112	Brunei	0.3851	Philippines	–0.0875
2002	Singapore	1.9583	Malaysia	1.0812	Thailand	0.7688	Brunei	0.3517	Philippines	–0.1576
2003	Singapore	1.9688	Malaysia	1.1016	Thailand	0.7234	Brunei	0.3638	Indonesia	–0.1557
2004	Singapore	1.9523	Malaysia	1.1292	Thailand	0.7011	Brunei	0.4030	Indonesia	–0.1152
2005	Singapore	2.0054	Malaysia	1.1140	Thailand	0.6922	Brunei	0.3512	Philippines	–0.1869
2006	Singapore	1.9707	Malaysia	1.1402	Thailand	0.6548	Brunei	0.3579	Indonesia	–0.0792
2007	Singapore	2.0348	Malaysia	1.1206	Thailand	0.5445	Brunei	0.2986	Indonesia	–0.0012
2008	Singapore	2.1125	Malaysia	1.0198	Thailand	0.5143	Brunei	0.3318	Indonesia	0.0342
2009	Singapore	2.1774	Malaysia	0.9289	Thailand	0.4893	Brunei	0.3507	Indonesia	0.0415
2010	Singapore	2.1097	Malaysia	1.0029	Thailand	0.4872	Brunei	0.3896	Indonesia	0.0842
2011	Singapore	2.0944	Malaysia	1.1388	Thailand	0.4710	Brunei	0.3683	Indonesia	0.0013
2012	Singapore	2.0806	Malaysia	1.1509	Thailand	0.4630	Brunei	0.3699	Indonesia	–0.0788
Overall										
Rank	**6**		**7**		**8**		**9**		**10**	
	Economy	**Score**	**Economy**	**Score**	**Economy**	**Score**	**Economy**	**Score**	**Economy**	**Score**
2000	Indonesia	–0.1641	Vietnam	–0.3582	Cambodia	–0.9110	Myanmar	–1.1786	Laos	–1.2981
2001	Indonesia	–0.1994	Vietnam	–0.3341	Cambodia	–0.9157	Laos	–1.2175	Myanmar	–1.2957
2002	Indonesia	–0.2266	Vietnam	–0.2931	Cambodia	–0.9383	Laos	–1.2438	Myanmar	–1.3006
2003	Philippines	–0.2447	Vietnam	–0.2738	Cambodia	–0.9813	Myanmar	–1.2111	Laos	–1.2908
2004	Philippines	–0.2419	Vietnam	–0.3663	Cambodia	–0.9640	Laos	–1.2053	Myanmar	–1.2930
2005	Indonesia	–0.2245	Vietnam	–0.3420	Cambodia	–0.9627	Laos	–1.1435	Myanmar	–1.3033
2006	Philippines	–0.2224	Vietnam	–0.3548	Cambodia	–0.9667	Laos	–1.1568	Myanmar	–1.3437
2007	Philippines	–0.2155	Vietnam	–0.3108	Cambodia	–0.9851	Laos	–1.1571	Myanmar	–1.3288
2008	Vietnam	–0.2904	Philippines	–0.2994	Cambodia	–0.9996	Laos	–1.1324	Myanmar	–1.2908
2009	Vietnam	–0.1959	Philippines	–0.4250	Cambodia	–1.0402	Laos	–1.0526	Myanmar	–1.2741
2010	Vietnam	–0.2230	Philippines	–0.4467	Cambodia	–1.0013	Laos	–1.1364	Myanmar	–1.2663
2011	Philippines	–0.3593	Vietnam	–0.4492	Cambodia	–0.8869	Laos	–1.1085	Myanmar	–1.2699
2012	Philippines	–0.2906	Vietnam	–0.4570	Cambodia	–0.7462	Laos	–1.1141	Myanmar	–1.3776

Source: Asia Competitiveness Institute.

Figure 2.2: Overall Competitiveness — 2000, 2011 and 2012

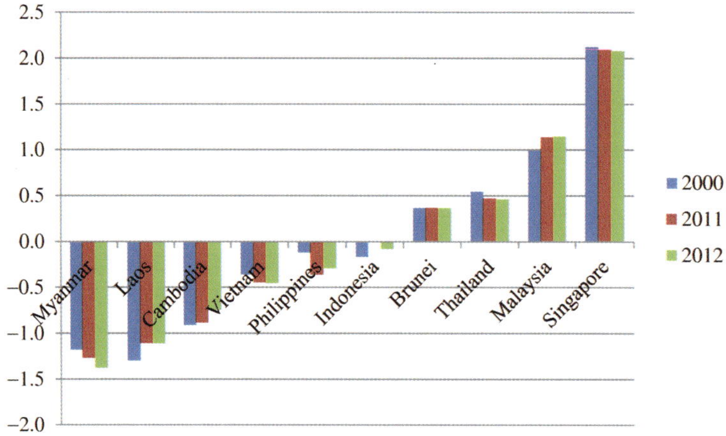

Source: Asia Competitiveness Institute.

remarkably over the years, it still came a distant second to its neighbor, Singapore. Thailand, Brunei and Cambodia maintained their third, fourth and eighth position, respectively, throughout the study period. Indonesia slipped below average, as indicated by its negative score, in 2012 although it continued to rank fifth. For those with consistently negative scores, Philippines's score improved slightly between 2011 and 2012 while Vietnam, Laos and Myanmar fared worse in 2012 compared to 2011.

As evident in Figure 2.2, while the gap between the top two most competitive economies has narrowed, albeit modestly, over the years, the remaining eight economies continue to considerably lag behind, with the Philippines, Vietnam and Myanmar plunging even more negatively in 2012 compared to 2000.

2.3.2 Macroeconomic Stability

Table 2.3 and Figure 2.3 show the 2012 rankings for the first environment, Macroeconomic Stability. The top three remained the same as the Overall Competitiveness ranking with Indonesia at fourth and Brunei at fifth, all with positive scores. However, unlike the Overall Competitiveness scores, the gap between the top two most competitiveness economies was as wide in 2012 as in 2000 with Singapore's scores remaining more than twice those of Malaysia. Indonesia, Brunei and Laos saw marked improvement in their scores while the Philippines, Vietnam, Cambodia and Myanmar's scores deteriorated in 2012 compared to 2000.

Among the three sub-environments constituting Macroeconomic Stability, Indonesia came as the top in Economic Vibrancy in 2012, followed by Singapore (see Figure 2.4). Thailand overtook Malaysia to become third. The Philippines's score turned positive while that of Brunei became negative in 2012. In the bottom four, Myanmar saw a marked improvement in its score while Vietnam, Laos and Cambodia performed more poorly in 2012 compared to 2011.

Table 2.3: Macroeconomic Stability Rankings and Standardised Scores, 2000–2012

Macroeconomic Stability

Rank	1		2		3		4		5	
	Economy	**Score**	**Economy**	**Score**	**Economy**	**Score**	**Economy**	**Score**	**Economy**	**Score**
2000	Singapore	2.1562	Malaysia	0.9077	Thailand	0.6735	Indonesia	0.2341	Brunei	–0.1823
2001	Singapore	2.0598	Malaysia	0.8756	Thailand	0.7058	Indonesia	0.1872	Brunei	0.0519
2002	Singapore	1.9859	Malaysia	0.9133	Thailand	0.7521	Indonesia	0.2821	Brunei	0.0130
2003	Singapore	2.1198	Malaysia	0.8732	Thailand	0.6850	Indonesia	0.3986	Brunei	–0.0977
2004	Singapore	2.1520	Malaysia	0.8813	Thailand	0.7722	Indonesia	0.3135	Brunei	–0.1465
2005	Singapore	2.1533	Malaysia	0.9081	Thailand	0.7493	Indonesia	0.2526	Brunei	–0.1120
2006	Singapore	2.0974	Malaysia	0.8844	Thailand	0.7543	Indonesia	0.4073	Brunei	–0.1111
2007	Singapore	2.0896	Malaysia	0.9295	Thailand	0.7381	Indonesia	0.4565	Philippines	–0.2148
2008	Singapore	2.1087	Malaysia	0.8279	Thailand	0.7582	Indonesia	0.6014	Brunei	–0.1712
2009	Singapore	2.2261	Thailand	0.7064	Malaysia	0.6693	Indonesia	0.6105	Brunei	–0.1573
2010	Singapore	2.2035	Malaysia	0.7255	Thailand	0.5937	Indonesia	0.5660	Brunei	0.0178
2011	Singapore	2.2375	Malaysia	0.7825	Indonesia	0.5680	Thailand	0.4867	Brunei	0.0409
2012	Singapore	2.1161	Malaysia	0.8426	Thailand	0.6299	Indonesia	0.4749	Brunei	0.0264

Macroeconomic Stability

Rank	6		7		8		9		10	
	Economy	**Score**	**Economy**	**Score**	**Economy**	**Score**	**Economy**	**Score**	**Economy**	**Score**
2000	Philippines	–0.2725	Vietnam	–0.3460	Cambodia	–0.5335	Myanmar	–1.1980	Laos	–1.4393
2001	Vietnam	–0.2157	Philippines	–0.3299	Cambodia	–0.5333	Laos	–1.1871	Myanmar	–1.6142
2002	Vietnam	–0.2474	Philippines	–0.2482	Cambodia	–0.6372	Laos	–1.1955	Myanmar	–1.6182
2003	Vietnam	–0.3144	Philippines	–0.3716	Cambodia	–0.6918	Myanmar	–1.2389	Laos	–1.3623
2004	Philippines	–0.3833	Vietnam	–0.5008	Cambodia	–0.5935	Laos	–1.1639	Myanmar	–1.3310
2005	Philippines	–0.3865	Vietnam	–0.4697	Cambodia	–0.6246	Laos	–1.0249	Myanmar	–1.4458
2006	Philippines	–0.3337	Vietnam	–0.5096	Cambodia	–0.7202	Laos	–0.9756	Myanmar	–1.4933
2007	Brunei	–0.2931	Vietnam	–0.4951	Cambodia	–0.8055	Laos	–0.9818	Myanmar	–1.4235
2008	Vietnam	–0.4876	Philippines	–0.5150	Cambodia	–0.7689	Laos	–1.0369	Myanmar	–1.3165
2009	Vietnam	–0.4063	Philippines	–0.6331	Cambodia	–0.8452	Laos	–0.9051	Myanmar	–1.2651
2010	Vietnam	–0.4257	Philippines	–0.4841	Cambodia	–0.7914	Laos	–1.1302	Myanmar	–1.2751
2011	Philippines	–0.4335	Vietnam	–0.6299	Cambodia	–0.7457	Laos	–1.0551	Myanmar	–1.2514
2012	Philippines	–0.3497	Vietnam	–0.4992	Cambodia	–0.6995	Laos	–1.0487	Myanmar	–1.4928

Source: Asia Competitiveness Institute

Figure 2.3: Macroeconomic Stability — 2000, 2011 and 2012

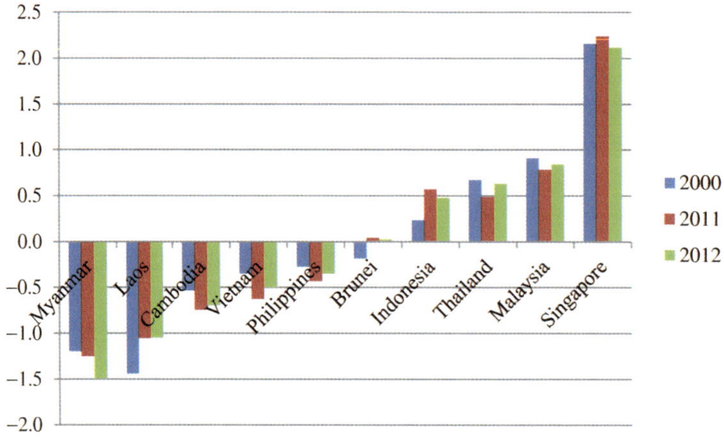

Source: Asia Competitiveness Institute.

Figure 2.4: Macroeconomic Stability (Economic Vibrancy) — 2000, 2011 and 2012

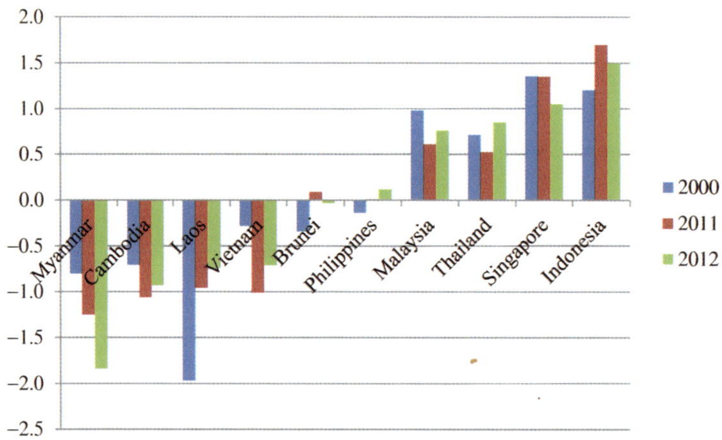

Source: Asia Competitiveness Institute.

In the ranking for Openness to Trade and Services, Singapore remained at the top in 2012, followed by Thailand and Malaysia at second and third, respectively (see Figure 2.5). Vietnam overtook Indonesia to become fourth as Indonesia's score plunged even more negatively in 2012. Brunei (sixth), the Philippines (seventh) and Laos (tenth) all saw a decline in their scores in 2012 compared to 2000. Cambodia and Myanmar, while remaining in the bottom three, performed better over the years.

Singapore and Malaysia were consistently the top two in the ranking for Attractiveness to Foreign Investors (see Figure 2.6). Brunei rose to become third as its score turned positive while those of Thailand and Cambodia turned negative in 2012 compared to 2000. Indonesia saw a remarkable increase in its score in 2012 compared to 2000 but it remained below average. In the bottom four, the Philippines and Vietnam fared worse while Laos and Myanmar improved over the study period.

Figure 2.5: Macroeconomic Stability (Openness to Trade and Services) — 2000, 2011 and 2012

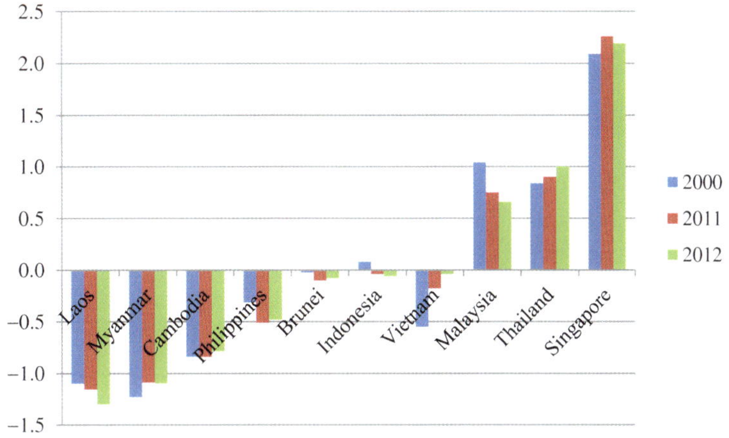

Source: Asia Competitiveness Institute.

Figure 2.6: Macroeconomic Stability (Attractiveness to Foreign Investors) — 2000, 2011 and 2012

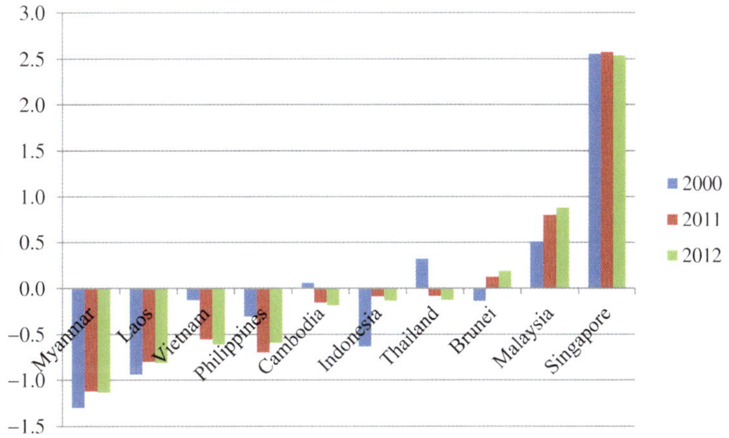

Source: Asia Competitiveness Institute.

2.3.3 Government and Institutional Setting

The same duo, Singapore and Malaysia, dominated the first and second place in the ranking for Government and Institutional Setting throughout the study period (see Table 2.4). Singapore is ahead of its neighbours as a developmental state with an able government that is not captured by private interests. Brunei overtook Thailand at third since 2009. Notwithstanding its political upheavals, Thailand remained above average and came as fourth, followed by the Philippines and Vietnam, both with lower scores in 2012 compared to 2000. In the bottom four, Indonesia, Cambodia and Laos improved, albeit marginally, over the years, while Myanmar fared worse.

Table 2.4: Government and Institutional Setting Rankings and Standardised Scores, 2000–2012

Government and Institutional Setting

Rank	1		2		3		4		5	
	Economy	**Score**	**Economy**	**Score**	**Economy**	**Score**	**Economy**	**Score**	**Economy**	**Score**
2000	Singapore	2.2693	Malaysia	0.7820	Thailand	0.5544	Brunei	0.4529	Philippines	–0.0825
2001	Singapore	2.2597	Malaysia	0.7858	Thailand	0.5783	Brunei	0.3796	Philippines	–0.0315
2002	Singapore	2.1408	Malaysia	0.9519	Thailand	0.7242	Brunei	0.2528	Philippines	–0.1287
2003	Singapore	2.0910	Malaysia	0.9197	Thailand	0.8536	Brunei	0.2530	Vietnam	–0.1458
2004	Singapore	2.0482	Malaysia	1.0397	Thailand	0.7397	Brunei	0.4226	Philippines	–0.2537
2005	Singapore	2.0332	Malaysia	1.1002	Thailand	0.7199	Brunei	0.4004	Philippines	–0.1866
2006	Singapore	2.1020	Malaysia	0.9913	Thailand	0.6583	Brunei	0.4256	Philippines	–0.2348
2007	Singapore	2.1706	Malaysia	0.9937	Thailand	0.5547	Brunei	0.4324	Philippines	–0.2147
2008	Singapore	2.3313	Malaysia	0.8253	Thailand	0.4646	Brunei	0.4439	Vietnam	–0.2743
2009	Singapore	2.3899	Malaysia	0.6473	Brunei	0.5178	Thailand	0.4600	Indonesia	–0.2614
2010	Singapore	2.3214	Malaysia	0.7887	Brunei	0.4853	Thailand	0.4054	Indonesia	–0.1357
2011	Singapore	2.2404	Malaysia	1.0061	Brunei	0.5430	Thailand	0.3730	Indonesia	–0.3216
2012	Singapore	2.2770	Malaysia	1.0132	Brunei	0.4883	Thailand	0.2377	Philippines	–0.2001

Government and Institutional Setting

Rank	6		7		8		9		10	
	Economy	**Score**	**Economy**	**Score**	**Economy**	**Score**	**Economy**	**Score**	**Economy**	**Score**
2000	Vietnam	–0.3243	Indonesia	–0.5557	Cambodia	–0.7959	Myanmar	–1.0355	Laos	–1.2647
2001	Vietnam	–0.2665	Indonesia	–0.5096	Cambodia	–0.8640	Myanmar	–1.1387	Laos	–1.1930
2002	Vietnam	–0.1978	Indonesia	–0.4630	Cambodia	–0.8837	Myanmar	–1.1935	Laos	–1.2030
2003	Philippines	–0.2891	Indonesia	–0.3601	Myanmar	–0.9579	Cambodia	–1.0387	Laos	–1.3259
2004	Vietnam	–0.2650	Indonesia	–0.4214	Cambodia	–0.9789	Laos	–1.0923	Myanmar	–1.2388
2005	Vietnam	–0.2802	Indonesia	–0.5244	Cambodia	–0.9321	Laos	–1.1148	Myanmar	–1.2155
2006	Indonesia	–0.2431	Vietnam	–0.3787	Cambodia	–0.9419	Laos	–1.1174	Myanmar	–1.2614
2007	Indonesia	–0.3442	Vietnam	–0.3722	Cambodia	–0.9003	Laos	–1.1335	Myanmar	–1.1865
2008	Philippines	–0.3074	Indonesia	–0.4274	Cambodia	–0.9678	Myanmar	–1.0213	Laos	–1.0670
2009	Vietnam	–0.2710	Philippines	–0.4816	Cambodia	–0.9052	Laos	–1.0353	Myanmar	–1.0606
2010	Vietnam	–0.3307	Philippines	–0.4000	Cambodia	–0.9332	Myanmar	–1.0384	Laos	–1.1629
2011	Philippines	–0.3356	Vietnam	–0.5225	Cambodia	–0.7659	Myanmar	–1.0832	Laos	–1.1339
2012	Vietnam	–0.4187	Indonesia	–0.4260	Cambodia	–0.6735	Laos	–1.1073	Myanmar	–1.1905

Source: Asia Competitiveness Institute.

Figure 2.7: Government and Institutional Setting — 2000, 2011 and 2012

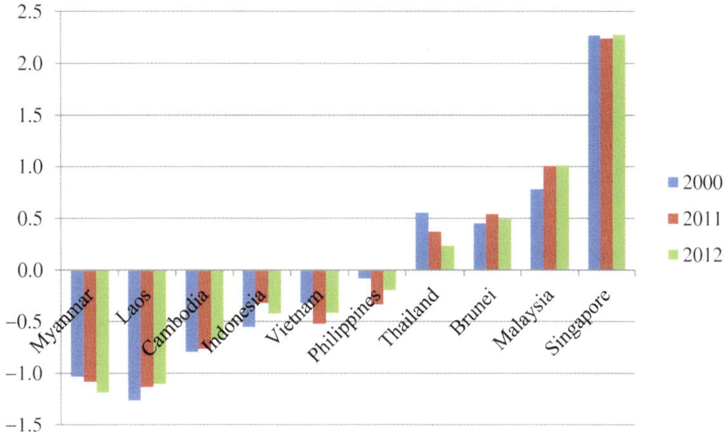

Source: Asia Competitiveness Institute.

Figure 2.8: Government and Institutional Setting (Government Policies and Fiscal Sustainability) — 2000, 2011 and 2012

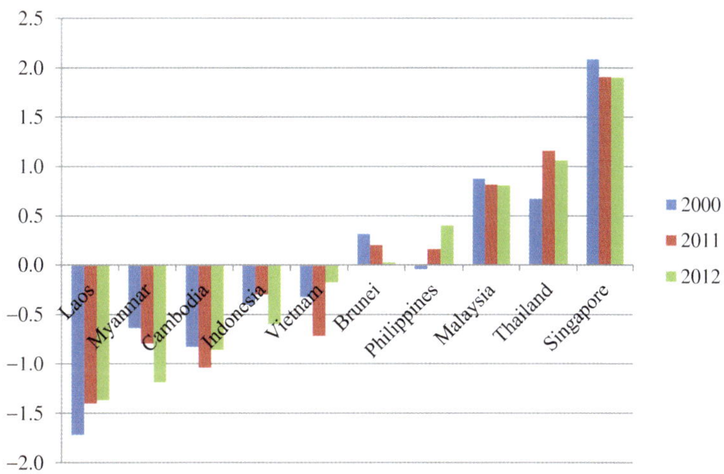

Source: Asia Competitiveness Institute.

Unsurprisingly, Singapore came at the top for all three sub-environments, namely Government Policies and Fiscal Sustainability (Figure 2.8), Institutions, Governance and Leadership (Figure 2.9), and Competition, Regulatory Standards and Rule of Law (Figure 2.10) throughout the selected years. Malaysia ranked at second for Institutions, Governance and Leadership and Competition, Regulatory Standards and Rule of Law. However, Malaysia lost its second place in the ranking for Government Policies and Fiscal Sustainability to Thailand by 2011. In the remaining two sub-environments, nonetheless, Thailand's score turned negative in 2012. The Philippines was above average for Government Policies and Fiscal Sustainability but below average for the other two sub-environments.

Figure 2.9: Government and Institutional Setting (Institutions, Governance and Leadership) — 2000, 2011 and 2012

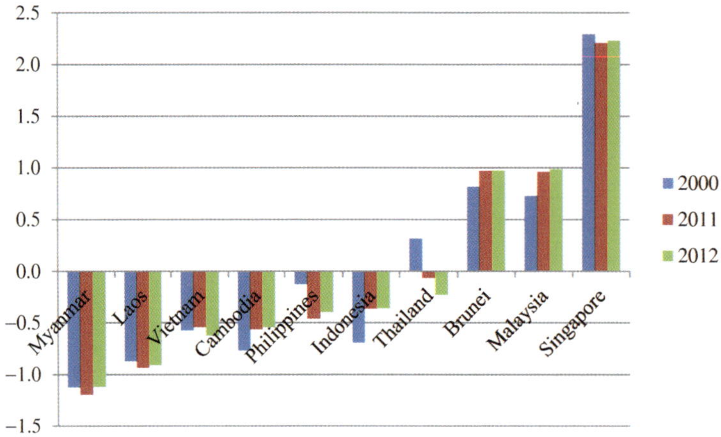

Source: Asia Competitiveness Institute.

Figure 2.10: Government and Institutional Setting (Competition, Regulatory Standards and Rule of Law) — 2000, 2011 and 2012

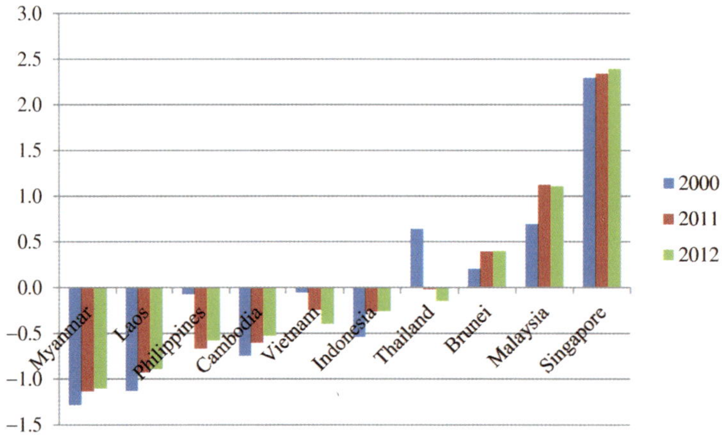

Source: Asia Competitiveness Institute.

2.3.4 Financial, Businesses and Manpower Conditions

Table 2.5 and Figure 2.11 show the rankings for Financial, Businesses and Manpower Conditions in the selected years. Malaysia overtook Singapore to become the top by 2011, followed by Thailand and Brunei consistently at third and fourth, respectively. Indonesia ranked fifth despite falling below average in 2012. Both the Philippines (sixth) and Vietnam (seventh) saw a marked deterioration in their scores in 2012 compared to 2000. In the bottom three, Laos and Myanmar are losing to Cambodia.

Table 2.5: Financial, Businesses and Manpower Conditions Rankings and Standardised Scores, 2000–2012

Financial, Businesses and Manpower Conditions

Rank	1		2		3		4		5	
	Economy	Score	Economy	Score	Economy	Score	Economy	Score	Economy	Score
2000	Singapore	1.8869	Malaysia	1.2310	Thailand	0.5368	Brunei	0.3147	Indonesia	–0.0369
2001	Singapore	1.9363	Malaysia	1.0842	Thailand	0.7104	Brunei	0.2913	Philippines	–0.0349
2002	Singapore	1.6130	Malaysia	1.3420	Thailand	0.9718	Brunei	0.3104	Vietnam	–0.1548
2003	Singapore	1.5874	Malaysia	1.4110	Thailand	0.8429	Brunei	0.4400	Vietnam	–0.0425
2004	Singapore	1.5480	Malaysia	1.4322	Thailand	0.8202	Brunei	0.4044	Indonesia	–0.0859
2005	Singapore	1.7594	Malaysia	1.2542	Thailand	0.8434	Brunei	0.2397	Vietnam	–0.0425
2006	Singapore	1.6762	Malaysia	1.4368	Thailand	0.7728	Brunei	0.3098	Vietnam	–0.1293
2007	Singapore	1.6784	Malaysia	1.4990	Thailand	0.5284	Brunei	0.3076	Indonesia	0.0258
2008	Singapore	1.7821	Malaysia	1.3812	Thailand	0.5387	Brunei	0.1980	Indonesia	0.1038
2009	Singapore	1.8066	Malaysia	1.3836	Thailand	0.5433	Brunei	0.1705	Vietnam	0.0838
2010	Singapore	1.6513	Malaysia	1.4975	Thailand	0.6865	Brunei	0.2125	Indonesia	0.0470
2011	Malaysia	1.6294	Singapore	1.6175	Thailand	0.7782	Brunei	0.0608	Indonesia	0.0347
2012	Malaysia	1.6369	Singapore	1.6190	Thailand	0.7753	Brunei	0.1742	Indonesia	–0.0486

Financial, Businesses and Manpower Conditions

Rank	6		7		8		9		10	
	Economy	Score	Economy	Score	Economy	Score	Economy	Score	Economy	Score
2000	Vietnam	–0.1058	Philippines	–0.1546	Laos	–1.1772	Myanmar	–1.2098	Cambodia	–1.2851
2001	Indonesia	–0.1551	Vietnam	–0.1942	Myanmar	–1.1363	Laos	–1.2167	Cambodia	–1.2850
2002	Philippines	–0.1730	Indonesia	–0.2728	Myanmar	–1.1025	Cambodia	–1.2353	Laos	–1.2988
2003	Philippines	–0.3015	Indonesia	–0.3105	Cambodia	–1.1277	Laos	–1.2223	Myanmar	–1.2768
2004	Vietnam	–0.0973	Philippines	–0.3100	Cambodia	–1.1730	Myanmar	–1.2214	Laos	–1.3174
2005	Philippines	–0.0745	Indonesia	–0.3262	Myanmar	–1.2151	Cambodia	–1.2159	Laos	–1.2224
2006	Philippines	–0.2417	Indonesia	–0.2446	Cambodia	–1.0677	Myanmar	–1.2262	Laos	–1.2862
2007	Vietnam	–0.0496	Philippines	–0.3526	Cambodia	–1.0834	Laos	–1.2674	Myanmar	–1.2863
2008	Vietnam	–0.0771	Philippines	–0.2737	Cambodia	–1.0748	Laos	–1.1988	Myanmar	–1.3794
2009	Indonesia	0.0279	Philippines	–0.4368	Laos	–1.0628	Cambodia	–1.2148	Myanmar	–1.3013
2010	Vietnam	0.0404	Philippines	–0.6348	Laos	–1.0776	Cambodia	–1.1366	Myanmar	–1.2862
2011	Vietnam	–0.3231	Philippines	–0.4007	Cambodia	–1.0941	Laos	–1.1028	Myanmar	–1.2002
2012	Philippines	–0.4083	Vietnam	–0.5101	Cambodia	–0.7775	Laos	–1.1766	Myanmar	–1.2844

Source: Asia Competitiveness Institute.

Figure 2.11: Financial, Businesses and Manpower Conditions — 2000, 2011 and 2012

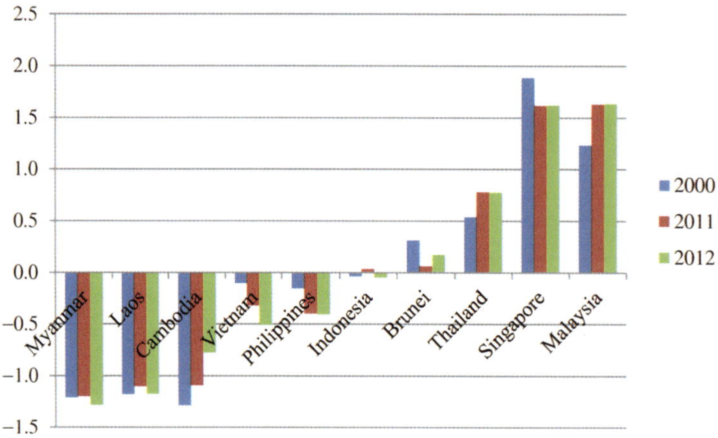

Source: Asia Competitiveness Institute.

Figure 2.12: Financial, Businesses and Manpower Conditions (Financial Deepening and Business Efficiency) — 2000, 2011 and 2012

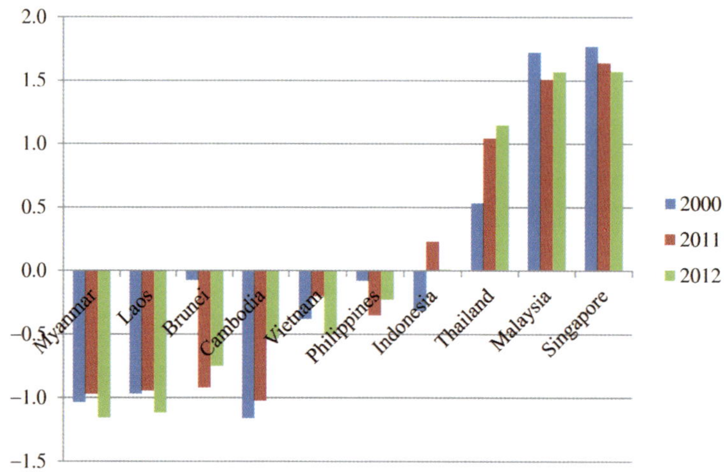

Source: Asia Competitiveness Institute.

The rankings for the three sub-environments are shown in Figures 2.12–2.14. Singapore remained at the top for Financial Deepening and Business Efficiency while Malaysia and Brunei took the top spot for Labour Market Flexibility and Productivity Performance, respectively. Indonesia rose above average for Financial Deepening and Business Efficiency, while Vietnam and the Philippines slipped below average for Labour Market Flexibility in 2012. In terms of Productivity Performance, only Singapore and Malaysia have improved over the years while the remaining eight economies have either stagnated or deteriorated.

Figure 2.13: Financial, Businesses and Manpower Conditions (Labour Market Flexibility) — 2000, 2011 and 2012

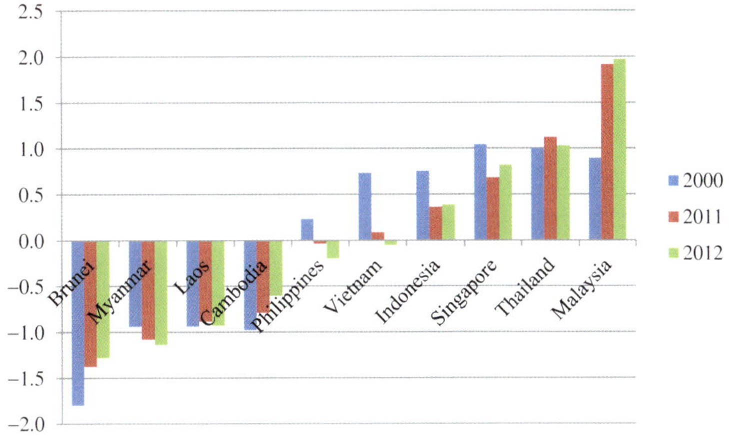

Source: Asia Competitiveness Institute.

Figure 2.14: Financial, Businesses and Manpower Conditions (Productivity Performance) — 2000, 2011 and 2012

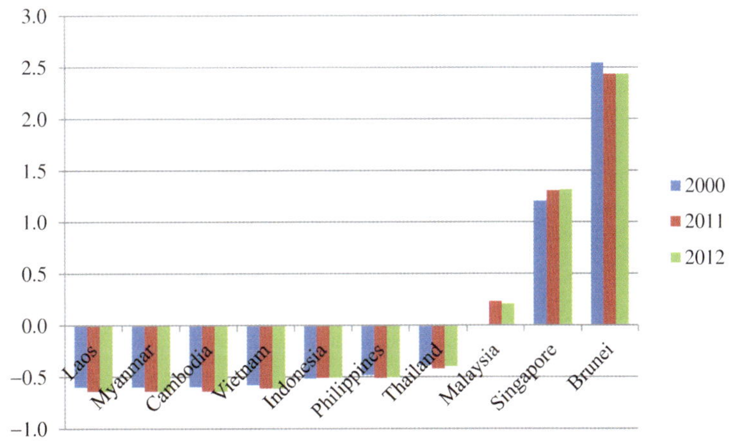

Source: Asia Competitiveness Institute.

2.3.5 Quality of Life and Infrastructure Development

Singapore, Malaysia and Brunei remained in the top three places in the ranking for Quality of Life and Infrastructure Development, followed by Thailand as the only four economies with positive scores throughout the study period (see Table 2.6). Philippines at fifth slipped below average since 2005. Vietnam, Cambodia and Laos fared better in 2012 compared to 2000, while Indonesia was stagnant and Myanmar was losing in competitiveness in this environment.

Table 2.6: Quality of Life and Infrastructure Development Rankings and Standardised Scores, 2000–2012

Quality of Life and Infrastructure Development

Rank	1		2		3		4		5	
	Economy	Score	Economy	Score	Economy	Score	Economy	Score	Economy	Score
2000	Singapore	2.0026	Malaysia	0.9585	Brunei	0.8591	Thailand	0.3722	Philippines	0.0417
2001	Singapore	1.9991	Malaysia	1.0029	Brunei	0.7909	Thailand	0.4080	Philippines	0.0522
2002	Singapore	1.9456	Malaysia	1.0357	Brunei	0.8039	Thailand	0.5690	Philippines	–0.0687
2003	Singapore	1.9009	Malaysia	1.1041	Brunei	0.8271	Thailand	0.4472	Philippines	0.0051
2004	Singapore	1.8869	Malaysia	1.0627	Brunei	0.8956	Thailand	0.4097	Philippines	0.0012
2005	Singapore	1.9222	Malaysia	1.1083	Brunei	0.8498	Thailand	0.4033	Philippines	–0.0857
2006	Singapore	1.8676	Malaysia	1.1674	Brunei	0.7818	Thailand	0.3875	Philippines	–0.0636
2007	Singapore	2.0343	Malaysia	0.9684	Brunei	0.7231	Thailand	0.3124	Philippines	–0.0624
2008	Singapore	2.0208	Malaysia	0.9446	Brunei	0.8241	Thailand	0.2453	Philippines	–0.0721
2009	Singapore	2.0509	Malaysia	0.9147	Brunei	0.8336	Thailand	0.1942	Philippines	–0.1023
2010	Singapore	2.0717	Malaysia	0.9092	Brunei	0.8075	Thailand	0.2192	Indonesia	–0.1483
2011	Singapore	2.0707	Malaysia	1.0225	Brunei	0.7913	Thailand	0.1986	Philippines	–0.2313
2012	Singapore	2.1199	Malaysia	1.0058	Brunei	0.7567	Thailand	0.1667	Philippines	–0.1776

Quality of Life and Infrastructure Development

Rank	6		7		8		9		10	
	Economy	Score	Economy	Score	Economy	Score	Economy	Score	Economy	Score
2000	Indonesia	–0.2839	Vietnam	–0.6262	Cambodia	–0.9520	Myanmar	–1.1708	Laos	–1.2011
2001	Indonesia	–0.3062	Vietnam	–0.6370	Cambodia	–0.9173	Laos	–1.1888	Myanmar	–1.2037
2002	Indonesia	–0.4356	Vietnam	–0.5501	Cambodia	–0.9260	Laos	–1.1838	Myanmar	–1.1900
2003	Indonesia	–0.3369	Vietnam	–0.5683	Cambodia	–0.9793	Laos	–1.1374	Myanmar	–1.2626
2004	Indonesia	–0.2568	Vietnam	–0.5694	Cambodia	–1.0245	Laos	–1.1400	Myanmar	–1.2653
2005	Indonesia	–0.2828	Vietnam	–0.5495	Cambodia	–1.0045	Laos	–1.1243	Myanmar	–1.2369
2006	Indonesia	–0.2308	Vietnam	–0.3763	Cambodia	–1.0685	Laos	–1.1661	Myanmar	–1.2990
2007	Indonesia	–0.1429	Vietnam	–0.3009	Cambodia	–1.0706	Laos	–1.1510	Myanmar	–1.3104
2008	Indonesia	–0.1443	Vietnam	–0.2942	Cambodia	–1.0888	Laos	–1.1160	Myanmar	–1.3195
2009	Vietnam	–0.1688	Indonesia	–0.2155	Cambodia	–1.0826	Laos	–1.0931	Myanmar	–1.3312
2010	Vietnam	–0.1558	Philippines	–0.2273	Cambodia	–1.0535	Laos	–1.0719	Myanmar	–1.3508
2011	Vietnam	–0.2762	Indonesia	–0.2763	Cambodia	–0.8524	Laos	–1.0304	Myanmar	–1.4165
2012	Indonesia	–0.3083	Vietnam	–0.3583	Cambodia	–0.7661	Laos	–1.0221	Myanmar	–1.4168

Source: Asia Competitiveness Institute

Figure 2.15: Quality of Life and Infrastructure Development — 2000, 2011 and 2012

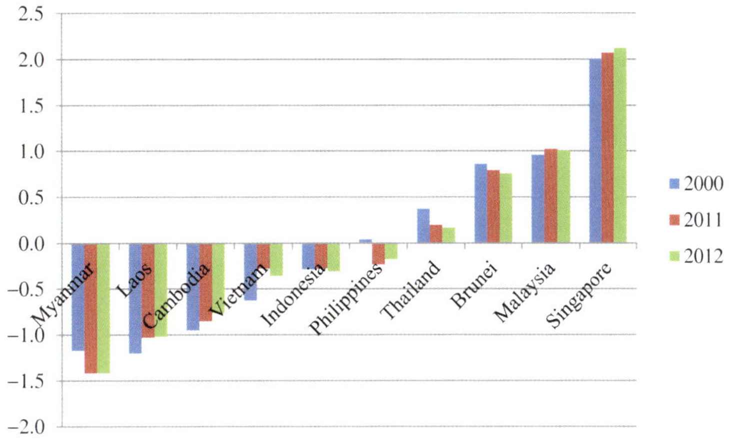

Source: Asia Competitiveness Institute.

Figure 2.16: Quality of Life and Infrastructure Development (Physical Infrastructure) — 2000, 2011 and 2012

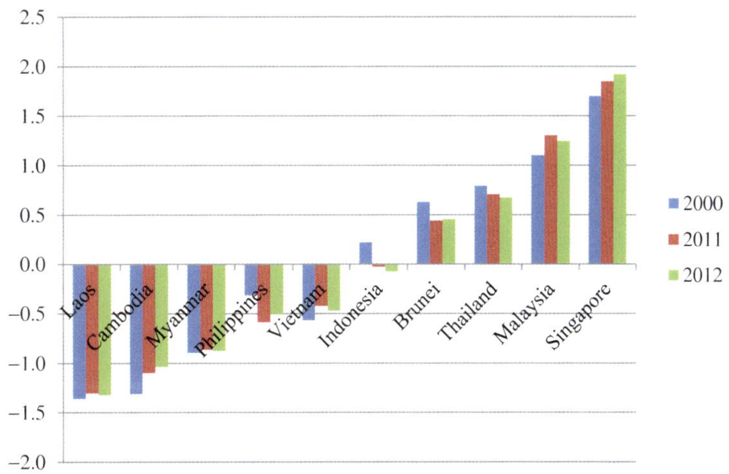

Source: Asia Competitiveness Institute.

For the three sub-environments, Singapore and Malaysia held the top two positions for Physical Infrastructure (Figure 2.16) and Technological Infrastructure (Figure 2.17), but Brunei has been ahead of Malaysia in terms of Standard of Living, Education and Social Stability. Indonesia and Thailand slipped below average for Physical Infrastructure and Standard of Living, Education and Social Stability, respectively, by 2011. Vietnam's score for Technological Infrastructure turned positive in 2011 but plunged back to negative in 2012. Cambodia overtook Indonesia in terms of Standard of Living, Education and Social Stability to rank sixth in 2011 and 2012.

Figure 2.17: Quality of Life and Infrastructure Development (Technological Infrastructure) — 2000, 2011 and 2012

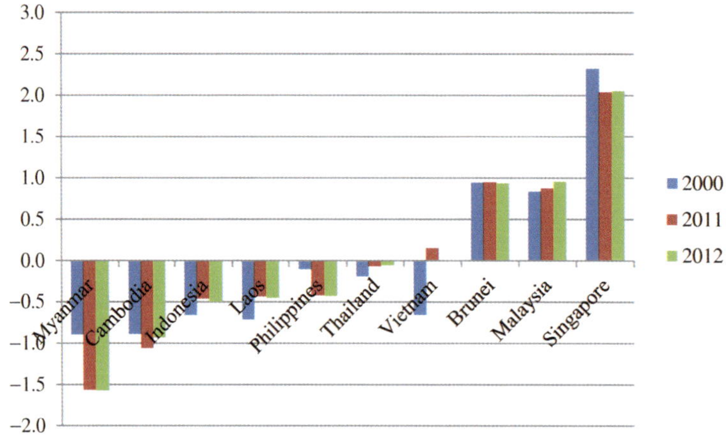

Source: Asia Competitiveness Institute.

Figure 2.18: Quality of Life and Infrastructure Development (Standard of Living, Education and Social Stability) — 2000, 2011 and 2012

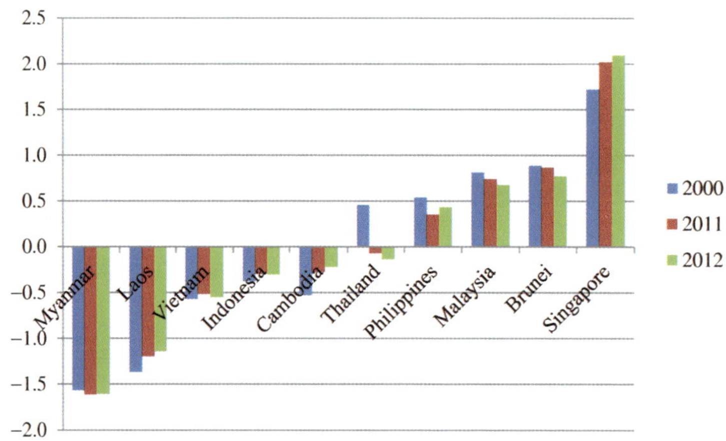

Source: Asia Competitiveness Institute.

2.3.6 Competitiveness Over Time Analysis

For a better comparison of the competitive performances over time, we split the study period into two sub-periods, 2000–2006 (Period A) and 2007–2012 (Period B). Our Competitiveness Over Time (COT) analysis is conducted for Overall Competitiveness as well as each of the 12 sub-environments, which is instructive for policy-making as it reveals where individual economies are losing competitiveness. For all COT Figures 2.19–2.31, countries above the 45° line improved their competitiveness in Period B relative to Period A; below the line

Figure 2.19: Competitiveness Over Time Analysis for Overall Competitiveness, Period B (2007–2012) Against Period A (2000–2006)

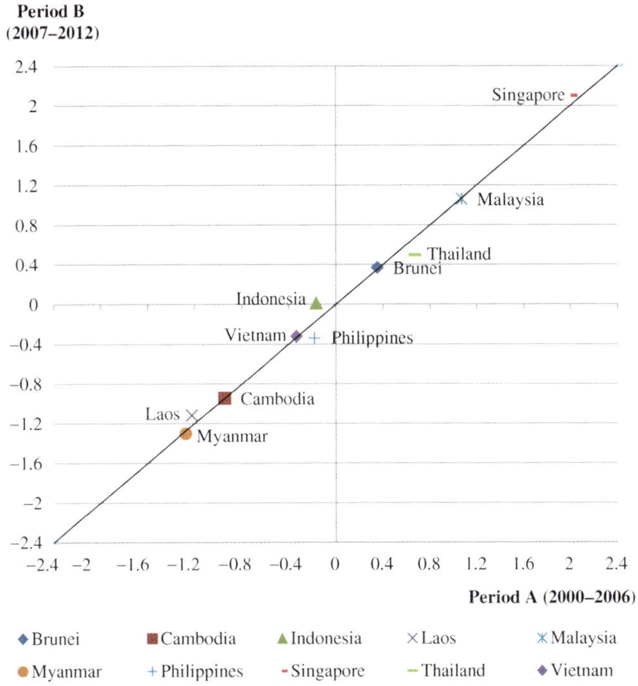

Source: Asia Competitiveness Institute.

means they performed better in the Period A; and on the line means no change over both sub-periods.

As shown in Figure 2.19, Singapore at the top of the 45° line is consistent with its top-notch performance in the Overall Competitiveness ranking detailed in Section 2.1. Malaysia, Brunei, Vietnam, Cambodia and Myanmar remained on the 45° line, indicating a lack of improvement over time. Thailand and the Philippines fell below the line, implying that their Overall Competitiveness has deteriorated. In contrast, Indonesia and Laos performed better in Period B than in Period A.

In Figure 2.20 for Economic Vibrancy, a sub-environment under MS, Indonesia rose to the top as well as fared better in Period B than in Period A. Laos and Myanmar also showed marked improvement while Vietnam and Cambodia are losing. Malaysia and Brunei fell slightly below the 45-degree line. Singapore, Thailand and the Philippines were on the 45° line.

In the other two sub-environments, Openness to Trade and Services (Figure 2.21) and Attractiveness to Foreign Investors (Figure 2.22), most ASEAN economies clustered along the 45° line with the exception of Vietnam and Indonesia as they fared better in Period B as compared to Period A in Figures 2.21 and 2.22, respectively.

Figure 2.20: Competitiveness Over Time Analysis for Economic Vibrancy, Period B (2007–2012) Against Period A (2000–2006)

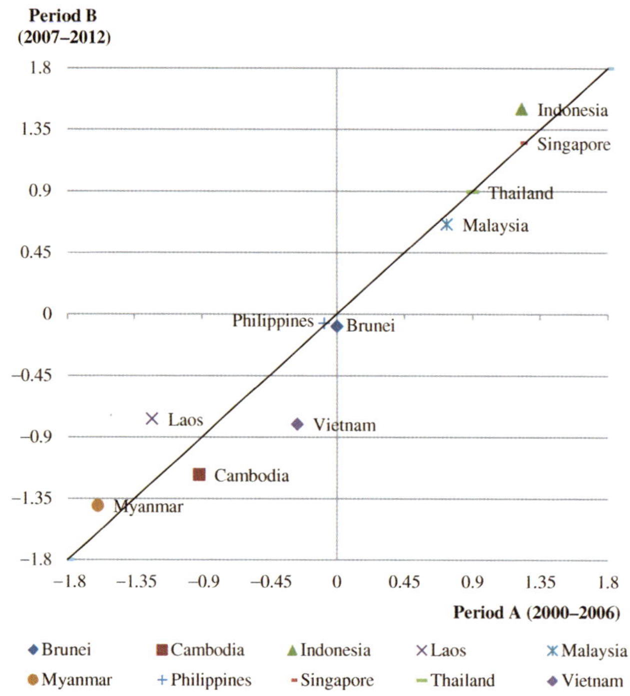

Source: Asia Competitiveness Institute.

Indonesia shows marked improvement in all three sub-environments under Government and Institutional Setting (Figures 2.23–2.25). Thailand and the Philippines, in contrast, fared worse in Period B as compared to Period in all three sub-environments. In Figure 2.23 for Government Policies and Fiscal Sustainability, all other countries remained on or very close to the 45° line. Nonetheless, the majority of ASEAN–10 (except Thailand the Philippines) see improvement in terms of Institutions, Governance and Leadership (Figure 2.24) and Competition, Regulatory Standards and Rule of Law (Figure 2.25) over time.

In Figure 2.26 for Financial Deepening and Business Efficiency, the first sub-environment under FBMC, Indonesia, Vietnam, Cambodia and Laos became more competitive in the second sub-period, whereas the Philippines, Brunei and Myanmar are losing. With regard to Labour Market Flexibility (Figure 2.27), Malaysia, Indonesia, Cambodia, Laos and Brunei show clear improvement but the remaining ASEAN economies, except for Singapore, fared worse in Period B as compared to Period A. In Figure 2.28 for Productivity Performance, Malaysia performed better over time while the rest of ASEAN clustered on the 45° line.

Among the three sub-environments under Quality of Life and Infrastructure Development, Vietnam has decidedly improved over time in terms of Technological Infrastructure (Figure 2.30). Indonesia and Singapore also show marked progress for Standard of Living, Education and Social Stability (Figure 2.31). In contrast, Thailand fared worse in the same sub-environment over time. Figure 2.29 for Physical Infrastructure shows the ASEAN–10 lying on or close to the 45° line.

Figure 2.21: Competitiveness Over Time Analysis for Openness to Trade and Services, Period B (2007–2012) Against Period A (2000–2006)

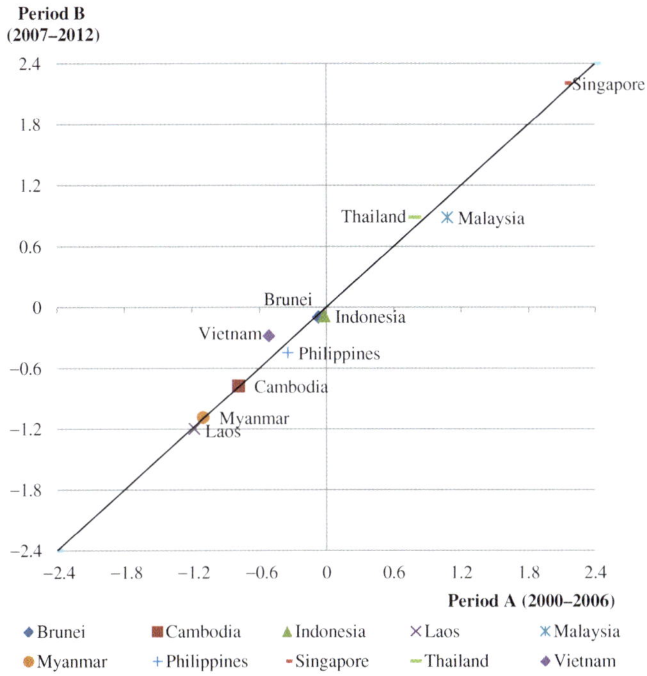

Source: Asia Competitiveness Institute.

Figure 2.22: Competitiveness Over Time Analysis for Attractiveness to Foreign Investors, Period B (2007–2012) Against Period A (2000–2006)

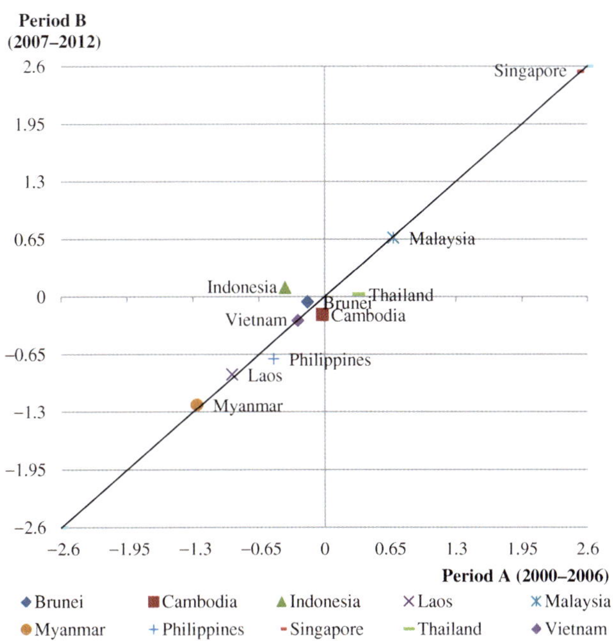

Source: Asia Competitiveness Institute.

Figure 2.23: Competitiveness Over Time Analysis for Government Policies and Fiscal Sustainability, Period B (2007–2012) Against Period A (2000–2006)

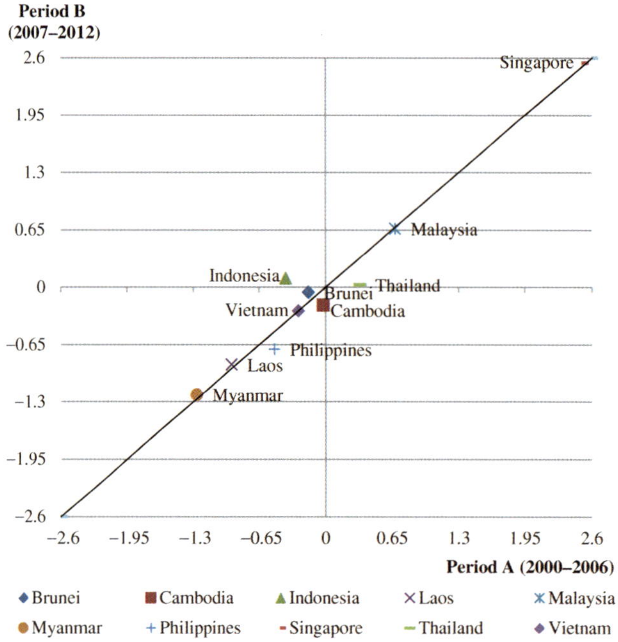

Source: Asia Competitiveness Institute.

Figure 2.24: Competitiveness Over Time Analysis for Institutions, Governance and Leadership, Period B (2007–2012) Against Period A (2000–2006)

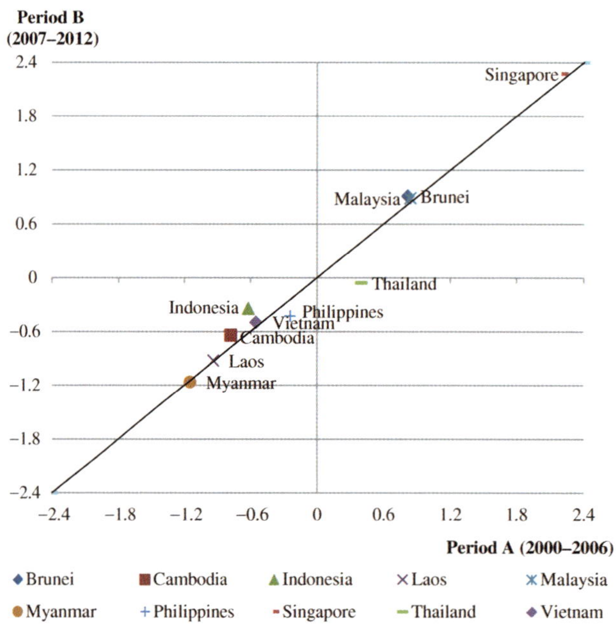

Source: Asia Competitiveness Institute.

Figure 2.25: Competitiveness Over Time Analysis for Competition, Regulatory Standards and Rule of Law, Period B (2007–2012) Against Period A (2000–2006)

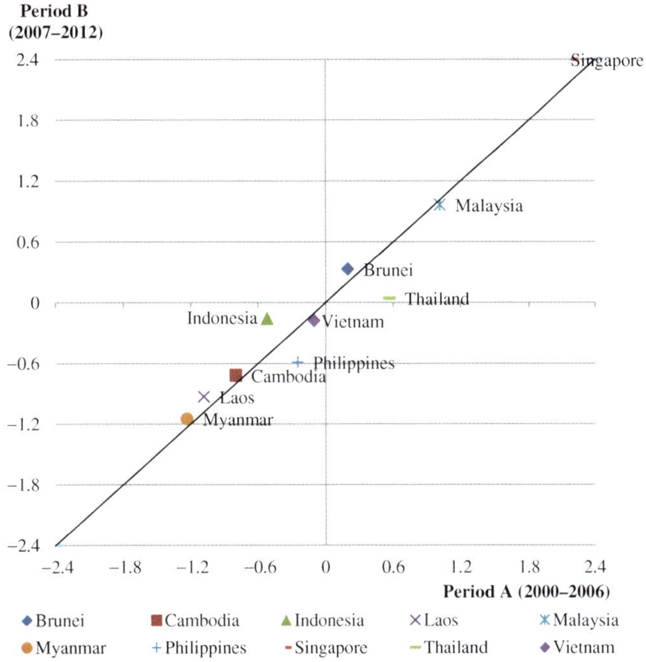

Source: Asia Competitiveness Institute.

Figure 2.26: Competitiveness Over Time Analysis for Financial Deepening and Business Efficiency, Period B (2007–2012) Against Period A (2000–2006)

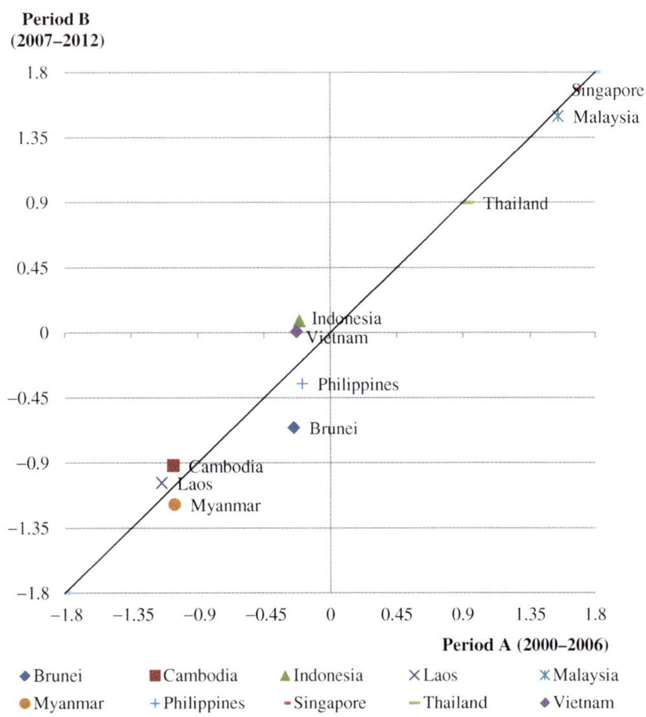

Source: Asia Competitiveness Institute.

Figure 2.27: Competitiveness Over Time Analysis for Labour Market Flexibility, Period B (2007–2012) Against Period A (2000–2006)

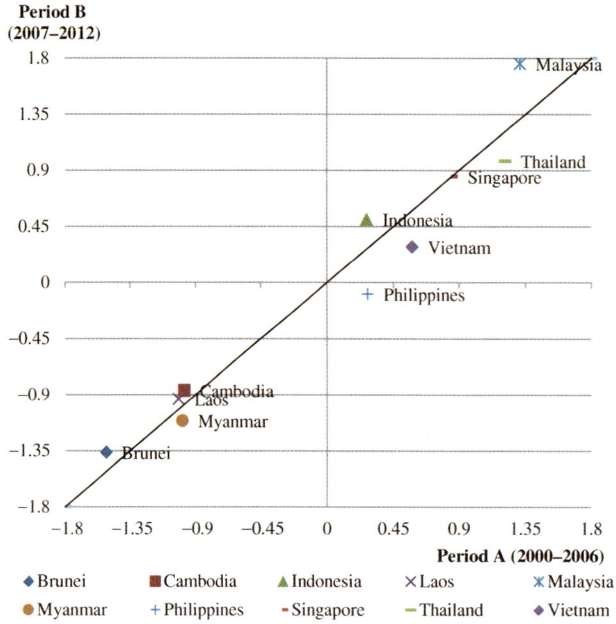

Source: Asia Competitiveness Institute.

Figure 2.28: Competitiveness Over Time Analysis for Productivity Performance, Period B (2007–2012) Against Period A (2000–2006)

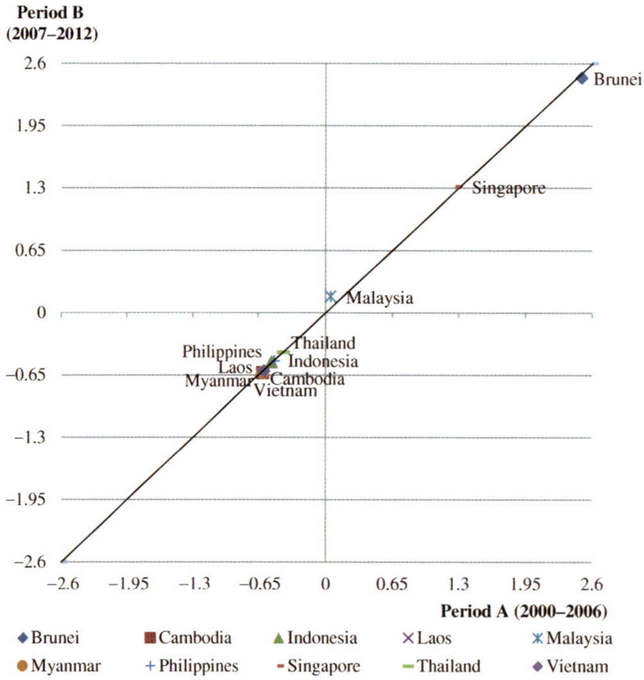

Source: Asia Competitiveness Institute.

Figure 2.29: Competitiveness Over Time Analysis for Physical Infrastructure, Period B (2007–2012) Against Period A (2000–2006)

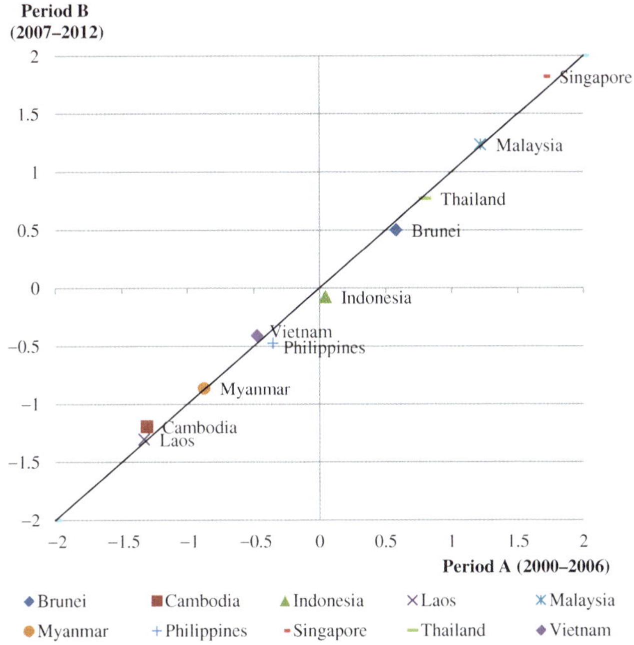

Source: Asia Competitiveness Institute.

Our COT analysis shows the majority of ASEAN-10 improving over time, albeit to various degrees. This indicates a notable heterogeneity within ASEAN-10 not only in stages of economic development but also paces of catch-up. It is clear from our analysis that economic reforms in the transitional CLMV have not been sufficiently deep and comprehensive. While CLV have made great strides in attracting foreign direct investment (FDI) and multinational corporations (MNCs) as part and parcel of the initial take-off stage, with Myanmar rising as the new frontier, keeping them remains a challenge as institutional capacity and productivity fail to come up to par.

Indonesia, in contrast, stands out as a clear outlier, showing impressive progress in the majority of sub-environments. Singapore is still the top performing economy by a large margin although it has been surpassed by Malaysia, Brunei and Indonesia in several sub-environments. Thailand and the Philippines, while maintaining most of their rankings, see their competitiveness erode over time. Self-competition is thus paramount regardless of where one stands in the rankings.

2.3.7 Maximising Strengths and Minimising Weaknesses

In this section, the top 20% strongest indicators and top 20% weakest indicators for each ASEAN economy will be first discussed. The analysis will bear in mind the contexts of high-income (Singapore and Brunei), middle-income (Indonesia, Malaysia, the Philippines and Thailand) and low-income transitional economies (CLMV). This is followed by analysis of

Figure 2.30: Competitiveness Over Time Analysis for Technological Infrastructure, Period B (2007–2012) Against Period A (2000–2006)

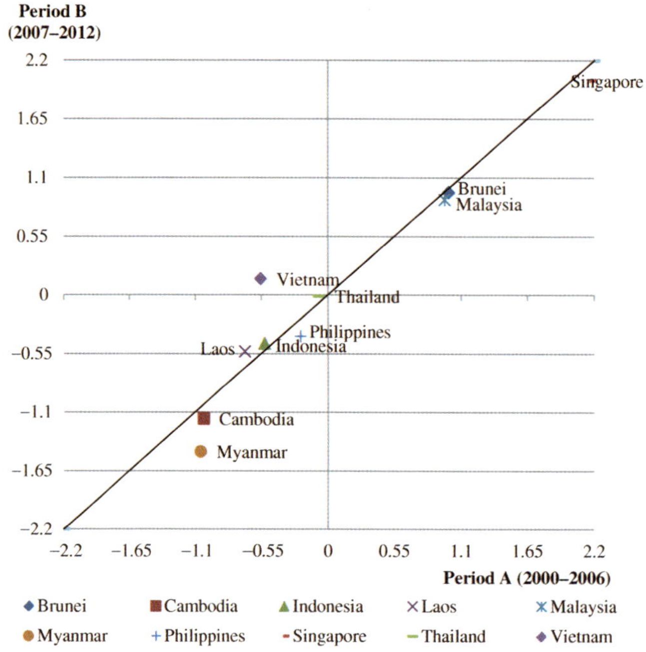

Source: Asia Competitiveness Institute.

Figure 2.31: Competitiveness Over Time Analysis for Standard of Living, Education and Social Stability, Period B (2007–2012) Against Period A (2000–2006)

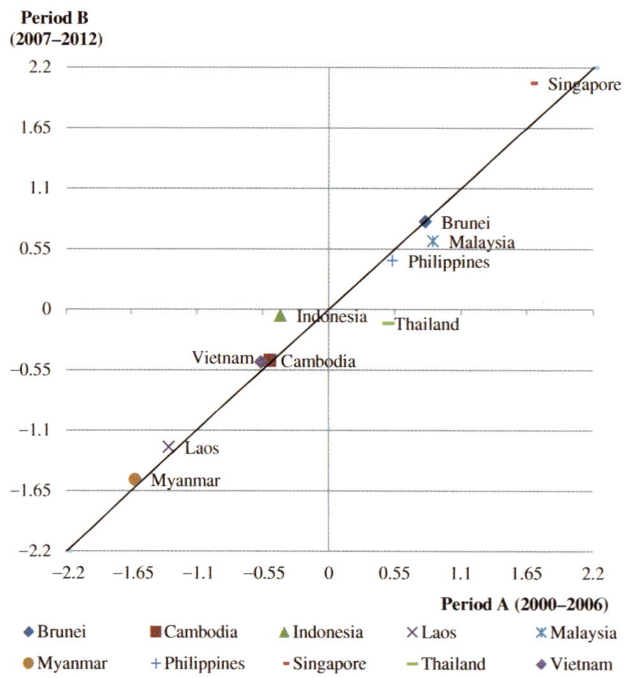

Source: Asia Competitiveness Institute.

spider web charts depicting each ASEAN economy's competitiveness *vis-à-vis* the ASEAN median (average ASEAN-10) and maximum competitiveness (set by top ASEAN economy).

Top 20% Strongest and Weakest Indicators

Starting alphabetically, Brunei's top 20% strongest and top 20% weakest indicators are presented in Tables 2.7 and 2.8. Labour productivity (#1 and #2 in Table 2.7) and fiscal balance (#3 and #4) are the two most notable strengths of the oil-rich state. On the other hand, the country's high labour costs can be prohibitive to potential foreign and local investors (#120 and #121 in Table 2.8). Among its top weaknesses is private sector (#107, #109) which is constrained by a relatively underdeveloped financial sector (#112, #118) as well as tax burden (#117).

Table 2.7: Top 20% Strongest Indicators List for Brunei, 2012

Rank	20% Strongest Indicator	Score	Environment
1	Agriculture, Value Added Per Worker	2.8857	FBMC
2	Industry, Value Added Per Worker	2.7061	FBMC
3	Budget Balance	2.4961	GIS
4	Current Account Balance	2.3667	MS
5	Gross Domestic Savings	2.0130	MS
6	Religious Unrest	2.0074	QLID
7	Government Environmental Protection Policies	1.9913	GIS
8	Access to Print Media	1.9250	QLID
9	Personal Computers	1.8999	QLID
10	Government Consumption Expenditure	1.8754	GIS
11	Sovereign Debt Risk	1.5245	FBMC
12	Corporate Income Tax	1.5244	MS
13	Secondary Schooling	1.5041	QLID
14	Information and Communication Technology (ICT) Expenditure	1.4678	QLID
15	Human Development Index	1.4606	QLID
16	Government Stability	1.4276	GIS
17	High Quality Management Staff	1.3648	FBMC
18	Inflation	1.3391	MS
19	Quality of Judiciary	1.3227	GIS
20	Political Change	1.3181	GIS
21	Government Responsiveness	1.2896	GIS
22	Public Security and Safety	1.2853	QLID
23	Quality of Regulatory Environment	1.1532	GIS
24	Access to Sanitation	1.1299	QLID

Source: Asia Competitiveness Institute.

Table 2.8: Top 20% Weakest Indicators List for Brunei, 2012

Rank	20% Weakest Indicator	Score	Environment
98	Output Industry (Value Added)	–0.8559	MS
99	Population	–0.8584	QLID
100	Government Revenue (Including Grants)	–0.8625	GIS
101	Urban Population Growth	–0.8709	QLID
102	Output Agriculture (Value Added)	–0.8986	MS
103	Labour Force	–0.9051	FBMC
104	Imports of Goods and Services	–0.9056	MS
105	International Reserves	–0.9120	GIS
106	Tax Revenue	–0.9757	GIS
107	Market Capitalisation	–1.0011	FBMC
108	Gross Domestic Product	–1.0027	MS
109	Listed Domestic Companies	–1.0108	FBMC
110	Output Services (Value Added)	–1.0145	MS
111	Deposit Interest Rate	–1.1037	GIS
112	Domestic Credit Provided by the Financial Sector	–1.1382	FBMC
113	Health Expenditure	–1.1681	QLID
114	Tax Revenue/Government Revenue	–1.2039	GIS
115	Regional Competitiveness	–1.2151	GIS
116	Gross Fixed Capital Formation	–1.3733	MS
117	Tax Burden	–1.3838	GIS
118	Claims on Private Sector	–1.3985	FBMC
119	Ease of Starting Business	–1.5885	GIS
120	Cost of Management Staff	–1.8786	FBMC
121	Cost of Production Labour	–1.9659	FBMC

Source: Asia Competitiveness Institute.

For Cambodia, labour cost emerges as one of its top strengths at #14 and #15 in Table 2.9 but somewhat dampened by the relatively high labour market risk (#115 in Table 2.10). Notable efforts have been made in strengthening the private sector, as shown in claims on private sector and corporate income tax at #1 and #7 in Table 2.9, but more heed needs to be paid to enterprise policy and ease of starting business (#99 and #102 in Table 2.10). Access to public services, particularly water, sanitation and primary education (#119, #120 and #121 in Table 2.10), remains a pressing challenge.

For Indonesia, its top 20% strongest indicators list (Table 2.11) has population, labour force, and agricultural value added as the top three, suggesting that labour-intensive agriculture in a dualistic economic structure is viable. It is also heartening that industry and services value added are among the top five strengths, which means Indonesia can leverage on its large workforce to

Table 2.9: Top 20% Strongest Indicators List for Cambodia, 2012

Rank	20% Strongest Indicator	Score	Environment
1	Claims on Private Sector	2.6650	FBMC
2	Freedom of Religion	2.2361	QLID
3	Freedom of Speech	1.5275	QLID
4	Government Environmental Protection Policies	1.2982	GIS
5	Daily Organic Water Pollutant Emissions	1.1308	QLID
6	Health Expenditure	1.0890	QLID
7	Corporate Income Tax	1.0723	MS
8	Real GDP Growth	1.0705	MS
9	Primary Schooling	0.8853	QLID
10	Unemployment Rate	0.8792	FBMC
11	Freedom of Press	0.8566	QLID
12	International Monetary Fund Credit	0.7738	GIS
13	External Debt	0.7272	GIS
14	Cost of Management Staff	0.7255	FBMC
15	Cost of Production Labour	0.6246	FBMC
16	Traffic Congestion	0.6240	QLID
17	Exchange Rate Appreciation	0.6166	GIS
18	Tax Burden	0.4703	GIS
19	Cellular Mobile Telephone Subscribers	0.4371	QLID
20	Legislature/Parliament (Functioning of Government System)	0.3384	GIS
21	Inflation	0.2454	MS
22	Tax Revenue/Government Revenue	0.1789	GIS
23	Monopolies and Cartel	0.1726	GIS
24	Business Impact of Rules on FDI	0.1434	MS

Source: Asia Competitiveness Institute.

Table 2.10: Top 20% Weakest Indicators List for Cambodia, 2012

Rank	20% Weakest Indicator	Score	Environment
98	Output Industry (Value Added)	−0.9197	MS
99	Policy towards Private Enterprise Rating	−0.9312	GIS
100	University and Post-Graduate Education	−0.9413	QLID
101	High Quality Management Staff	−0.9522	FBMC
102	Ease of Starting Business	−0.9593	GIS
103	Availability of Scientists and Engineers	−0.9636	FBMC
104	Human Development Index	−0.9934	QLID
105	Market Capitalisation	−1.0011	FBMC
106	Output Services (Value Added)	−1.0096	MS
107	Listed Domestic Companies	−1.0108	FBMC
108	Gross Domestic Product	−1.0135	MS
109	Interest Rate Spread	−1.0139	GIS
110	Regional Competitiveness	−1.0306	GIS
111	Government Revenue (Including Grants)	−1.0352	GIS
112	Information and Communication Technology (ICT) Expenditure	−1.0386	QLID
113	Tax Revenue	−1.0390	GIS
114	Sovereign Debt Risk	−1.0401	FBMC
115	Labour Activism	−1.0404	QLID
116	Internet Subscribers	−1.1456	QLID
117	Gross Domestic Savings	−1.1986	MS
118	Urban Population	−1.2701	QLID
119	Access to Water	−1.8661	QLID
120	Literacy Rate	−1.9631	QLID
121	Access to Sanitation	−2.1378	QLID

Source: Asia Competitiveness Institute.

Table 2.11: Top 20% Strongest Indicators List for Indonesia, 2012

Rank	20% Strongest Indicator	Score	Environment
1	Population	2.6425	QLID
2	Output Agriculture (Value Added)	2.5761	MS
3	Labour Force	2.5603	FBMC
4	Output Industry (Value Added)	2.3978	MS
5	Output Services (Value Added)	2.0612	MS
6	Freedom of Speech	1.5275	QLID
7	Gross Domestic Product	1.4270	MS
8	Gross Fixed Capital Formation	1.3507	MS
9	Political System Risk	1.0350	GIS
10	Market Capitalisation	1.0244	FBMC
11	Tax Revenue	0.9467	GIS
12	Government Revenue (Including Grants)	0.9427	GIS
13	Cost of Management Staff	0.7255	FBMC
14	Regional Competitiveness	0.7152	GIS
15	Deposit Interest Rate	0.6788	GIS
16	Tax Burden	0.6262	GIS
17	Listed Domestic Companies	0.5710	FBMC
18	Imports of Goods and Services	0.5413	MS
19	Foreign Direct Investment Inflows	0.5383	MS
20	Secondary Schooling	0.5321	QLID
21	Cost of Production Labour	0.5221	FBMC
22	Availability of Scientists and Engineers	0.4825	FBMC
23	Levelness of Playing Field	0.4394	MS
24	Claims on Private Sector	0.4123	FBMC

Source: Asia Competitiveness Institute.

Table 2.12: Top 20% Weakest Indicators List for Indonesia, 2012

Rank	20% Weakest Indicator	Score	Environment
98	Domestic Credit to Private Sector	–0.7077	FBMC
99	Institutional Sophistication	–0.7156	GIS
100	Personal Computers	–0.7346	QLID
101	Religious Unrest	–0.7381	QLID
102	Freedom of Religion	–0.7454	QLID
103	Internet Subscribers	–0.7496	QLID
104	Openness to Trade	–0.7571	MS
105	Public Security and Safety	–0.8440	QLID
106	Government Stability	–0.8566	GIS
107	Lack of Discrimination Against Investors	–0.8866	MS
108	Access to Sanitation	–0.8876	QLID
109	Quality of Central Bank/Monetary Authority	–0.8879	GIS
110	Air Quality	–0.8984	QLID
111	Labour Market Rating	–0.9082	FBMC
112	Lack of Market Distortion Due to State-Ownership	–0.9145	GIS
113	Policy Towards Private Enterprise Rating	–0.9312	GIS
114	High Quality Management Staff	–0.9522	FBMC
115	Money Supply (M2)	–1.0034	GIS
116	Information and Communication Technology (ICT) Expenditure	–1.0386	QLID
117	Real GDP Growth	–1.1562	MS
118	Exchange Rate Appreciation	–1.3039	GIS
119	Unemployment Rate	–1.6258	FBMC
120	International Monetary Fund Credit	–2.1697	GIS
121	External Debt	–2.4458	GIS

Source: Asia Competitiveness Institute.

develop resource-based industries and services, for instance agro-processing and simultaneously tackle its unemployment problem. What it needs is an enabling macroeconomic environment which remains one of its top weaknesses (#109, #115, and #118 in Table 2.12). Equally imperative is prudent fiscal management as high external debt (#121) exposes the country to solvency and liquidity risk.

Among Indonesia's top strengths are entrepreneurial growth (#10 and #17) and foreign direct investment (#19), supported by not only relatively cheap and educated labour force (#13. #20 and #21) but also a level playing field (#23) although improvement is required in terms of credit to the private sector (#98 in Table 2.12).

Laos is clearly accelerating its catching-up to join the more mature ASEAN members as evident in its top three strengths that include ICT infrastructure (#1) and investment (#3) (see

Table 2.13). Among its top 20% strengths are manpower cost (#8 and #10), political stability (#15) and a relatively conducive business environment (#17, #18 and #21).

On the other hand, human capital remains a challenge for Laos to keep up with the demands of industrialisation (#114, #116 and #119 in Table 2.14). While Laos seems to have made great strides in improving its technological infrastructure, more heed should be paid to physical infrastructure (#112 and #117) as a prerequisite for attracting and retaining FDI and MNCs (#102).

Malaysia is a credible challenger to Singapore in Overall Competitiveness and even surpassed Singapore as the top in Labour Market Flexibility by 2011. Its top strengths include a thriving private sector supported by a strong legislature and financial system (#1 to #4 in Table 2.15). Its high-quality manpower (#6, #8, #13) poses a threat to Singapore's overall top spot in Financial,

Table 2.13: Top 20% Strongest Indicators List for Laos, 2012

Rank	20% Strongest Indicator	Score	Environment
1	Telecommunications System	2.7618	QLID
2	Urban Population Growth	2.6091	QLID
3	Gross Fixed Capital Formation	1.2351	MS
4	Daily Organic Water Pollutant Emissions	1.1308	QLID
5	Real GDP Growth	1.0999	MS
6	Unemployment Rate	0.9775	FBMC
7	International Monetary Fund Credit	0.8240	GIS
8	Cost of Management Staff	0.7255	FBMC
9	External Debt	0.7189	GIS
10	Cost of Production Labour	0.6246	FBMC
11	Traffic Congestion	0.6098	QLID
12	Exchange Rate Appreciation	0.5063	GIS
13	Government Environmental Protection Policies	0.4555	GIS
14	Government Consumption Expenditure	0.3266	GIS
15	Political Change	0.2274	GIS
16	Primary Schooling	0.1271	QLID
17	Corporate Income Tax	0.0403	MS
18	Tax Revenue/Government Revenue	–0.0551	GIS
19	Claims on Private Sector	–0.0763	FBMC
20	Tax Burden	–0.1685	GIS
21	Ease of Starting Business	–0.2813	GIS
22	Budget Balance	–0.2830	GIS
23	Inflation	–0.3413	MS
24	Deposit Interest Rate	–0.3651	GIS

Source: Asia Competitiveness Institute.

Table 2.14: Top 20% Weakest Indicators List for Laos, 2012

Rank	20% Weakest Indicator	Score	Environment
98	Listed Domestic Companies	–1.0108	FBMC
99	Information and Communication Technology (ICT) Expenditure	–1.0386	QLID
100	Sovereign Debt Risk	–1.0401	FBMC
101	Labour Activism	–1.0404	QLID
102	Foreign Owned Companies	–1.0410	MS
103	Output Services (Value Added)	–1.0453	MS
104	Religious Unrest	–1.0478	QLID
105	Gross Domestic Product	–1.0581	MS
106	Government Revenue (Including Grants)	–1.0650	GIS
107	Barriers to Trade	–1.0664	GIS
108	Corporate Governance	–1.0699	GIS
109	Tax Revenue	–1.0707	GIS
110	Quality of Stock Market Regulatory Authority	–1.0793	FBMC
111	Cellular Mobile Telephone Subscribers	–1.0878	QLID
112	Local Road System	–1.0973	QLID
113	Strength of Auditing and Reporting Standards	–1.1823	GIS
114	Extent of Staff Training	–1.2035	FBMC
115	Regional Competitiveness	–1.2151	GIS
116	Secondary Schooling	–1.2793	QLID
117	Access to Water	–1.6535	QLID
118	Current Account Balance	–1.7436	MS
119	Literacy Rate	–1.9761	QLID
120	Lending Rate	–2.2852	GIS
121	Interest Rate Spread	–2.5951	GIS

Source: Asia Competitiveness Institute.

Table 2.15: Top 20% Strongest Indicators List for Malaysia, 2012

Rank	20% Strongest Indicator	Score	Environment
1	Listed Domestic Companies	2.1632	FBMC
2	Monopolies and Cartel	1.6582	GIS
3	Quality of Banking and Financial System	1.5828	FBMC
4	Legislature/Parliament (Functioning of Government System)	1.5524	GIS
5	Rail Facilities	1.5446	QLID
6	Availability of Scientists and Engineers	1.5327	FBMC
7	Air Quality	1.5087	QLID
8	Extent of Staff Training	1.5020	FBMC
9	Money Supply (M2)	1.4998	GIS
10	Ease of Starting Business	1.4964	GIS
11	Government Stability	1.4276	GIS
12	Market Capitalisation	1.4251	FBMC
13	High Quality Management Staff	1.3648	FBMC
14	Domestic Credit to Private Sector	1.3492	FBMC
15	Corporate Governance	1.3400	GIS
16	Internet Subscribers	1.3241	QLID
17	Government Responsiveness	1.2896	GIS
18	Domestic Credit Provided by the Financial Sector	1.2760	FBMC
19	Quality of Stock Market Regulatory Authority	1.2034	FBMC
20	Airport Facilities	1.1672	QLID
21	Intellectual Property Rights Protection	1.1643	GIS
22	Port Facilities	1.1599	QLID
23	Transparency of Processes for Investors	1.1545	MS
24	Transparency of Government	1.1545	GIS

Source: Asia Competitiveness Institute.

Table 2.16: Top 20% Weakest Indicators List for Malaysia, 2012

Rank	20% Weakest Indicator	Score	Environment
98	Output Agriculture (Value Added)	0.1326	MS
99	Current Account Balance	0.1321	MS
100	Government Environmental Protection Policies	0.1276	GIS
101	Unemployment Rate	0.0933	FBMC
102	Access to Print Media	0.0891	QLID
103	Health Expenditure	0.0256	QLID
104	Secondary Schooling	0.0222	QLID
105	Cost of Production Labour	–0.0125	FBMC
106	Foreign Direct Investment Inflows	–0.0651	MS
107	Urban Population Growth	–0.0936	QLID
108	GDP Per Capita	–0.1183	MS
109	Industry, Value Added Per Worker	–0.2044	FBMC
110	Deposit Interest Rate	–0.2497	GIS
111	Traffic Congestion	–0.3448	QLID
112	Claims on Private Sector	–0.3484	FBMC
113	Tax Burden	–0.4355	GIS
114	Population	–0.4489	QLID
115	External Debt	–0.5237	GIS
116	Labour Force	–0.5379	FBMC
117	Freedom of Speech	–0.6547	QLID
118	Budget Balance	–0.7336	GIS
119	Freedom of Religion	–0.7454	QLID
120	Daily Organic Water Pollutant Emissions	–1.0652	QLID
121	International Monetary Fund Credit	–1.1853	GIS

Source: Asia Competitiveness Institute.

Businesses and Manpower Conditions. Also highly competitive is its transport infrastructure including rail (#5), airport (#20) and seaport (#22).

Nonetheless, maintaining FDI inflows has been challenging for Malaysia (#106 in Table 2.16) amid rising labour cost (#105) as well as declining labour pool (#114, #116). While its controversial capital controls proved effective during the Asian Financial Crisis, as admitted belatedly by IMF, Malaysia's fiscal vulnerability remains stark as shown in its external debt (#115) and budget balance (#118).

Myanmar as a latecomer to competitiveness is not without its strengths among which are taxes (#1 in Table 2.17), macroeconomic stability (#4, #5) and labour cost (#6, #7). Myanmar's immense potential as a food basket is evident in its agricultural output and productivity (#18 and

Table 2.17: Top 20% Strongest Indicators List for Myanmar, 2012

Rank	20% Strongest Indicator	Score	Environment
1	Tax Burden	2.2266	GIS
2	Daily Organic Water Pollutant Emissions	1.3608	QLID
3	Deposit Interest Rate	1.3191	GIS
4	Inflation	0.8945	MS
5	External Debt	0.7674	GIS
6	Cost of Management Staff	0.7255	FBMC
7	Cost of Production Labour	0.6246	FBMC
8	Traffic Congestion	0.5896	QLID
9	International Monetary Fund Credit	0.5322	GIS
10	Interest Rate Spread	0.2773	GIS
11	Air Quality	0.2570	QLID
12	Literacy Rate	0.2512	QLID
13	Claims on Private Sector	0.0784	FBMC
14	Access to Sanitation	0.0217	QLID
15	Labour Force	−0.0016	FBMC
16	Urban Population Growth	−0.1033	QLID
17	Population	−0.1143	QLID
18	Output Agriculture (Value Added)	−0.3091	MS
19	Unemployment Rate	−0.3979	FBMC
20	Access to Water	−0.4039	QLID
21	Current Account Balance	−0.4061	MS
22	Agriculture, Value Added Per Worker	−0.4649	FBMC
23	Lending Rate	−0.5303	GIS
24	Foreign Direct Investment Inflows	−0.5483	MS

Source: Asia Competitiveness Institute.

Table 2.18: Top 20% Weakest Indicators List for Myanmar, 2012

Rank	20% Weakest Indicator	Score	Environment
98	Local Road System	−1.0973	QLID
99	Telephone Main Lines	−1.1239	QLID
100	Strength of Auditing and Reporting Standards	−1.1823	GIS
101	Extent of Staff Training	−1.2035	FBMC
102	Regional Competitiveness	−1.2151	GIS
103	Domestic Credit to Private Sector	−1.2636	FBMC
104	Openness to Trade	−1.2741	MS
105	Internet Subscribers	−1.3027	QLID
106	Human Development Index	−1.3474	QLID
107	University and Post-Graduate Education	−1.3720	QLID
108	Quality of Government Policies	−1.5030	GIS
109	Health Expenditure	−1.5260	QLID
110	Government Effectiveness	−1.5494	GIS
111	Gross Fixed Capital Formation	−1.7171	MS
112	Government Consumption Expenditure	−1.7227	GIS
113	Freedom of Press	−1.8147	QLID
114	Quality of Regulatory Environment	−1.8159	GIS
115	Primary Schooling	−1.8931	QLID
116	Real GDP Growth	−1.9094	MS
117	Corporate Income Tax	−2.0041	MS
118	Tax Revenue/Government Revenue	−2.2154	GIS
119	Exchange Rate Appreciation	−2.3949	GIS
120	Cellular Mobile Telephone Subscribers	−2.4649	QLID
121	Money Supply Growth (M2)	−2.5570	MS

Source: Asia Competitiveness Institute.

#22), high water and air quality (#2 and #11) and labour force (#15). In fact, the country seems poised to become the next investment hotspot with increasing FDI inflows (#24). That said, its governance (#108, #110, #114 in Table 2.18) and infrastructure (#98, #99) leave much to be desired. Much awaits the political scene to be resolved for the country to move forward.

For Philippines, talent pool is an enormous strength to tap (#8 in Table 2.19) if the lack of productive employment can be resolved at home (#120 in Table 2.20), which will translate into a transfusion of skills and capital, particularly from the professionally skilled diaspora around the world. Much is dependent on improving the domestic business environment that is plagued with cronyism (#108) and bureaucratic inertia (#99, #100, #109). Democracy may enable voices to be heard (#2, #21) but will hinder growth without political stability (#119) and good governance (#111, #112, #114). Infrastructure, both physical and technological, is another major impediment as shown in Table 2.20 (#103, #106, #107, #116).

Table 2.19: Top 20% Strongest Indicators List for Philippines, 2012

Rank	20% Strongest Indicator	Score	Environment
1	Air Quality	1.6346	QLID
2	Freedom of Speech	1.5275	QLID
3	Labour Activism	1.4848	QLID
4	Real GDP Growth	1.4548	MS
5	Exchange Rate Appreciation	1.0690	GIS
6	Tax Revenue/Government Revenue	1.0498	GIS
7	Daily Organic Water Pollutant Emissions	0.9654	QLID
8	High Quality Management Staff	0.9149	FBMC
9	Lack of Market Distortion Due to State-Ownership	0.8555	GIS
10	Lending Rate	0.8511	GIS
11	Literacy Rate	0.7821	QLID
12	Interest Rate Spread	0.7612	GIS
13	Freedom of Religion	0.7454	QLID
14	Strength of Auditing and Reporting Standards	0.7195	GIS
15	Extent of Staff Training	0.5798	FBMC
16	Traffic Congestion	0.5375	QLID
17	Cost of Production Labour	0.5341	FBMC
18	Population	0.5095	QLID
19	Health Expenditure	0.4891	QLID
20	Money Supply Growth (M2)	0.4384	MS
21	Freedom of Press	0.4265	QLID
22	Quality of Stock Market Regulatory Authority	0.4015	FBMC
23	Market Capitalisation	0.3457	FBMC
24	Social Unrest	0.3159	QLID

Source: Asia Competitiveness Institute.

Table 2.20: Top 20% Weakest Indicators List for Philippines, 2012

Rank	20% Weakest Indicator	Score	Environment
98	Access to Print Media	−0.6903	QLID
99	Government Responsiveness	−0.6959	GIS
100	Quality of Police	−0.7091	GIS
101	Institutional Sophistication	−0.7156	GIS
102	Personal Computers	−0.7199	QLID
103	Rail Facilities	−0.7821	QLID
104	Money Supply (M2)	−0.8066	GIS
105	Levelness of Playing Field	−0.8381	MS
106	Telephone Main Lines	−0.8609	QLID
107	Port Facilities	−0.8654	QLID
108	Lack of Cronyism	−0.8712	GIS
109	Bureaucracy/Red Tape	−0.8772	GIS
110	Government Environmental Protection Policies	−0.9284	GIS
111	Quality of Judiciary	−0.9328	GIS
112	Effectiveness of Legal System	−0.9342	GIS
113	Gross Domestic Savings	−0.9523	MS
114	Legislature/Parliament (Functioning of Government System)	−0.9704	GIS
115	Monopolies and Cartel	−0.9724	GIS
116	Airport Facilities	−1.0050	QLID
117	Religious Unrest	−1.0478	QLID
118	Corporate Income Tax	−1.1981	MS
119	Political Change	−1.2485	GIS
120	Unemployment Rate	−1.8223	FBMC
121	Primary Schooling	−1.8931	QLID

Source: Asia Competitiveness Institute.

Singapore is *sui generis* as one of the four Asian Tigers, evolving from a port city in the *entrepôt* trade to an economic powerhouse within less than half a century. Its globalised way is reflected in its strengths in attracting FDI (#1 in Table 2.21) and high productivity (#2), both reinforced by quality of political leadership (#3). Its top-notch ICT infrastructure (#10, #11) further strengthens its position as host to many MNC headquarters in Southeast Asia.

Despite dominating the top spot in the rankings of Overall Competitiveness as well as the majority of sub-environments for over a decade, Singapore is not without its weaknesses (Table 2.22). Its small population (#113, #115) and rising manpower costs (#119, #120) threaten to erode its attractiveness to MNCs as newcomers with large, young and increasingly educated workforce emerge. Local small and medium-sized companies need an extra boost as well (#98, #99, #105) for Singapore to remain competitive with less reliance on FDI.

Table 2.21: Top 20% Strongest Indicators List for Singapore, 2012

Rank	20% Strongest Indicator	Score	Environment
1	Foreign Direct Investment Inflows	2.8088	MS
2	Services, Value Added Per Worker	2.7582	FBMC
3	Quality of Political Leadership	2.7574	GIS
4	GDP Per Capita	2.6316	MS
5	Transparency of Processes for Investors	2.5930	MS
6	Transparency of Government	2.5930	GIS
7	Social Unrest	2.5724	QLID
8	Quality of Central Bank/ Monetary Authority	2.5650	GIS
9	Openness to Trade	2.5428	MS
10	Telephone Main Lines	2.5401	QLID
11	Networked Readiness Index	2.5122	QLID
12	Lack of Corruption	2.4739	GIS
13	Foreign Owned Companies	2.4283	MS
14	Lack of Discrimination Against Investors	2.4124	MS
15	Integrity of Legal System	2.4121	GIS
16	Barriers to Trade	2.3786	GIS
17	Exports of Goods and Services	2.3756	MS
18	Institutional Sophistication	2.3585	GIS
19	Effectiveness of Legal System	2.3561	GIS
20	Labour Unrest	2.3504	QLID
21	Policy Towards Private Enterprise Rating	2.3425	GIS
22	Intellectual Property Rights Protection	2.3231	GIS
23	Levelness of Playing Field	2.3212	MS
24	Port Facilities	2.2998	QLID

Source: Asia Competitiveness Institute.

Table 2.22: Top 20% Weakest Indicators List for Singapore, 2012

Rank	20% Weakest Indicator	Score	Environment
98	Domestic Credit Provided by the Financial Sector	0.5132	FBMC
99	Corporate Income Tax	0.4629	MS
100	Unemployment Rate	0.2407	FBMC
101	Interest Rate Spread	0.2297	GIS
102	Tax Burden	0.2281	GIS
103	Output Industry (Value Added)	0.2062	MS
104	Telecommunications System	−0.2010	QLID
105	Claims on Private Sector	−0.2112	FBMC
106	Government Consumption Expenditure	−0.2324	GIS
107	Urban Population Growth	−0.2408	QLID
108	Real GDP Growth	−0.3534	MS
109	Agriculture, Value Added Per Worker	−0.3574	FBMC
110	Tax Revenue/Government Revenue	−0.4422	GIS
111	Inflation	−0.4617	MS
112	Freedom of Speech	−0.6547	QLID
113	Population	−0.7888	QLID
114	Daily Organic Water Pollutant Emissions	−0.8100	QLID
115	Labour Force	−0.8223	FBMC
116	Output Agriculture (Value Added)	−0.8999	MS
117	Government Environmental Protection Policies	−1.1266	GIS
118	Deposit Interest Rate	−1.1336	GIS
119	Cost of Management Staff	−1.8786	FBMC
120	Cost of Production Labour	−1.9659	FBMC
121	Traffic Congestion	−2.8372	QLID

Source: Asia Competitiveness Institute.

Notwithstanding its political tumult since the 2006 *coup d'etat*, Thailand continues to be one of the top tourism destinations (#1 in Table 2.23). Its private sector remains strong amid political uncertainty (#6, #14, #16), contributing to its stable Overall Competitiveness scores as shown in Section 2.1. Fiscal balance is also not compromised (#5, #10). That said, prolonged political instability (#117, #120 in Table 2.24), coupled with weak governance (#111, #112), will eventually undermine the confidence of both domestic and foreign investors. A level playing field (#110) free of cronyism (#115) is vital to restore trust in the country renowned for its hospitality.

Table 2.23: Top 20% Strongest Indicators List for Thailand, 2012

Rank	20% Strongest Indicator	Score	Environment
1	International Tourism Receipts	2.3835	MS
2	Domestic Credit Provided by the Financial Sector	1.9899	FBMC
3	Total Value of Stocks Traded	1.8916	FBMC
4	University and Post-Graduate Education	1.6929	QLID
5	Tax Revenue	1.5936	GIS
6	Domestic Credit to Private Sector	1.2983	FBMC
7	Unemployment Rate	1.2722	FBMC
8	Money Supply (M2)	1.2509	GIS
9	Gross Domestic Product	1.2332	MS
10	Government Revenue (Including Grants)	1.1453	GIS
11	International Reserves	1.1385	GIS
12	Net Foreign Assets	1.1111	GIS
13	Regional Competitiveness	1.0383	GIS
14	Ease of Starting Business	0.9955	GIS
15	Airport Facilities	0.9528	QLID
16	Market Capitalisation	0.9517	FBMC
17	Tax Revenue/Government Revenue	0.9474	GIS
18	Government Consumption Expenditure	0.8883	GIS
19	Output Industry (Value Added)	0.8627	MS
20	Electric Power	0.8431	QLID
21	Imports of Goods and Services	0.8239	MS
22	Access to Sanitation	0.8096	QLID
23	Gross Fixed Capital Formation	0.7598	MS
24	Freedom of Religion	0.7454	QLID

Source: Asia Competitiveness Institute.

Table 2.24: Top 20% Weakest Indicators List for Thailand, 2012

Rank	20% Weakest Indicator	Score	Environment
98	Transparency of Government	–0.4349	GIS
99	Tax Burden	–0.4350	GIS
100	Legislature/Parliament (Functioning of Government System)	–0.4515	GIS
101	Telecommunications System	–0.4526	QLID
102	Religious Unrest	–0.4621	QLID
103	Corporate Income Tax	–0.4806	MS
104	Intellectual Property Rights Protection	–0.5301	GIS
105	Monopolies and Cartel	–0.5379	GIS
106	International Monetary Fund Credit	–0.6019	GIS
107	Integrity of Legal System	–0.6030	GIS
108	Freedom of Speech	–0.6547	QLID
109	Urban Population	–0.6712	QLID
110	Levelness of Playing Field	–0.7057	MS
111	Quality of Police	–0.7122	GIS
112	Government Responsiveness	–0.7184	GIS
113	Government Stability	–0.7614	GIS
114	Social Unrest	–0.8123	QLID
115	Lack of Cronyism	–0.8499	GIS
116	Air Quality	–0.8688	QLID
117	Political System Risk	–0.9056	GIS
118	External Debt	–0.9092	GIS
119	Daily Organic Water Pollutant Emissions	–1.2865	QLID
120	Political Change	–1.3039	GIS
121	Urban Population Growth	–1.3674	QLID

Source: Asia Competitiveness Institute.

Vietnam, often dubbed "Asia's next tiger," has broad-based strengths (see Table 2.25), particularly in public services including health care (#2) and education (#5) in the top five. Stable government (#18, #22) and low-cost, educated workforce (#5, #7, #13) reinforce the country's attractiveness as a host of FDI.

However, Vietnam risks becoming a "paper tiger" if it does not completely shed the legacy of the central planning era. While it has stepped up its efforts in privatisation and equitisation of state-owned corporations, market distortion at the expense of the private sector remains (#101 in Table 2.26). Inadequate banking system (#99), underdeveloped equity market (#98) and weak corporate governance (#111) threaten to decimate local small- and medium-sized enterprises.

Table 2.25: Top 20% Strongest Indicators List for Vietnam, 2012

Rank	20% Strongest Indicator	Score	Environment
1	Deposit Interest Rate	2.0998	GIS
2	Health Expenditure	1.9111	QLID
3	Tax Revenue/Government Revenue	1.0980	GIS
4	Cellular Mobile Telephone Subscribers	0.8942	QLID
5	Primary Schooling	0.8304	QLID
6	Domestic Credit to Private Sector	0.8050	FBMC
7	Cost of Management Staff	0.7255	FBMC
8	Domestic Credit Provided by the Financial Sector	0.6935	FBMC
9	Money Supply (M2)	0.6805	GIS
10	Interest Rate Spread	0.6742	GIS
11	Labour Force	0.6391	FBMC
12	Unemployment Rate	0.6336	FBMC
13	Cost of Production Labour	0.6246	FBMC
14	Regional Competitiveness	0.6244	GIS
15	Traffic Congestion	0.6227	QLID
16	Religious Unrest	0.5873	QLID
17	Access to Water	0.5776	QLID
18	Political Change	0.4943	GIS
19	Urban Population Growth	0.4736	QLID
20	International Monetary Fund Credit	0.4154	GIS
21	Population	0.3968	QLID
22	Government Stability	0.3807	GIS
23	Quality of Police	0.3548	GIS
24	Literacy Rate	0.3419	QLID

Source: Asia Competitiveness Institute.

Table 2.26: Top 20% Weakest Indicators List for Vietnam, 2012

Rank	20% Weakest Indicator	Score	Environment
98	Market Capitalisation	–0.8335	FBMC
99	Quality of Banking and Financial System	–0.8361	FBMC
100	Bureaucracy/Red Tape	–0.9100	GIS
101	Lack of Market Distortion Due to State-Ownership	–0.9145	GIS
102	Intellectual Property Rights Protection	–0.9305	GIS
103	Business Impact of Rules on FDI	–0.9445	MS
104	High Quality Management Staff	–0.9522	FBMC
105	Government Environmental Protection Policies	–0.9689	GIS
106	Electric Power	–1.0089	QLID
107	Sovereign Debt Risk	–1.0401	FBMC
108	Labour Activism	–1.0404	QLID
109	Foreign Owned Companies	–1.0410	MS
110	Barriers to Trade	–1.0664	GIS
111	Corporate Governance	–1.0699	GIS
112	Quality of Stock Market Regulatory Authority	–1.0793	FBMC
113	Local Road System	–1.0973	QLID
114	Air Quality	–1.1428	QLID
115	Government Consumption Expenditure	–1.1546	GIS
116	Strength of Auditing and Reporting Standards	–1.1823	GIS
117	Extent of Staff Training	–1.2035	FBMC
118	Secondary Schooling	–1.2793	QLID
119	Tax Burden	–1.4063	GIS
120	Freedom of Press	–1.8147	QLID
121	Inflation	–2.4844	MS

Source: Asia Competitiveness Institute.

Perhaps a deeper and more comprehensive *Doi Moi* is what it takes for Vietnam to propel itself into the rank of the more mature ASEAN members. Poor infrastructure (106, #113) and lack of high quality manpower (#108, #114) are the other major obstacles in Vietnam's path to prosperity.

Median Competitiveness Web Analysis

As another guide for policy-makers, we employ spider web chart analysis to show how competitive individual economies are in each of the 12 sub-environments relative to ASEAN–10 average and to ASEAN–10 maximum.

Figure 2.32: Median Competitiveness Web Analysis for Brunei, 2012

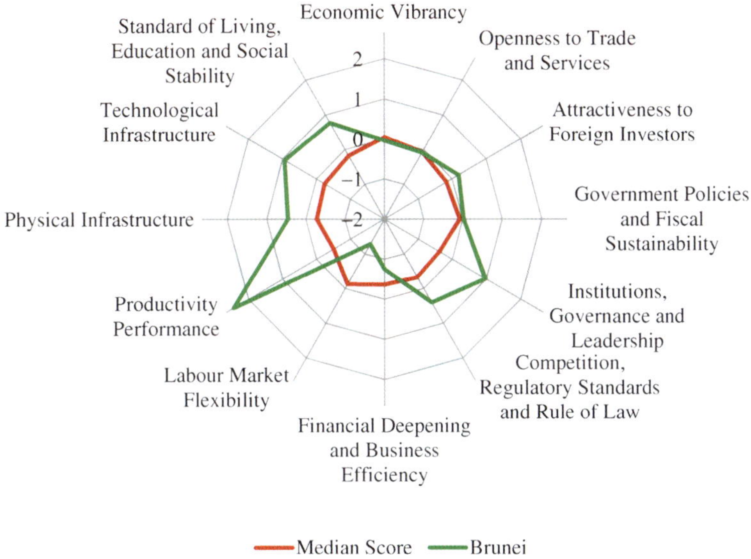

Source: Asia Competitiveness Institute.

Starting alphabetically with Brunei in Figure 2.32 (in green), its Productivity Performance, Technological Infrastructure and Standard of Living, Education and Social Stability exceed the ASEAN-10 median (in red) by a large margin, while its Labour Market Flexibility and Financial Deepening and Business Efficiency are oddly below the regional average. For Economic Vibrancy, Openness to Trade and Services and Attractiveness to Foreign Investors, the oil-rich state is either at, or close to, the median level, suggesting more can be done in these areas, leveraging on its strong Institutions, Governance and Leadership and Competition, Regulatory Standards and Rule of Law.

For Cambodia (Figure 2.33), its Attractiveness to Foreign Investors and Standard of Living, Education and Social Stability are about the regional median while Economic Vibrancy is way–off. Improvement on Government Policies and Fiscal Sustainability, Openness to Trade and Services and Physical Infrastructure toward median level is imperative for Cambodia to catch up with ASEAN median in terms of Economic Vibrancy and Financial Deepening and Business Efficiency.

In Figure 2.34, Indonesia has performed exceptionally well above the ASEAN-10 median for Economic Vibrancy, as consistent with Sections 2.2 and 2.6. Its Labour Market Flexibility, Financial Deepening and Business Efficiency and Physical Infrastructure also exceed the regional median, albeit to a lesser extent. Given such strengths as well as Institutions, Governance and Leadership and Competition, Regulatory Standards and Rule of Law at the median level, sound policies are vital to turn potential into reality. In this regard, ACI's studies on Indonesia's 33 provinces can be helpful in narrowing such gaps.[11]

[11] See Tan *et al*. (2013a, 2015).

Figure 2.33: Median Competitiveness Web Analysis for Cambodia, 2012

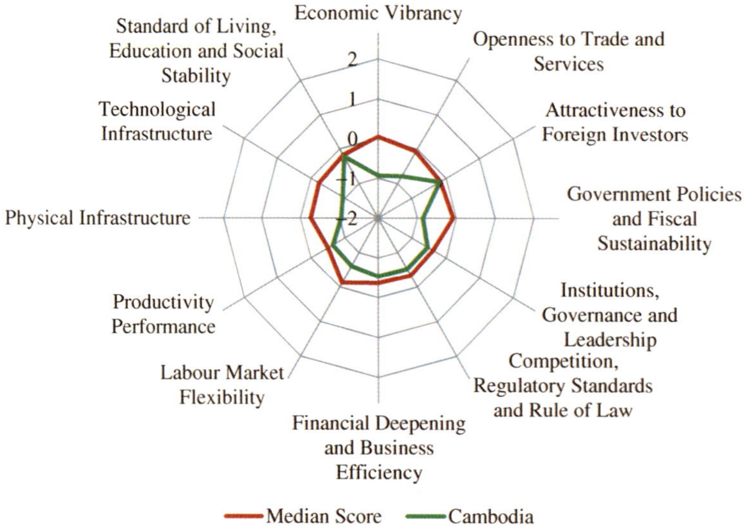

Source: Asia Competitiveness Institute.

Figure 2.34: Median Competitiveness Web Analysis for Indonesia, 2012

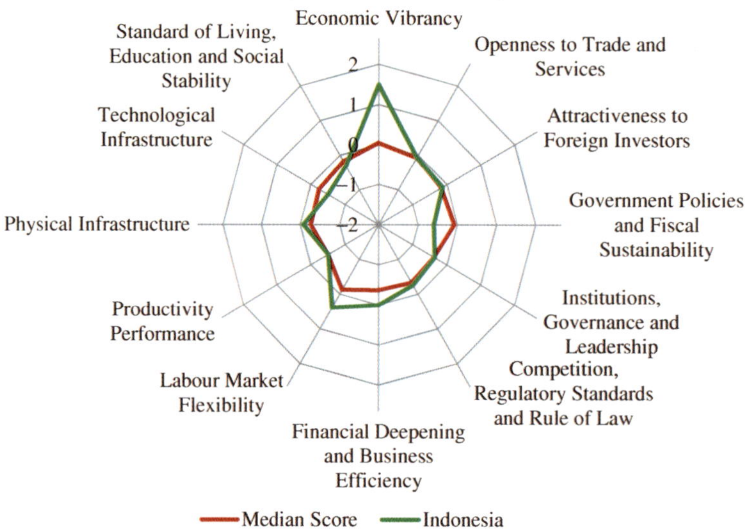

Source: Asia Competitiveness Institute.

Laos in Figure 2.35 remains far below the ASEAN-10 median but not without potential. Its Productivity Performance and Technological Infrastructure are very close to the median level, suggesting the country can harness these strengths to attract foreign investors, promote trade and bring the remaining sub-environments up to par.

It is no surprise that Malaysia, as always second to Singapore in Overall Competitiveness rankings, is well above the ASEAN median in all sub-environments (Figure 2.36). In line with

Figure 2.35: Median Competitiveness Web Analysis for Laos, 2012

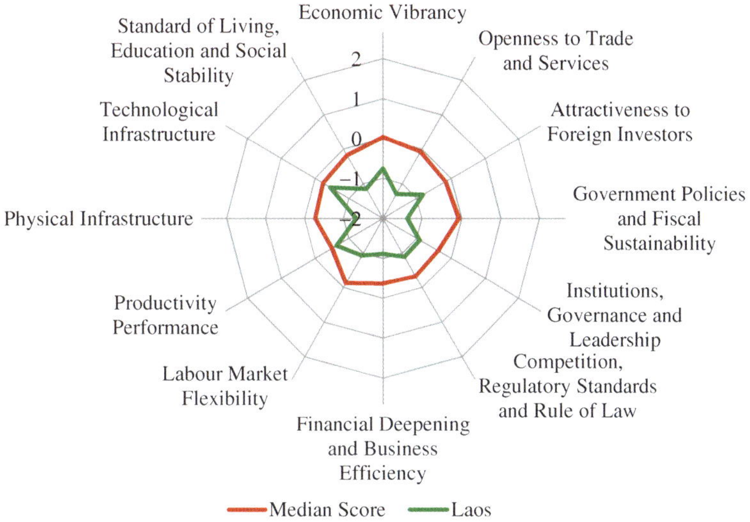

Source: Asia Competitiveness Institute.

Figure 2.36: Median Competitiveness Web Analysis for Malaysia, 2012

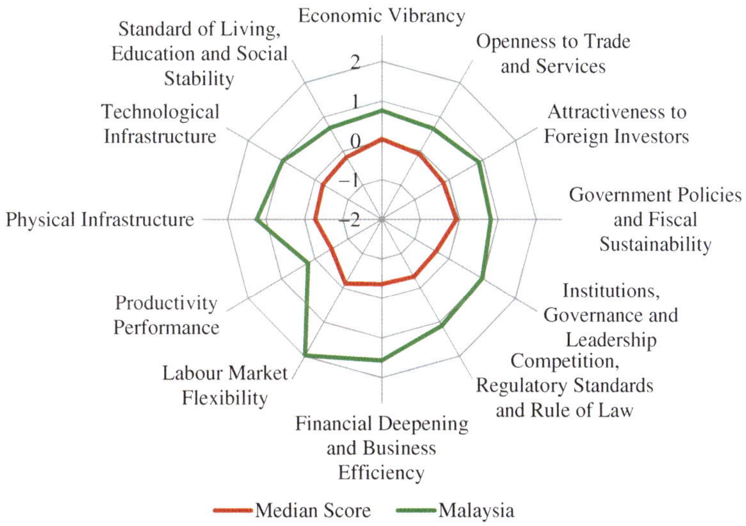

Source: Asia Competitiveness Institute.

its top spot in the 2012 ranking of Labour Market Flexibility, Malaysia exceeds the regional average by the largest margin. With apparent strengths in Financial Deepening and Business Efficiency and Physical Infrastructure, more can be done to bring Productivity Performance and Standard of Living, Education and Social Stability up to similar levels.

Myanmar, being tenth in most sub-environments, is unsurprisingly below ASEAN median in most sub-environments (Figure 2.37). However, its Productivity Performance being coincident

Figure 2.37: Median Competitiveness Web Analysis for Myanmar, 2012

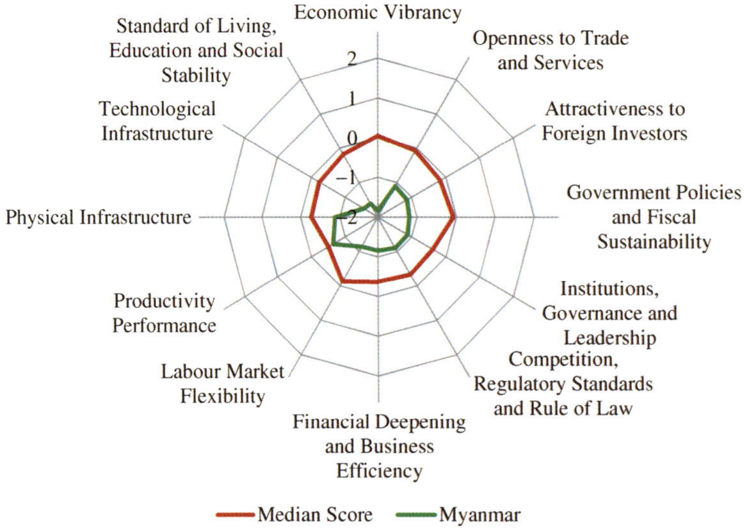

Source: Asia Competitiveness Institute.

Figure 2.38: Median Competitiveness Web Analysis for Philippines, 2012

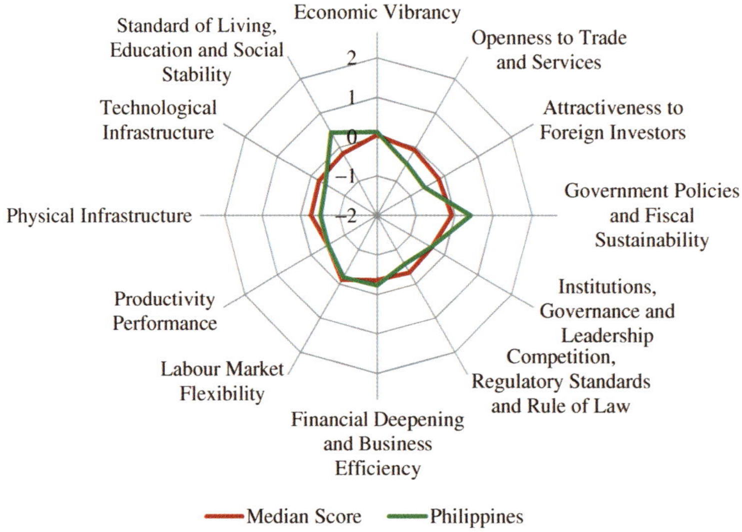

Source: Asia Competitiveness Institute.

with the median, and, to a lesser extent, Physical Infrastructure provide ample grounds for hope. Working on Economic Vibrancy, which is the farthest from regional median, should be the top priority.

The Philippines (Figure 2.38) as one of the more mature economies in ASEAN-10 exceeds the regional average in Standard of Living, Education and Social Stability as well as Government Policies and Fiscal Sustainability. Nonetheless, it is below median for Openness to Trade and

Services, Attractiveness to Foreign Investors, Physical Infrastructure and Competition, Regulatory Standards and Rule of Law. Improvement on these sub-environments would lead to more job creation, which in turn will unleash the country's enormous manpower potential through a backflow of Filipino professionals working abroad, enhancing the country's competitiveness which has been lagging.

Consistently ranked at the top in Overall Competitiveness, Singapore (Figure 2.39) is way above ASEAN median in all sub-environments. Nonetheless, it has to watch Labour Market Flexibility. Reliance on foreign labour is perhaps too restrictive. Maintaining competitiveness on the back of value add should be the Singaporean way going forward.

As the fourth most competitive economy in ASEAN-10 throughout the period 2000–2012, Thailand (Figure 2.40) well exceeds the regional median in most aspects of competitiveness, notably Openness to Trade and Services, Financial Deepening and Business Efficiency, Physical Infrastructure and Government Policies and Fiscal Sustainability. More heed should be paid to bring Standard of Living, Education and Social Stability up to standard as the aforementioned sub-environments. Improvement on Institutions, Governance and Leadership, Competition, Regulatory Standards and Rule of Law, Productivity Performance and Technological Infrastructure is required if Thailand is to credibly challenge Malaysia, or even Singapore, for the top two spots.

Vietnam (Figure 2.41) is a tad above ASEAN median for Technological Infrastructure and Labour Market Flexibility. It is, however, losing competitiveness in Attractiveness to Foreign Investors and Economic Vibrancy and lagging below regional average in Standard of Living, Education and Social Stability, suggesting its growth momentum is running out of steam. More work can be done in terms of Institutions, Governance and Leadership as the primary force driving the country forward or it risks falling behind its neighbours Cambodia and Laos.

Figure 2.39: Median Competitiveness Web Analysis for Singapore, 2012

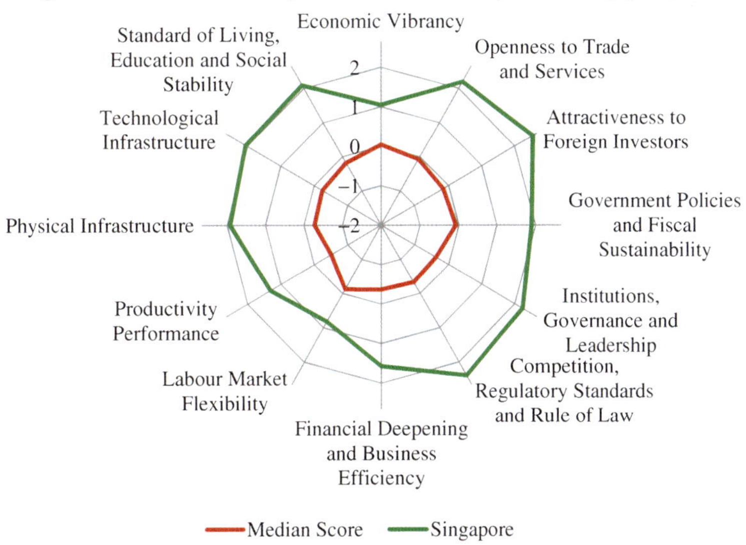

Source: Asia Competitiveness Institute.

Figure 2.40: Median Competitiveness Web Analysis for Thailand, 2012

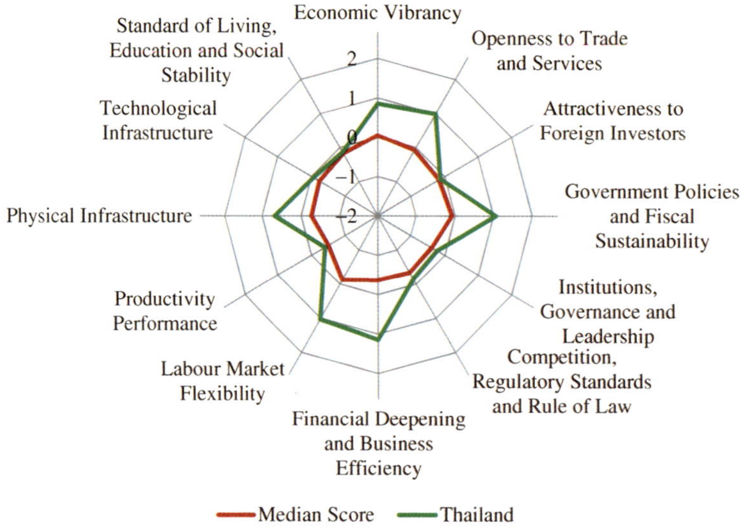

Source: Asia Competitiveness Institute.

Figure 2.41: Median Competitiveness Web Analysis for Vietnam, 2012

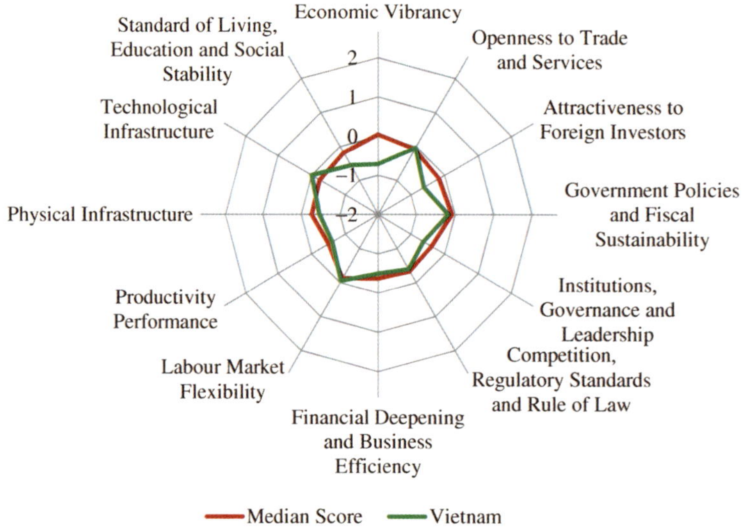

Source: Asia Competitiveness Institute.

Maximum Competitiveness Web Analysis

Maximum competitiveness web analysis presented in this section reveals how far individual ASEAN economies are from the highest scores for each of the 12 sub-environments, thereby serving as a rudimentary roadmap for catching-up.

Brunei (Figure 2.42) is well below the ASEAN-10 maximum in most aspects of competitiveness despite being in the high-income band with Singapore. Its Labour Market Flexibility is the furthest from the regional maximum. Its Productivity Performance at the regional top is encouraging if it looks to diversify its economy toward a sustainable future.

Cambodia (Figure 2.43) is well below the ASEAN maximum for all sub-environments. Given its gaps relative to the top performer, Attractiveness to Foreign Investors could be prioritised for

Figure 2.42: Maximum Competitiveness Web Analysis for Brunei, 2012

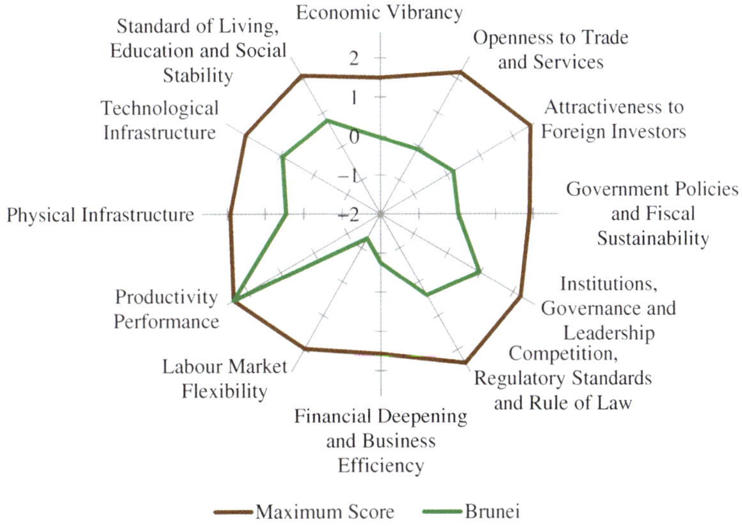

Source: Asia Competitiveness Institute.

Figure 2.43: Maximum Competitiveness Web Analysis for Cambodia, 2012

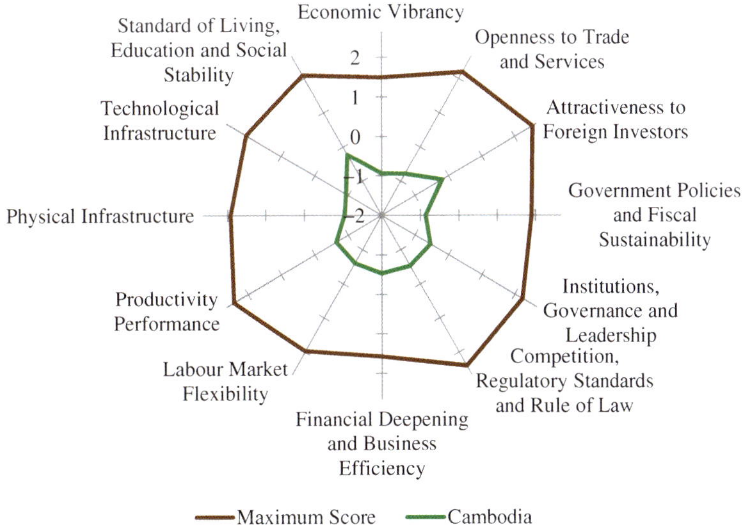

Source: Asia Competitiveness Institute.

employment creation to improve Standard of Living, Education and Social Stability. Foreign capital can also be harnessed for infrastructure upgrade as well as facilitate development in other aspects of competitiveness.

For Indonesia (Figure 2.44) as the top in Economic Vibrancy, all other sub-environments leave much to be desired, especially Government Policies and Fiscal Sustainability being the furthest from the regional maximum. Openness to Trade and Services and Financial Deepening and Business Efficiency could be Indonesia's starting point to narrow the gap with the top four economies (Singapore, Malaysia, Thailand and Brunei).

Laos (Figure 2.45) remains far distant from ASEAN maximum in all sub-environments. With relatively high scores in Technological Infrastructure and Productivity Performance, it could pursue a dualistic economic structure as a food basket as well as taking over labour-intensive manufacturing activities from more developed economies. More work on the institutional front would be essential to expedite the catching-up process.

While Malaysia (Figure 2.46) has done exceptionally well for Labour Market Flexibility and Financial Deepening and Business Efficiency, its distance to ASEAN maximum for the remaining sub-environments remains large, with the widest being Productivity Performance. Equally pressing is bringing Standard of Living, Education and Social Stability up to par with the other components of competitiveness.

Myanmar (Figure 2.47) remains well below the ASEAN maximum for all sub-environments, notably Economic Vibrancy. Expediting its opening-up would certainly help in this regard. Coupled with improvement in Openness to Trade and Services and Attractiveness to Foreign Investors, Myanmar will be able to narrow the development gap with, first, Cambodia, Laos and

Figure 2.44: Maximum Competitiveness Web Analysis for Indonesia, 2012

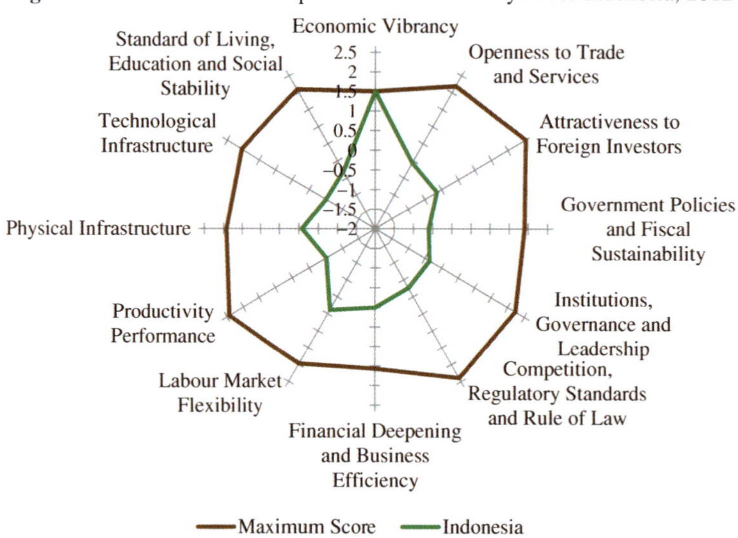

Source: Asia Competitiveness Institute.

Figure 2.45: Maximum Competitiveness Web Analysis for Laos, 2012

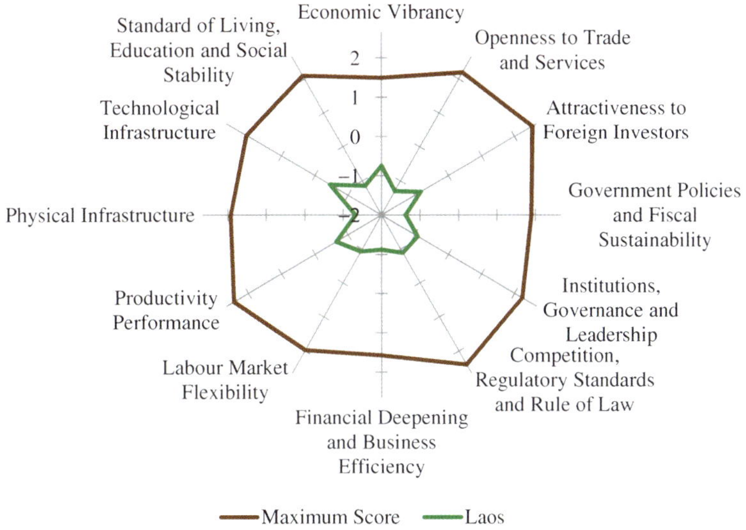

Source: Asia Competitiveness Institute.

Figure 2.46: Maximum Competitiveness Web Analysis for Malaysia, 2012

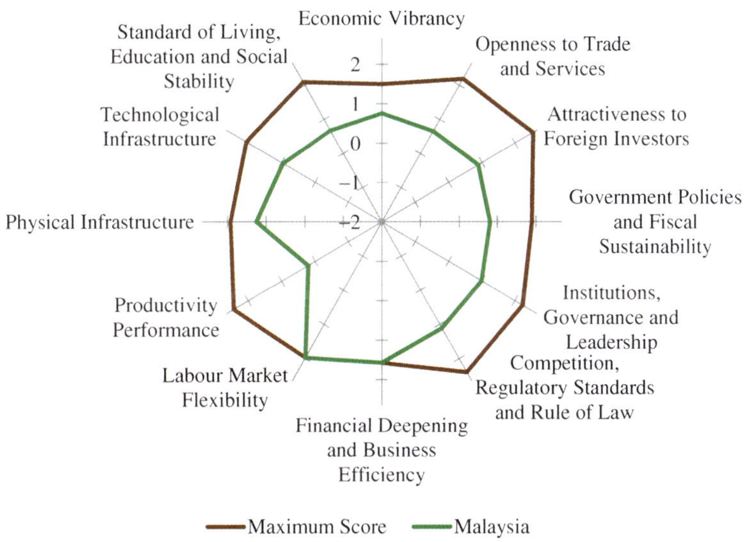

Source: Asia Competitiveness Institute.

Vietnam, and subsequently the rest of ASEAN-10. Socio-political stability is inarguably a pre-requisite for the country to eventually take-off.

The Philippines (Figure 2.48), despite being one of the more mature economies in ASEAN-10, remains remarkably distant from the regional maximum for Openness to Trade and Services,

Figure 2.47: Maximum Competitiveness Web Analysis for Myanmar, 2012

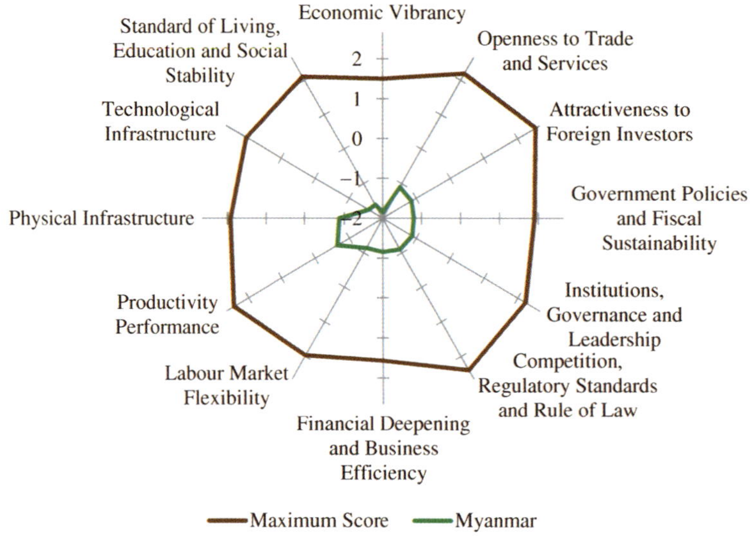

Source: Asia Competitiveness Institute.

Figure 2.48 Maximum Competitiveness Web Analysis for Philippines, 2012

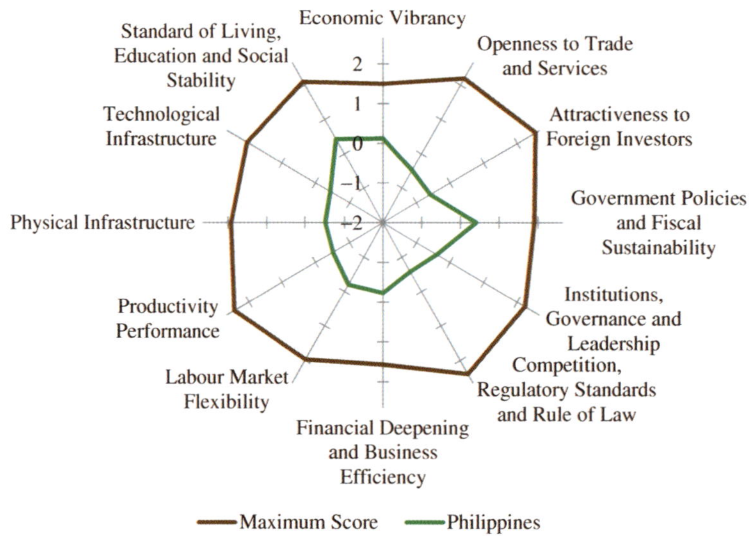

Source: Asia Competitiveness Institute.

Attractiveness to Foreign Investors and Infrastructure. As noted earlier, productive employment is vital for the country to regain its competitiveness. Leveraging on its strength in Government Policies and Fiscal Sustainability, reform could be kick-started in Competition, Regulatory Standards and Rule of Law, paving the way for a thriving private sector.

Notwithstanding setting the regional maximum in virtually all sub-environments, Singapore (Figure 2.49) has been disappointingly lagging in Productivity Performance. Cultivating innovation

Figure 2.49: Maximum Competitiveness Web Analysis for Singapore, 2012

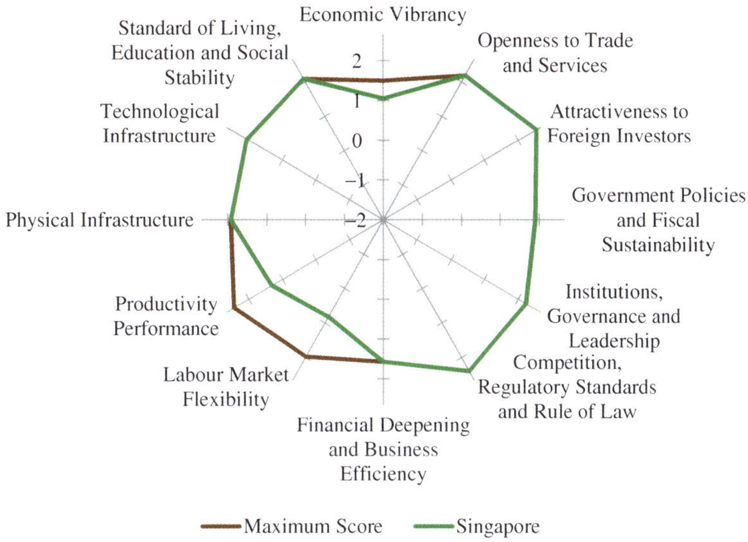

Source: Asia Competitiveness Institute.

Figure 2.50: Maximum Competitiveness Web Analysis for Thailand, 2012

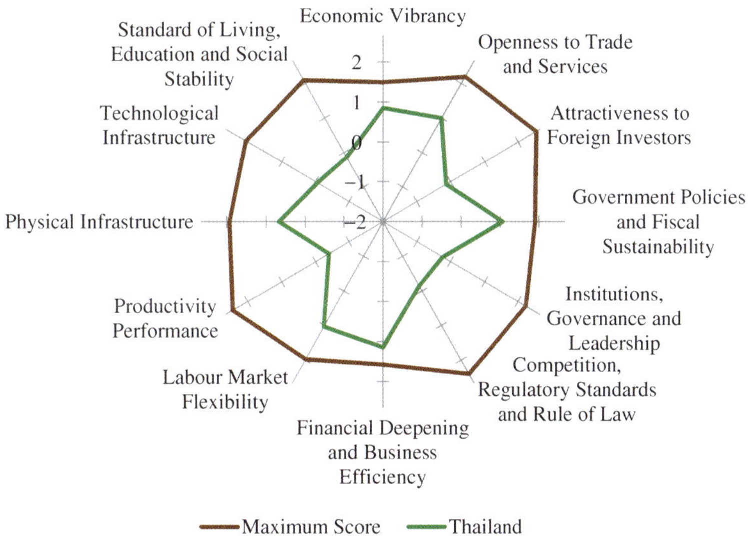

Source: Asia Competitiveness Institute.

and entrepreneurial spirit would be the crucial first step towards the knowledge-based economy it is aspiring to become.

Thailand (Figure 2.50) is close to the regional maximum for Economic Vibrancy, Government Policies and Fiscal Sustainability and Financial Deepening and Business Efficiency. Nonetheless, it is oddly well below the maximum for Standard of Living, Education and Social Stability given its strong performances in the aforementioned aspects. Socio-political stability needs to be

Figure 2.51: Maximum Competitiveness Web Analysis for Vietnam, 2012

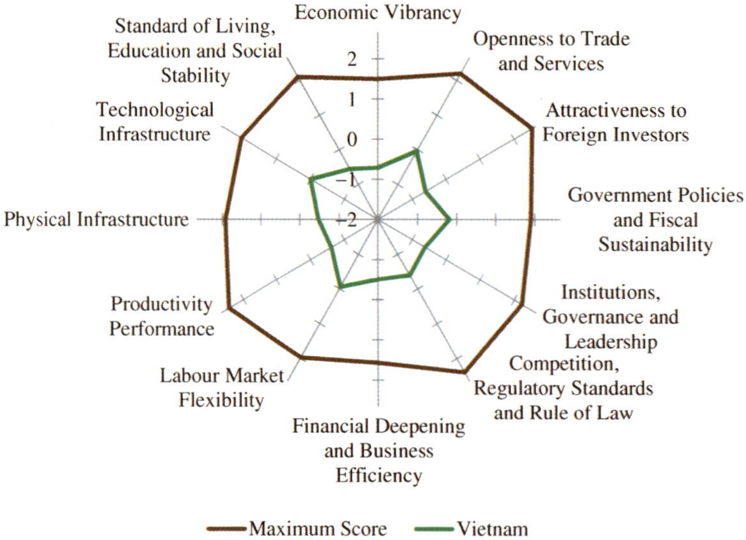

Source: Asia Competitiveness Institute.

restored for maintaining its Attractiveness to Foreign Investors as well as improving Competition, Regulatory Standards and Rule of Law.

Vietnam (Figure 2.51) is well below ASEAN maximum in all sub-environments although it has narrowed the distance for Openness to Trade and Services, Technological Infrastructure and Labour Market Flexibility. With growing competition from Cambodia, Laos and Myanmar, Vietnam cannot be complacent in regaining its reform momentum which seems to be fizzling, particularly for Economic Vibrancy and Attractiveness to Foreign Investors.

2.3.8 Upgrading Competitiveness through *What-if* Simulation Analysis

This section discusses the *what-if* simulation analysis by the methodology detailed in Section 2.5. The simulation is conducted for the Overall Competitiveness as well as each of the four environments. Tables 2.27–2.31 show the standardised scores and rankings of individual ASEAN-10 economies before and after simulation.

In Table 2.27 for Overall Competitiveness, most impressive is Vietnam, jumping from seventh to fifth after the simulation. The first, second, third and fifth ranking remain unchanged but with improved scores. For instance, Indonesia, while staying at fifth, attains a positive score of 0.2775 as compared to the negative –0.0788 before the simulation. The remaining economies climb one place after having the scores of their top 20% weakest indicators raised to zero (the average score).

Likewise for Macroeconomic Stability, the first, second, third and fifth rankings stay the same but with higher scores (see Table 2.28). Philippines at six also remains unchanged and with improved score. The bottom four, Vietnam, Cambodia, Laos and Myanmar, all move up one spot after the policy simulation.

Table 2.27: *What-if* Competitiveness Simulation Analysis on Overall Competitiveness, 2012

Economy	Rank		Score	
	Before	After	Before	After
Brunei	4	3	0.3699	0.8088
Cambodia	8	7	–0.7462	–0.3691
Indonesia	5	5	–0.0788	0.2775
Laos	9	8	–1.1141	–0.7211
Malaysia	2	2	1.1509	1.2375
Myanmar	10	9	–1.3776	–0.9256
Philippines	6	5	–0.2906	–0.0211
Singapore	1	1	2.0806	2.2384
Thailand	3	3	0.4630	0.6097
Vietnam	7	5	–0.4570	–0.1090

Source: Asia Competitiveness Institute.

Table 2.28: *What-if* Competitiveness Simulation Analysis on Macroeconomic Stability, 2012

Economy	Rank		Score	
	Before	After	Before	After
Brunei	5	5	0.0264	0.3907
Cambodia	8	7	–0.6995	–0.4507
Indonesia	4	3	0.4749	0.6332
Laos	9	8	–1.0487	–0.7411
Malaysia	2	2	0.8426	0.8524
Myanmar	10	9	–1.4928	–0.9941
Philippines	6	6	–0.3497	–0.1830
Singapore	1	1	2.1161	2.1738
Thailand	3	3	0.6299	0.6910
Vietnam	7	6	–0.4992	–0.2585

Source: Asia Competitiveness Institute.

In Table 2.29 for Government and Institutional Setting, Indonesia, Cambodia and Myanmar see impressive jumps from seventh to fifth, eighth to sixth and tenth to eighth, respectively, after the simulation. Thailand joins Brunei at third. With the exception of the top three and the fifth ranking with no change, the remaining economies have an incremental change of one place up, with Vietnam on par with Indonesia and the Philippines at fifth.

Table 2.29: *What-if* Competitiveness Simulation Analysis on Government and Institutional Setting, 2012

Economy	Rank		Score	
	Before	After	Before	After
Brunei	3	3	0.4883	0.8598
Cambodia	8	6	–0.6735	–0.4404
Indonesia	7	5	–0.4260	0.0326
Laos	9	8	–1.1073	–0.6556
Malaysia	2	2	1.0132	1.1495
Myanmar	10	8	–1.1905	–0.7190
Philippines	5	5	–0.2001	0.0848
Singapore	1	1	2.2770	2.3295
Thailand	4	3	0.2377	0.5155
Vietnam	6	5	–0.4187	–0.1116

Source: Asia Competitiveness Institute.

Table 2.30: *What-if* Competitiveness Simulation Analysis on Financial, Businesses and Manpower Conditions, 2012

Economy	Rank		Score	
	Before	After	Before	After
Brunei	4	3	0.1742	1.0384
Cambodia	8	6	–0.7775	–0.3262
Indonesia	5	4	–0.0486	0.4191
Laos	9	8	–1.1766	–0.8366
Malaysia	1	1	1.6369	1.7153
Myanmar	10	9	–1.2844	–1.0552
Philippines	6	5	–0.4083	–0.1545
Singapore	2	1	1.6190	2.0623
Thailand	3	3	0.7753	0.7753
Vietnam	7	5	–0.5101	0.0060

Source: Asia Competitiveness Institute.

For Financial, Businesses and Manpower Conditions, Vietnam and Cambodia step up from seventh to fifth and eight to sixth after simulation (see Table 2.30). Singapore becomes on par with Malaysia at the top and Brunei joins Thailand at third. The rest of ASEAN-10 move up one place.

In Table 2.31 for Quality of Life and Infrastructure Development, the top four remains unchanged after policy simulation but with improved scores. Cambodia remarkably leaps from

Table 2.31: *What-if* Competitiveness Simulation Analysis on Quality of Life and Infrastructure Development, 2012

Economy	Rank		Score	
	Before	After	Before	After
Brunei	3	3	0.7567	0.8642
Cambodia	8	5	–0.7661	–0.2227
Indonesia	6	5	–0.3083	0.0028
Laos	9	8	–1.0221	–0.5804
Malaysia	2	2	1.0058	1.1182
Myanmar	10	8	–1.4168	–0.8370
Philippines	5	4	–0.1776	0.1702
Singapore	1	1	2.1199	2.2400
Thailand	4	4	0.1667	0.4080
Vietnam	7	5	–0.3583	–0.0625

Source: Asia Competitiveness Institute.

eighth to fifth. Vietnam and Myanmar climb two places from seventh to fifth and tenth to eighth, respectively. The CLMV grouping moving up as much certainly makes complacency not an option for the more mature ASEAN economies.

2.4. CONCLUDING REMARKS AND POLICY IMPLICATIONS

In general, the rankings of the Overall Competitiveness do not see considerable changes over the period 2000–2012. The development divide in the ASEAN–10 is evident in our annual rankings where Singapore, Malaysia, Thailand and Brunei remain in the top four for 13 consecutive years. Indonesia has decidedly surpassed the Philippines and Vietnam at 5th. The bottom four are Vietnam, Cambodia, Laos and Myanmar. With the exception of the top four, the standardised scores of the remaining six economies were negative in 2012, indicating their below average performances. The four environments and 12 sub-environments, however, are more diverse in rankings with Malaysia overtaking Singapore at the top for Financial, Businesses and Manpower Conditions and Brunei leading the region in Productivity Performance.

Overall, our competitiveness analysis of ASEAN-10 indicates wide variations not only in levels of economic development but also paces of catching-up within the bloc. Indonesia, for instance, has been surging ahead, particularly in sub-environments such as Economic Vibrancy and Financial Deepening and Business Efficiency, as shown in our results of its top 20% strongest indicators and median spider web chart. In contrast, for Vietnam, the catching-up momentum seems to be running out of steam as signified in its below-median Economic Vibrancy and Attractiveness to Foreign Investors. Cambodia and Laos seems are quietly rising with marked improvement in many aspects of competitiveness. The Philippines risks being outstripped by the newer members as its standardised scores plunged more negatively over time. Thailand, too, cannot be complacent on the socio-political front if it is to restore trust and confidence in both domestic and foreign investors.

As noted in Section 2.4, a competitiveness ranking in itself is like a beauty contest that stops short of offering insights on how to improve the rankings. The top 20% strongest and weakest indicators are thus revealed as a roadmap for policy-makers to harness their strengths and resolve their weaknesses on the path towards a more competitive and prosperous economy. Furthermore, our policy simulation analysis shows how much individual countries improve in terms of standardised scores and rankings if their scores of the top 20% weaknesses are raised to the average. The post-simulation results for the CLMV grouping are particularly impressive, proving that improved competitiveness is totally feasible with the right solutions.

While economic convergence is certainly desirable, diversity in development stages as well as competitiveness can be a powerful source of synergy, complementing and supplementing one another for increased overall regional competitiveness. A dualistic economic structure with agriculture and agro-based industrialisation-cum-servicisation is totally viable, leveraging on the strengths of individual member states. Thailand, for instance, has been a successful dualistic economy. Malaysia and Indonesia are endowed with abundant fertile soil conducive for a wide variety of agricultural products from rubber to palm oil. CLMV are currently net exporters of rice. The more developed ASEAN-6 and the newer, transitional CLMV form a pair like two hands clapping in terms of agricultural resource endowments and potential dualism. For this strategy to materialise, regional integration efforts need to be more earnest. Whether it is ASEAN Free Trade Area (AFTA), ASEAN Trade in Goods Agreement or AEC, ASEAN has many platforms to realise its potential.

Part and parcel in such dualistic economic structure, as well as for more seamless flows of trade and investment, is enhanced intra-regional physical connectivity which is one of the three pillars[12] of the Master Plan on ASEAN Connectivity (MPAC) officially adopted by all ten member states in 2010. Fundamentally MPAC shares important similarities with China's 21st Maritime Silk Road supported by the newly established Asian Infrastructure Investment Bank (AIIB). Both aim to fill the gaps in infrastructure for enhanced connectivity and economic cooperation, albeit on different scales. As the entire ASEAN bloc are founding members of AIIB, efforts to align AIIB and the Maritime Silk Road with MPAC are critical for narrowing the development divide within ASEAN. In other words, ASEAN centrality must remain as the primary strategic thrust in the ongoing Asianisation of global institutional arrangements, be it the Maritime Silk Road or AIIB.

In fact, in view of the aforementioned China-led global initiatives, ACI has proposed the Asia Economic Connectivity Vision 2030 (AECV2030) which also aims to facilitate wider physical and information connectivity across Asia in a multi-modal way.[13] In contrast with China's multi-

---

[12]The remaining two pillars are institutional connectivity and people-to-people connectivity. See ASEAN Secretariat (2010)

[13]See Tan *et al.* (2015).

tracked approach, ACI's vision is ASEAN-centric and led by ASEAN through open, consultative and consensual decision-making, thus allowing respective countries to position its best interests, in accordance with its development stage and needs, so as to benefit from the wide array of opportunities offered by increased connectivity and trade.

While regional integration can play an important part in narrowing the development gap, the fruit of deeper integration is more likely to be equally distributed if individual countries systematically address their weaknesses, in addition to enhancing their strengths. In this regard, ACI's annual competitiveness analysis for ASEAN-10 can be instructive as action plans. Building institutional capacity remains the first and foremost prerequisite to ensure greater competitiveness and sustainable growth, as evident in the experience of Singapore. In all, competition in parallel with cooperation is the cornerstone of the vision for a highly competitive and resilient ASEAN.

REFERENCES

ASEAN Secretariat. 2010. *Master Plan on ASEAN Connectivity*. Jakarta: ASEAN Secretariat. http://www.asean.org/resources/publications/asean-publications/item/master-plan-on-asean-connectivity-2.

Basu Das, Sanchita, Jayant Menon, Rodolfo Severino and Omkar Lal Shrestha. 2013. *The ASEAN Economic Community: A Work in Progress*. Singapore: Asian Development Bank and Institute of Southeast Asian Studies.

Bristow, Gillian. 2010. *Critical Reflections on Regional Competitiveness: Theory, Policy, Practice (Routledge Studies in Human Geography)*. Oxon: Routledge.

Camagni, Roberto. 2002. "On the Concept of Territorial Competitiveness: Sound or Misleading?" *Urban Studies* 39(13): 2395–2411.

IMD World Competitiveness Center. 2014. *IMD World Competitiveness Yearbook*. Lausanne: International Institute for Management Development.

Kitson, Michael, Ronald Martin and Peter Tyler. 2004. "Regional Competitiveness: An Elusive yet Key Concept?" *Regional Studies* 38(9): 991–99.

Krugman, Paul. 1994. "Competitiveness: A Dangerous Obsession." *Foreign Affairs* 73(2): 28–44.

McKinsey Global Institute. 2014. *Southeast Asia at the crossroads: Three paths to prosperity*. http://www.mckinsey.com/insights/energy_resources_materials/three_paths_to_sustained_economic_growth_in_southeast_asia.

Porter, Michael E. 1980. *Competitive Strategy: Techniques for Analyzing Industries and Competitors*. New York: Free Press.

Tan, Khee Giap, Linda Low, Tan Kong Yam and Lim Lijuan Amanda. 2013. Annual Analysis of Competitiveness, Development Strategies and Public Policies on ASEAN-10: 2000–2010. Singapore: Pearson Education.

Tan, Khee Giap, Sangiita Yoong Wei Cher, Sasidaran Gopalan and Nguyen Le Phuong Anh. 2015. *2014 Annual Competitiveness Ranking and Simulation Study for ASEAN-10 and Development Strategies to Enhance Asia Economic Connectivity*. Singapore: World Scientific.

World Economic Forum. 2014. *Global Competitiveness Report 2013–2014*. Geneva: World Economic Forum.

Discussant Notes 2.1

Mrs Somhatai Panichewa

Chief Executive Officer,
Amata VN Public Company Limited, Thailand

Mr Chartchai Panichewa

President,
ThaiSri Insurance Public Company Limited, Thailand

Deputy Minister Nguyen Cam Tu, Dr Nguyen Dinh Cung, Professor Tan Kong Yam, Associate Professor Tan Khee Giap, ladies and gentlemen, a very good morning to all of you. Thank you for inviting me to speak at the ASEAN Review Seminar today.

Firstly, I would like to congratulate you on Singapore's General Election which is to be held in September. As a Thai citizen, I am still waiting the election. It may take place, at the earliest, in February next year.

Amata Corporation is currently investing a total of US$1.2 billion in two countries. We are an industrial estate developer with 1,000 manufacturing factories, 80% of which are foreign direct investment (FDI). At the moment, we are implementing three projects in two countries which will generate an equivalent of 10% of Thailand's GDP. We have been operating in Vietnam for 20 years. In 1995 when we first came to Vietnam, 80% of FDI flows into Vietnam were from Japan. With our experience in bringing FDI to Vietnam, we are very proud that the Vietnamese government has given the green light to our four projects in both the North and the South of Vietnam. We are proud to be the creator of jobs, GDP and tax revenues for the countries we invest in. Myanmar will be our next investment destination after Vietnam.

I just came back to Vietnam two weeks ago and I am very impressed with the economic development of Vietnam. Vietnam is aiming to achieve 6.2% GDP growth this year, which I think is not too high for the country to attain. Why? The new Investment Law became effective in July this year, allowing FDI to operate the real estate business. This is particularly impactful as in Thailand, investment in real estate must be in the form of joint venture with Thai companies owning at least 51%. In Vietnam, foreigners are now permitted to buy properties without working in the country, which is very attractive for investors. However, the effects may only be felt in the microeconomic sector, particularly among speculators. Vietnam needs to focus more on the real sector. Otherwise, how can Vietnam jump from seventh to fifth according to ACI's *what-if* simulation? How can Vietnam strengthen the real sector, not the real estate sector?

The real sector in Vietnam is manufacturing. If you look at the top ten manufacturing companies in the country, nine are state-owned enterprises and the other is Samsung. Where are the small and medium-sized enterprises (SMEs)? I would like to share that Amata Corporation has been cooperating with one district in Tokyo named Ota. We built Ota Technopark in Thailand which is a hub for Japanese SMEs that relocate in Thailand. How has the Japanese government helped? The Japanese government subsidised the administrative costs. Although the SMEs only occupy 300 square metres of the Technopark, the government sponsored the costs of building meeting rooms, reception and accounting costs. I would like to see this type of support extended to the Vietnamese SMEs to make them more competitive. Otherwise, how can they grow and create real demand in the economy?

Vietnam's policies, in my view, have been well thought out. Nonetheless, execution has been problematic, particularly in terms of transparency and corporate governance. Policies from the central government are very clear, but they become grey at the regional and provincial level and are not properly executed. Foreign investors thus feel insecure. For example, my company issued interim dividends for 15 years. However five years ago, a foreign company restated its financial statements, turning profits to losses, which led the government to ban interim dividends. This is not fair to us who have been truthful in financial reporting and have never restated financial figures. Furthermore, when tax inspectors come to inspect us, they do so for the most recent five years and usually impose penalties and fines. The perception among investors is that if tax collectors cannot find faults, they cannot go back. They are supposed to help the private sector, not to take from the private sector.

I love Vietnam. Vietnam has been my second home for 20 years. I believe if Vietnam can overcome its weaknesses, the country will grow even faster with more FDI inflows. In the next five years, we will triple our investment in three countries. In terms of land, we will expand from 100 square kilometres to 500 square kilometres. If you are looking to go to Vietnam and Myanmar, we would love to accompany you as a consultant, as a friend and as a partner. Thank you very much.

Chapter 3

How Productive is the Agricultural Sector in ASEAN?

3.1 INTRODUCTION

Notwithstanding the dramatic rise in manufacturing output that anchors the structural transformation towards industry in ASEAN, agriculture remains a vital economic sector in the region. As home to over 600 million people or 9% of the world population, sustaining agricultural productivity is paramount to maintain food security. Equally important is the fact that the less developed ASEAN member states are still heavily reliant on agriculture as a primary source of growth, as reflected in the shares of their agricultural sector in national output and employment (see Table 3.1).

In particular, agriculture accounts for more than 30% of Gross Domestic Product (GDP) in Myanmar, Laos and Cambodia. In Cambodia, Laos, Myanmar and Vietnam (CLMV), agricultural share of total employment remains high at more than 60%. Such a wide gap between agricultural contribution to GDP and its share in employment in the less developed economies in ASEAN signifies the lag between industrialisation efforts and the absorption of rural labour into non-agricultural sectors, which in turn highlights the significance of agricultural development to economic growth and people's livelihoods. Widespread rural poverty, as indicated in the perpetually higher rural poverty headcount ratios compared to national averages, further underscores the importance of agricultural growth in the regionwide aspiration for inclusive prosperity (see Figure 3.1).

Decades of industrialisation efforts present both opportunities and challenges to the region's agricultural sector. On the one hand, the combination of industrialisation, urbanisation and improvement in urban living standard has driven up the demand for not only food products but also agricultural inputs for industrial production, benefiting farmers with both new markets and higher prices. On the other hand, reallocation of land and labour out of agriculture intensifies the risk of food shortages, as evident in the experience of China which has turned from a net food exporter to a net food importer.[1] In fact, the global food crisis between 2007 and 2008 highlights the region's vulnerability to agricultural shocks as ASEAN member states, comprising both net food importers and net food exporters, resorted to concurrent, uncoordinated export bans and extreme import orders, exacerbating the situation.[2]

[1] See Tan, Suo and Rajan (2016b, forthcoming).

[2] Falling global stocks of rice and other grains, adverse weather and soaring oil prices in 2007 and 2008 led to a surge in prices of rice and other foodstuffs in many parts of the world including ASEAN. The initial policy responses to the global food crisis in ASEAN varied according to whether the country in question was a net exporter or a net importer of food. While the former, notably Thailand and Vietnam, raised export taxes and placed restrictions on exports, the

Table 3.1: Share of Agriculture in GDP and Employment in ASEAN, 2000 and 2011

Country	Agricultural Share of GDP (%)		Agricultural Employment (% of Total Employment)	
	2000	**2011**	**2000**	**2011**
Brunei	1.02	0.63	0.65	0.51
Cambodia	37.84	36.68	69.92	65.46
Indonesia	15.60	13.79	48.26	40.73
Laos	45.17	29.52	76.68	74.76
Malaysia	8.60	11.78	18.88	12.06
Myanmar	57.24	n.a.	70.36	66.78
Philippines	13.97	12.72	39.66	33.11
Singapore	0.10	0.04	0.15	0.0
Thailand	9.02	13.34	56.46	47.74
Vietnam	22.73	20.08	67.35	62.73

Source: World Bank and Food and Agriculture Organisation (FAO).

Figure 3.1: Rural Poverty Incidence in ASEAN, 2012

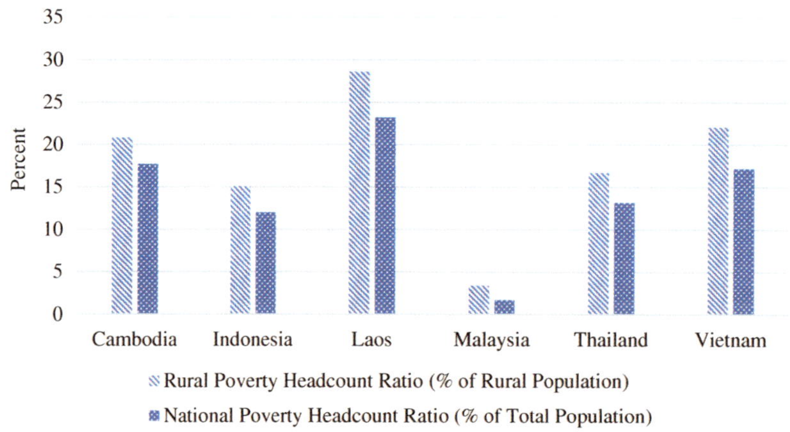

Note: Data as of 2012, except for Thailand (2011). Brunei, Myanmar, Philippines and Singapore are excluded due to data unavailability.
Source: World Bank.

In view of these realities, agriculture has received renewed focus among the ASEAN members. In the aftermath of the 2007–2008 global food crisis, the ASEAN-10 countries officially

latter such as Malaysia, Indonesia and Philippines reduced import restrictions and tariffs while trying to secure food supplies in the international market. In 2008, Vietnam, Cambodia and Laos banned exports of rice although Cambodia subsequently abolished the policy to take advantage of high international food prices. Malaysia and Philippines, in contrast, resorted to "panic" buying from Thailand and Vietnam, respectively, at the peak of the crisis. See Asian Development Bank (ADB) (2008) for analysis on the underlying causes of the 2007–2008 global food crisis and Chandra and Lontoh (2010) for details on policy responses in ASEAN countries.

launched the ASEAN Integrated Food Security (AIFS) Framework, supported by a Strategic Plan of Action on ASEAN Cooperation in Food, Agriculture and Forestry in 2009, which crystallised a concrete set of strategies across four major components:[3]

(i) Food security and emergency and shortage relief;
(ii) Sustainable food trade development;
(iii) Integrated food security information system and
(iv) Agricultural innovation.

This framework was accompanied by the establishment of the ASEAN Plus Three Emergency Rice Reserve (APTERR)[4] which came into force in 2012, serving as a regional cooperative mechanism to reduce the high vulnerability of the region to fluctuations in the production of basic foodstuffs and thus to the instability of the region's food supply.

APTERR defines two types of reserve: the initial earmark of stocks, which is 787,000 tonnes, and the stockpiles which comprised both rice stocks and cash contributions. A governing council is tasked with facilitating multilateral decision-making, and a secretariat oversees day-to-day operations of the initiative which are funded by member countries. The council also acts as an intermediary that evaluates the degree of emergency in food-deficit countries and matches them with another country with food surplus.

Nonetheless, while being strong on principles, APTERR lacks specifics on operational details.[5] For instance, the procedures, terms, prices and conditions with which food-deficit countries may obtain a release of the reserve remain opaque. In addition, APTERR's current volume of earmarked reserve at 787,000 tonnes is inadequate to contain a food crisis in relatively larger countries.[6] Releases of the reserve have been, in some instances, hampered by unclear or conflicting national regulations.[7] Furthermore, APTERR is centred on food accessibility rather than the supply side of food security, which is typically determined by food production growth. In fact, while the 2009 Strategic Plan of Action on ASEAN Cooperation on Food, Agriculture and Forestry outlines strategic thrusts for augmenting agricultural supply capacity in the region, concrete policy measures and operational specifics remain lacking.

ACI thus seeks to contribute to the policy discourse on agricultural development and food security in the region with this chapter on agricultural productivity in ASEAN, which is a relatively underdeveloped topic in the literature. As a region endowed with vast fertile land, diverse topography and abundant labour, there is ample scope for ASEAN to enhance productivity in agriculture, thereby fulfilling its potential to not only achieve food self-sufficiency but also become the food basket of Asia. Fostering agricultural productivity growth is of central importance to not only food security but also competitiveness and inclusive economic growth, particularly in light of the accelerating depletion of natural resources available for agricultural production.

[3]See ASEAN Secretariat (2009).

[4]Member countries are ASEAN-10, Japan, South Korea and China.

[5]See Briones (2011) and Trethewie (2013).

[6]See Briones *et al*. (2012).

[7]See Briones (2011).

As with our other flagship projects on competitiveness in Greater China, India and ASEAN, ACI applies evidence-based, quantitative research methodologies in this chapter to estimate the annual change of agricultural Total Factor Productivity (TFP) — broadly defined as the portion of output not explained by the amount of inputs used in the production — in eight ASEAN countries (Brunei and Singapore excluded) over the period from 2000 to 2011. Specifically, we employ both the growth accounting methodology and the Malmquist index data envelopment analysis (DEA).

Following this Introduction, Section 3.2 summarises the key trends, opportunities and challenges facing ASEAN's agricultural sector. Section 3.3 reviews the existing literature on agricultural productivity in ASEAN at the regional and national level. Section 3.4 presents the empirical methodologies for measuring TFP growth and data sources. Section 3.5 discusses the results of the study. Section 3.6 focuses on the case of Malaysia which attained the highest annual TFP growth over the study period. Particular emphasis is placed on Malaysia's successful development of the agribusiness industry. Section 3.7 concludes with policy implications for enhancing agricultural productivity.

3.2 AGRICULTURE IN ASEAN: TRENDS, OPPORTUNITIES AND CHALLENGES

Southeast Asia has a total land area of 4.5 million square kilometres of which about half is continental and the other half is archipelagic or insular.[8] Most of the region has a tropical humid climate with ample rainfall to keep the soil from drying. The rest of Southeast Asia, including central Myanmar, central plain of Cambodia, Northeast Thailand and the Philippines, have tropical sub-humid climate with less than 1,500 millimetres of annual rainfall, which causes the soil to dry out for part of the year. The region is also endowed with abundant fertile soils, particularly in the lowlands conducive for rice production. Nonetheless, the upland areas with heavy rainfalls have suffered from soil erosion.

Figure 3.2 shows the percentage of land area that is arable in individual ASEAN-8 countries in 2000 and 2011. Malaysia has the lowest amount of arable land which appears to be stagnant at around 2.9% of total land area over the decade between 2000 and 2011. Thailand is endowed with the highest share of arable land of more than 30.5%, followed by Cambodia and Vietnam. Laos witnessed the most remarkable increase in the percentage of arable land from merely 4% in 2000 to 6.2% in 2011. Cambodia and Indonesia also registered a marked rise in their proportions of arable land by 1.7 and 1.6%, respectively, over the same period.

Small-scale, traditional farming has been the dominant form of agricultural production in the region for thousands of years. Rice is both the staple food and primary agricultural commodity in the region, dominating the lowland areas. The region alone produces about 35% of global rice output.[9] Recent decades have seen commercial cash crops flourishing with remarkable increases in the number of plantation crops adopted by countries in the region such as palm oil, rubber,

[8]See Capistrano and Marten (1986).

[9]See Baldwin *et al.* (2012, 3).

Figure 3.2: Arable Land as a Percentage of Total Land Area in ASEAN-8, 2000 and 2011

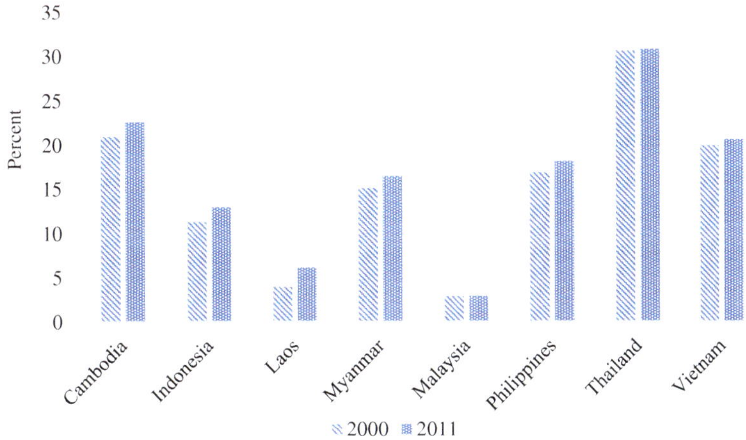

Source: World Bank.

sugar, tea, coffee, coconut. While plantations of perennial crops are typically owned and managed by state-owned or multinational firms, production of annual crops is still primarily undertaken by smallholder farmers. Contract farming between farmers and agro-processing companies has increasingly become common, whereby farmers provide land and labour while the contracting companies furnish seeds, technical expertise and market.[10]

ASEAN's shift toward commercial cash crops is evident in the changes in the proportions of major crops in regional agricultural output (see Figure 3.3). The share of rice in regional output declined from 35.7% to 31.7%, while palm oil saw a considerable increase from 10.1% to 15.3% between 2000 and 2011. The shares of rubber and maize rose from 8.2% and 4.6% to 8.5% and 5.4%, respectively, over the same period. Cassava has replaced coconuts as the sixth most important crop, contributing 5.3% to total regional output. Bananas saw a decline in its share of regional output although it remains as the fourth most important crop in the region.

The share of commercial cash crops in ASEAN's total agricultural exports also increased markedly between 2000 and 2011 (see Figure 3.4). For instance, the shares of palm oil and rubber, the region's top-2 agricultural exports, rose from 15.2% to 25.1% and 12.1% to 21.3%, respectively, from 2000 to 2011. Although the shares of coffee and coconut oil decreased over the study period, raw sugar has grown in significance and became the sixth most major crop in regional agricultural exports in 2011. Meanwhile, rice's share declined somewhat from 9.5% in 2000 to 7.6% in 2011.

While ASEAN as a whole experienced an annual growth rate of 2.7% in agricultural output between 1985 and 2010, higher than the global agricultural growth of 2.3%, the region lags behind China (4.3%) and India (3.1%) in the same period.[11] This trend is even more pronounced

[10]See, for instance, Sriboonchitta and Wiboonpoongse (2008), Manorom *et al.* (2011), Cahyadi and Waibel (2013) and Woods (2015) for case studies of contract farming in ASEAN.

[11]See Golan and Kohli (2013, 5). Also see Tan *et al.* (2016b, forthcoming) and Tan *et al.* (2015) for provincial and state-level details.

Figure 3.3: Share of Major Crops in ASEAN's Gross Agricultural Production by Value (US$), 2000 and 2011

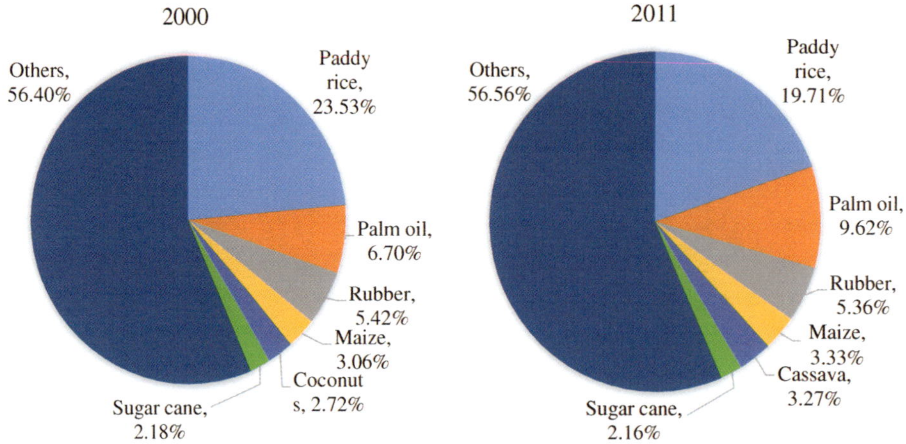

Note: Calculated based on gross production value in constant 2004–2006 million US$ obtained from FAO.
Source: FAO.

Figure 3.4: Share of Major Crops in ASEAN's Agricultural Exports, 2000 and 2011

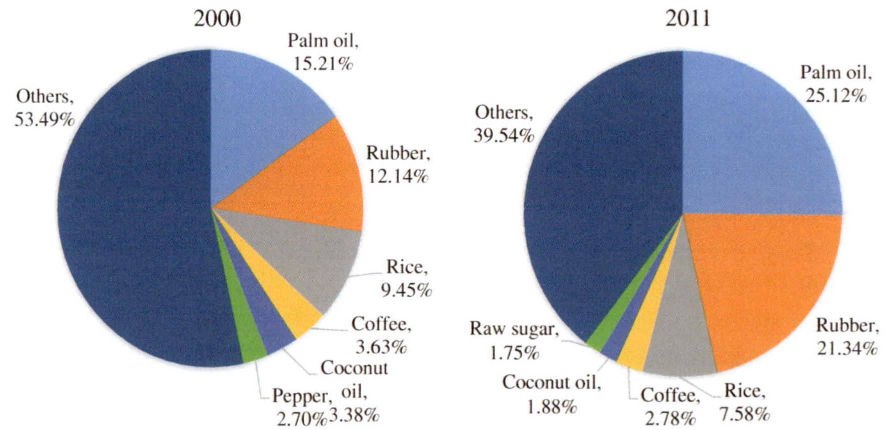

Source: FAO.

in the decade between 2000 and 2011 as China's annual growth rate of agricultural output (6.6%) exceeded that of ASEAN (4.10%) by a wide margin (see Figure 3.5). India's annual agricultural output growth rate (4.15%) was also slightly higher than ASEAN's average. Notable heterogeneity is also observed within ASEAN over the same period, with Cambodia having the highest annual growth rate of 8.11%, more than three times that of Thailand being the lowest (2.46%).

Wide variations among the ASEAN member states are also prevalent in agricultural output (see Figure 3.6). In particular, Indonesia is the biggest contributor, constituting more than 30% of total regional agricultural output over the period 2000–2011. Thailand accounted for about one-fifth of ASEAN's output in the same period. In contrast, Cambodia and Laos together only

Figure 3.5: Annual Agricultural Output Growth Rates of ASEAN-8, China and India, 2000–2011

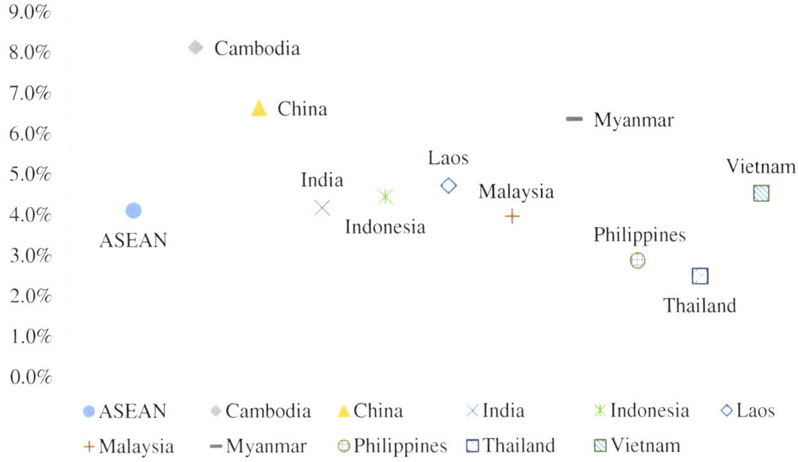

Note: The study period for China and India are 2000–2012 and 2000–2010, respectively. ASEAN average annual growth rate is measured in international US$ from FAO. Share of agricultural output to total regional agricultural output is used to calculate the ASEAN average growth rate. The growth rates for China and India are derived by ACI.
Source: FAO.

Figure 3.6: ASEAN-8 Countries' Shares of Regional Agricultural Output, 2000 and 2011

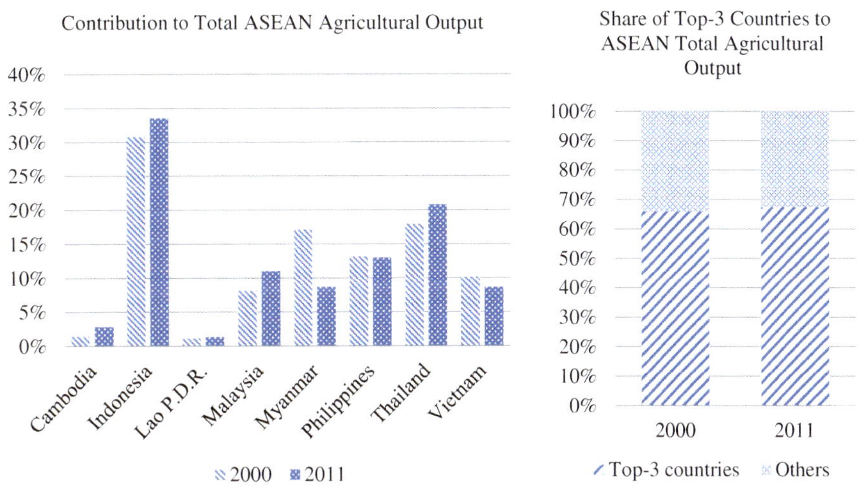

Source: FAO.

made up less than 5% of the regional agricultural output. Also notable is the sharp drop in Myanmar's share in regional output from 17.2% in 2000 to 8.7% in 2011.

Growth rates of agricultural inputs exhibit diversity among the ASEAN-8 economies (see Table 3.2). Between 2000 and 2011, Cambodia registered impressive growth in labour productivity as well as mechanisation and usage of fertiliser. Malaysia experienced the highest growth rate of labour productivity in the region over the same period with moderate increases in

Table 3.2: Annual Growth of Labour Productivity and Non-Labour Inputs in ASEAN-8, 2000–2011

Country	Average Annual Growth Rates (%)			
	Output/ Worker	Land/ Worker	Machinery/ Worker	Fertiliser/ Worker
Cambodia	7.80	2.82	11.74	19.49
Indonesia	3.97	1.74	0.52	6.19
Laos	2.28	0.45	−0.91	9.75
Malaysia	7.96	3.02	2.96	7.73
Myanmar	−1.45	1.81	11.17	−3.57
Philippines	2.54	0.33	4.64	−2.75
Thailand	5.52	2.18	3.73	4.26
Vietnam	0.49	−0.02	−1.21	0.004

Source: US Department of Agriculture (USDA) and FAO.

other inputs. Myanmar, in contrast, saw a decline in labour productivity but remarkable growth in mechanisation. On the other hand, Vietnam, being a net agricultural exporter, appears to stagnate in terms of labour productivity growth and growth of other inputs per worker.

Patterns of agricultural trade are also varied among the ASEAN-8 member states. In particular, Malaysia, Indonesia, Thailand and Vietnam are net agricultural exporters while Cambodia, Laos, Myanmar and the Philippines are net agricultural importers. Agricultural trade balances have important implications not only on national economic performance but also on food security. Net agricultural exporters are more likely to achieve food security through self-reliance or by exchanging their exports for food. Net agricultural importers are at risk when food shortages occur, as evident in the 2007–2008 global food crisis when exporters introduced export bans or cut down on their exports. While regional cooperation in food emergency relief may play a critical part in ensuring food security, improving agricultural production remains paramount, particularly for the less developed net agricultural importers, namely Cambodia, Laos and Myanmar.

3.2.1 Overview of the Agricultural Sectors in ASEAN-8

As seen in the foregoing discussion, ASEAN-8 as a bloc is notable for its pronounced variations in agricultural output growth, productivity as well as trade performance. It is thus instructive to examine the different natural endowments, key agricultural trends as well as binding constraints facing the sector in the individual member states, which is the focus of this section.

Cambodia

Cambodia, one of the world's fastest growing economies in the past two decades, is endowed with vast flat, arable land and large rivers conducive for agriculture which has played a critical role in the country's development in addition to its booming garment industry and tourism sector.

The country has the most favourable agricultural land-to-population ratio in ASEAN. Together with Myanmar, Cambodia has the highest growth rate in per capita agricultural production from the mid-1990s to the late 2000s.[12] The country also has one of the most liberal policy pertaining to foreign ownership in most sectors. As a result, there has been a surge in private local and foreign investment in agriculture, particularly in large-scale plantations including rubber, sugar cane and palm oil.

Notwithstanding these advantages, Cambodia remains a net agricultural importer. This is due to underinvestment in agricultural infrastructure, notably irrigation, rural roads, extension services and rural credit. Much of the country's agriculture is dependent on an inadequate subsistence rain-fed irrigation system, resulting in a low level of productivity in terms of both labour and land. Another major constraint facing Cambodia's agricultural sector is its small workforce relative to the size of its agricultural land, which is further aggravated by the rapid development of the labour-intensive garment industry.

Indonesia[13]

Indonesia's agricultural sector can be divided into two groups, food agriculture, which concentrates in Java and Bali, and commodity agriculture, which is dominated by Sumatera and Kalimantan. The palm oil boom in the past decade, primarily driven by demands from China and India, has propelled Indonesia into the rank of the largest palm oil exporters in the world. The country's oil palm areas rose six-fold from around 1 million hectares in 1994 to 6 million hectares in 2007. The share of palm oil in Indonesia's agricultural exports increased from 14% in 1995 to 50% in 2008.[14] The country's spectacular growth of the palm oil industry has enabled traditionally lagging non-Java provinces to catch up with the populous and more developed provinces in Java and Bali. Indonesia is also one of the world's largest producers of rubber and spices.

Nonetheless, while Indonesia is a net agricultural exporter, it is also a net food importer and thus vulnerable to food insecurity. Java, the centre of the country's rice sector, is facing binding land constraints, which highlights the importance of raising agricultural productivity. It is also worth noting that Indonesia's agricultural research intensity, measured by the share of public investment in agricultural R&D in total agricultural value added, has declined by about two-thirds since the late 1970s.[15]

Laos

Laos has the second lowest ratio of arable land to total land in ASEAN, only after Malaysia. Correspondingly, it has the lowest arable land to rural population ratio in the region which poses

---

[12] See Intal *et al.* (2011, 13–14).

[13] See Tan *et al.* (2016c, forthcoming) for details at the provincial level.

[14] See Intal *et al.* (2011, 32).

[15] See Warr (2011, 4).

significant problems to the country's labour-intensive agricultural sector. Furthermore, Laos has diverse topography comprising: the relatively flat lands of the Mekong Corridor conducive for rice, the rich soils of the Bolovens plateau suited to horticulture and coffee, the mountains of Vientiane Plain and Northern Lowlands which are experiencing severe erosion, and the poor soils of the remote Northern and Central-Southern highlands. Therefore, agricultural development must be customised to specific areas and their respective topography. The country's land-locked nature and inadequate infrastructure further constrains the country's agricultural sector, particularly when agricultural exports are heavily dependent on sea transport.[16]

Laos's food agriculture remains largely on a subsistence basis and reliant on traditional production practices. The country's agricultural yield is hence significantly lower than other countries in ASEAN. However, Laos's recent efforts to shift toward agricultural commercialisation have started to bear fruits. For instance, commercial maize cultivation has been thriving in the poorly endowed Northern lowlands. Cross-border contract farming in cabbage with Thailand and sugar cane with China has proved to be beneficial to Laotian farmers with higher incomes and access to improved inputs, technical support and credit.[17] The rubber sector is also growing as a result of foreign investment from China and Vietnam.

Malaysia

As a relatively more industrialised economy compared to the remaining ASEAN-8 countries focused in this chapter, agriculture has long existed in duality with Malaysia's manufacturing sector. The country's primary crops are geared toward industrial use, including oil palm, rubber, timber and cocoa, accounting for about 60% of agricultural GDP.[18] Despite having the lowest ratio of arable land to total land in ASEAN, Malaysia has been the leading agricultural exporter in the region. The country's agricultural development efforts have been geared towards modernisation and commercialisation of agriculture through promoting private sector participation.

A competitive agribusiness industry, particularly in palm oil, underpins the country's success in agricultural development. In fact, Malaysia is one of the few middle-income developing countries with a relatively developed agro-based sector. Malaysia's agricultural strategies and policies as well as constraints and challenges will be elaborated upon in greater detail in Section 4.6.

Myanmar

Myanmar has a distinct comparative advantage in agriculture, spanning a wide range of elevation, latitude, temperature and rainfall, as well as endowed with favourable soils. It also has abundant water resources with four major river basins and largely untouched groundwater. The country's agricultural landscape can be divided into three broad zones, namely the Delta, the Dry Zone and the hilly areas. The densely populated Delta zone is the national centre of rice

---

[16]See Intal *et al.* (2011, 21).

[17]See Manorom *et al.* (2011).

[18]See Ahmad and Suntharalingam (2009, 2).

production. The Dry zone is home to rainfed crops including rice where water permits. The hilly areas are dominated by tree crops, horticulture along with rice, maize and pulses.

Despite the aforementioned considerable potential and large public investments in irrigation, reservoirs and road transport, the country's agricultural sector remains underperforming. While Myanmar has a wide variety of exportable agricultural crops, its actual exports are confined to primarily pulses, rice and maize. With the exception of pulses, export volumes of the other products are often minimal and volatile. It is found that most agricultural output gains have been from land expansion rather than increased yields.[19] Myanmar's crop yields are generally lower than those in ASEAN. Low levels of input use, particularly improved seeds and fertiliser, weak agricultural research and extension services and limited access to rural credit also characterise Myanmar's agricultural sector.

Philippines

Philippines's agricultural comparative advantage lies in tropical fruits, coconut oil and fishery, which constitute the bulk of the country's agricultural exports. Nonetheless, while the country experienced robust growth in the agricultural sector until the early 1980s, output growth rates of virtually all crops decelerated throughout the 1980s and early 1990s owing mainly to a reduction in cultivated area and natural calamities.[20] Consequently, Philippines has become one of the world's top rice importers, and its agricultural imports have increasingly outstripped exports in recent decades. Furthermore, compared to other ASEAN member states, the Philippines has a relatively low land-to-population ratio but high population growth, which pose significant challenges to national food security.

The Philippines government has since stepped up its efforts in improving agricultural growth. For instance, public investment in irrigation and agricultural R&D, particularly in rice, increased in the 1990s. However, the reversion of price policies during the same period towards import substitution of less competitive subsectors such as rice, corn and poultry rather than export-oriented crops such as coconut and banana has hampered the development of the agricultural sector.[21]

Thailand

Endowed with the highest ratio of arable land to total land area in ASEAN, Thailand has one of the most developed agricultural sectors in the region. The country has successfully maintained its position as one of the world's top exporters in rice, rubber, cassava, canned tuna, shrimp, chicken and sugar, among others. Agribusiness firms have been the driver of agricultural diversification and R&D, resulting in the country having the highest ratio of agricultural R&D to agricultural GDP in ASEAN. They have also developed strong linkages with smallholder

---

[19] See Haggblade *et al.* (2013).

[20] See Mundlak *et al.* (2002, 33).

[21] See Tiongco (2011, 367).

farmers through the adoption of contract farming. A number of export-oriented agribusiness companies also invest in meeting the sanitary and phytosanitary standards of the European Union, the United States and Japan, paving the way for Thai agricultural products to enter these lucrative yet demanding markets.[22]

Steadfast government support also plays a vital role in the stellar growth of Thailand's agriculture. Extensive infrastructural investments have been made in irrigation, roads, electricity and telecommunications. In recent years, the Thai government has focused on developing postharvest facilities including refrigeration, food-testing laboratories and equipment. Other supportive policies include fiscal incentives and tariff elimination on important inputs.[23]

Vietnam

Having the third largest ratio of arable land to total land area in ASEAN, Vietnam is also bestowed with diverse topography with high latitude range which is conducive for a variety of crops. In particular, the mountainous and hilly soils are suitable for afforestation and cultivation of perennial crops and fruit crops. The delta soils form the country's centre of food production. These regions are typically amply moisturised and less prone to soil degradation. Institutional reforms as part of *Doi Moi* (renovation) programme in 1986 in rural areas unleashed the country's potential in agricultural production, enabling its remarkable transformation from a recipient of emergency food assistance to a net agricultural exporter since the early 1990s. This was followed by major public investments in irrigation and fertiliser which contributed to higher crop intensity and yield. As a result, Vietnam's agricultural exports increased substantially at an average annual growth rate of 12.6% between 1995–2010,[24] ranging from rice and coffee to fishery, pepper, cashew nuts, rubber, fruits and wood products.

Notwithstanding these impressive achievements, Vietnam's agricultural growth, which has largely been based on expansion of irrigation and physical inputs, has seen signs of slowdown, falling from over 4% annually in the late 1990s to below 4% in 2006–2007.[25] Land fragmentation, outdated technology, weak supply chains and unregulated quality have eroded the sector's competitiveness. While liberalisation of the sector has taken place, state-owned enterprises remain the dominant actor in both agricultural input and output markets, imposing barriers to growth as well as breeding mismanagement and corruption.

3.2.2 Opportunities and Challenges from a Regional Perspective

While each of the eight ASEAN member states is facing unique binding constraints that could potentially impede agricultural growth as noted, the agricultural sector in ASEAN as a whole also shares a common set of opportunities and challenges. On the demand side, rising per capita incomes and increasing access to international markets through intra- and inter-regional trade

[22] See Poapongsakorn (2011).

[23] See Intal *et al.* (2011, 29).

[24] See The Ministry of Agriculture and Rural Development of Vietnam and FAO (2013, 10).

[25] See Coxhead *et al.* (2010, 39).

mean growing demand for the region's agricultural products. On the supply side, however, climate change and land degradation pose serious threats to agricultural production growth in ASEAN. While technological advances may help to overcome supply-side issues, maintaining yield growth has been a challenge for ASEAN.

Economic growth, technological innovations and increased global and regional trade present important opportunities for the development of the agricultural sector in ASEAN. First, urbanisation and a growing middle class resultant from the region's remarkable economic development has led to a shift in food preferences toward high-value agricultural commodities such as fruits and vegetables. Demand for industrial crops has also increased as industrialisation gathers pace in the region. Concomitantly, the proliferation of supermarkets has provided another important outlet of agricultural produce, particularly high-value commercial crops, in addition to traditional wet markets. It has been estimated between 2001 and 2009, the annual compound sales growth rate averaged at 17.4% for Malaysia, Indonesia, Philippines and Thailand, while that of Vietnam reached a remarkable 45.4%.[26] Furthermore, innovations in institutional arrangements such as the establishment of farmer–producer organisations and contract farming have helped to integrate farmers into modern supply chains and benefit from them.

Second, ASEAN countries have made important breakthroughs in agricultural research and development (R&D) which contribute to higher yields and more sustainable agricultural practices. Notable examples include palm oil in Malaysia and Indonesia, organic agriculture in Indonesia, tilapia culture in the Philippines, baby corn and orchids in Thailand, among others. Agricultural R&D expenditure in less developed countries such as Vietnam and Cambodia quadrupled between 1996 and 2008.[27] While R&D spending in low-income and lower-middle-income countries in the region remains low compared to their GDP, it has produced remarkable impacts on these countries' agricultural production, as evident in the case of Vietnam being the second largest rice exporter in the world. Additionally, the spread of contract farming across the region has also promoted technological transfer between agro-processing firms and farmers, particularly in the introduction of new varieties, eco-friendly agricultural practices as well as yield-enhancing production technologies. Advances in agricultural technology have opened the door for smallholder farmers in ASEAN to high-value commodity markets, thereby raising their incomes and ensuring a more sustainable livelihood.

Last but not least, trade liberalisation has increased the region's access to both high-income and other developing countries. Lower import barriers in ASEAN's trading partners have facilitated the growth of rice as well as high-value agricultural exports such as fruits, vegetables and seafood. Impressive economic growth within Asia has led to the expansion of intra-regional agricultural trade. Intra-regional agricultural trade in Asia-Pacific has in fact grown faster than trade with the rest of the world. In 2008, for instance, more than half of agricultural exports from Asia were within the region.[28] China's transformation into a net importer of food commodities, a trend that is likely to continue in the coming decades, also has important implications for the

---

[26] See Reardon *et al.* (2012, 12333).

[27] See Flaherty *et al.* (2013, 4).

[28] See Ong (2014, 40).

agricultural sector in ASEAN. Within ASEAN, the enactment of the ASEAN Free Trade Area (AFTA) in 1993 and subsequently the ASEAN Trade in Goods Agreement (ATIGA) in 2010 has facilitated the reduction in tariff and non-tariff trade barriers and resultantly an improved flow of goods across the member states. Regional trade initiatives, notably the ASEAN+*x* free trade agreements with Australia and New Zealand, China, India, Japan and Korea, have further unlocked the export potential for the region's agricultural sector.

Nonetheless, the agricultural sector in ASEAN is facing immense challenges in terms of climate change, land degradation and maintaining yield growth. Southeast Asia is one of the world's most vulnerable regions to climate change due to a combination of long coastlines, high concentration of population and heavy reliance on agriculture and natural resources. Decreasing precipitation, increasing frequency and intensity of heat waves, droughts, floods and tropical cyclones pose tremendous threats to the region's agricultural production and food security. Over the past fifty years, the region's average temperature has increased at the rate of 0.1–0.3°C per decade and the sea level has risen by 1–3 millimetre annually.[29] It is found that in tropical countries, even moderate warming can significantly reduce yields.[30] Climate change has also exacerbated water shortages, further constraining the irrigation systems which are already inadequate in the less developed ASEAN member states. Small-scale farming, which characterises the majority of the regional agricultural production, is particularly vulnerable as farmers are likely to suffer from local climate change impacts which may be hard to predict. Small farm sizes, low capitalisation and crop diversification restrict their capability to adapt and hence further amplify their vulnerability.

Another formidable challenge for ASEAN's agriculture is land degradation which has reduced the potential for yield increases in approximately 50% of the region's cultivated areas.[31] Many of Southeast Asia's soils tend to have low moisture-holding capacity and are thus prone to drought and desertification. Better soils are abundant in lowland areas such as the Mekong but population pressure as well as high concentration of economic activity have accelerated soil degradation. Soil erosion will also increase the likelihood of landslides in hilly areas. Much of land degradation stems from a combination of intensive cultivation and excessive use of chemical fertilisers. In many poor rural areas, farmers have been forced into cultivating in ecologically fragile areas such as steep hillside slopes, thereby exacerbating land erosion. Oil palm plantations have caused degradation of rainforests in Indonesia, particularly in Sumatera and Kalimantan. While there has been a shift toward more sustainable cultivation practices such as increasing use of organic fertilisers, changing crop schedules, changing timing of crop activities, soil degradation remains a critical threat to agricultural productivity and rural livelihoods in ASEAN.

Low crop yields and yield growth, particularly for major food crops, are also important constraints to the development of the agricultural sector in ASEAN. Table 3.3 shows the average annual yield (in tonnes/hectare) and yield growth over the period 2000–2011 for ASEAN's

[29] See Ong (2014, 44).

[30] See Thapa and Gaiha (2011).

[31] See Squires (2009, 5).

Table 3.3: Average Annual Yield and Yield Growth of Selected Crops in ASEAN, China and India, 2000–2011

	Average Yield (Tonnes/Hectare)			**Average Yield Growth (%)**		
	China	**India**	**ASEAN**	**China**	**India**	**ASEAN**
Rice	6.35	3.14	3.89	0.47	1.91	1.75
Rubber	1.18	1.74	1.07	0.87	1.26	2.49
Maize	5.19	2.09	3.22	1.41	3.59	4.15
Coconuts	10.36	5.00	5.31	4.47	0.38	0.77
Cassava	16.14	30.67	16.77	0.24	3.67	3.47
Bananas	23.80	32.65	19.27	2.11	0.59	2.93

Source: FAO.

major crops, namely rice, rubber, maize, coconuts, cassava and bananas, in ASEAN, China and India. In general, annual yields of most crops in ASEAN remain lower than those of China, especially for rice, maize, coconuts and bananas, with the exception of cassava where ASEAN posts higher yield. Compared to India, ASEAN's yields are lower for rubber, cassava and bananas while only modestly higher for rice, maize and coconuts. In terms of annual yield growth, ASEAN exhibits higher growth rates for rubber, maize and bananas. Nonetheless, for rice, ASEAN's yield growth lags behind that of India; whereas for coconuts and cassava, ASEAN's yield growth is lower than that of China and India, respectively.

3.3 LITERATURE REVIEW

While there have been several studies on agricultural TFP for Asia as a region and selected Asian countries, there are far fewer studies on agricultural TFP for ASEAN specifically. The literature on agricultural TFP growth in ASEAN can be classified into two broad categories: global or regional studies, which usually provide TFP growth rates at both regional and national level, and national case studies, which offer estimates at the country-specific level (see Table 3.4). The majority of studies use data from FAO Agrostat database with input series typically including land, labour, livestock, fertiliser and farm machinery.

In particular, Suhariyanto and Thirtle (2001) estimated agricultural TFP growth for eighteen Asian economies from 1965 to 1996 by adopting the Malmquist index with respect to the sequential frontier. With regard to ASEAN which comprised Cambodia, Indonesia, Laos, Malaysia, Myanmar, the Philippines, Thailand and Vietnam in the study, the authors found that regional TFP growth rate over the entire study period was 0.32% annually which was less than half the growth rate of 0.73% posted by East Asia (China, Japan, South Korea, North Korea and Mongolia). National TFP growth exhibited wide variations with Cambodia experiencing a negative growth rate of −1.86% and Malaysia having the highest growth rate of 3.55% between 1965–1996. Thailand, Vietnam and Myanmar saw a decline in agricultural TFP growth, while Laos, the Philippines and Indonesia registered healthy growth rates of 1.75%, 1.33% and 0.17%,

Table 3.4: Existing Literature on Agricultural Total Factor Productivity (TFP) in ASEAN

Author/Year	Study Period	Research Scope	Methodology	Annual TFP Growth (%)
Avila and Evenson (2010)	1961–2001	78 developing economies	Accounting Relationship Derivation	ASEAN: 1.49; Cambodia: −1.89; Indonesia: 2.02; Laos: 1.86; Malaysia: 2.51; Philippines: 0.79; Thailand: 1.13; Vietnam: 1.45
Coelli and Rao (2003)	1980–2000	93 developed and developing economies	Malmquist Productivity Indices	Cambodia: 5.7; Indonesia: −1.9; Laos: 3.4; Malaysia: 0.4; Myanmar: 1.8; Philippines: 0.8; Thailand: –0.5; Vietnam: 2.4
Fuglie (2012)	1961–1970; 1971–1980; 1981–1990; 1991–2000; 2001–2009	87 developed and developing countries	Growth Accounting	ASEAN: 0.57 (1961–1970), 2.10 (1971–1980), 0.54 (1981–1990), 1.69 (1991–2000), 3.29 (2001–2009)
Mundlak, Larson and Butzer (2002)	1961–1998	Indonesia, Philippines and Thailand	Growth Accounting	Philippines: 0.25 (1961–1998); Indonesia:1.49 (1971–1998); Thailand:1.08 (1971–1995)
Rungsuriyawiboon and Wang (2009)	1980–2004	5 Asian regions and 10 Asian transitional economies	Malmquist Productivity Indices	ASEAN: 1.132 (1980–2004); Laos: –0.091 (1980–2004); Myanmar: 2.862 (1989–2004); Vietnam: 0.672 (1986–2004)
Son Nghiem and Coelli (2010)	1976–1997	Vietnam	Malmquist Productivity Indices	Vietnam: 3.3–3.5
Suhariyanto and Thirtle (2001)	1965–1996	18 Asian economies	Malmquist Productivity Indices	ASEAN: 0.32; Cambodia: −1.86; Indonesia: 0.17; Laos: 1.75; Malaysia: 3.55; Myanmar: –0.02; Philippines: 1.33: Thailand: −1.00; Vietnam: –0.18
Umetsu, Lekprichakul and Chakravorty (2003)	1971–1990	Philippines	Malmquist Productivity Indices	Philippines: 0.7

respectively, over the same period. By dividing the study period into four sub-periods, the authors also showed that agricultural TFP growth in ASEAN reached its peak at 0.80% over the period 1971–1980 before plunging to its lowest level of –0.18% between 1981–1990 and recovering at 0.69% between 1991–1996. Within ASEAN, only Malaysia, Laos and the Philippines showed positive TFP growth over all four sub-periods.

Coelli and Rao (2003) arguably offer the most ambitious study on the topic, measuring TFP growth for 93 developed and developing countries for the period 1980–2000 using the DEA-based Malmquist productivity indices. According to their estimation, Asia was the best performer with an annual TFP growth of 2.9%. In terms of individual country performance, China led the world with an average annual growth of 6.0% in TFP, followed by Cambodia with 5.7%. Within ASEAN, Laos and Vietnam rounded out the top-3 countries with highest TFP growth rates at 3.4 and 2.4%, respectively. Thailand and Indonesia, nevertheless, posted negative TFP growth while the rest showed positive growth in TFP over the study period.

Employing a generalised Malmquist productivity index and a parametric output distance function approach to decompose TFP growth into its associated components, Rungsuriyawiboon and Wang (2009) measured TFP growth in six Asian regions across 27 countries, namely Central Asia, Eastern Asia, Southern Asia, Southeast Asia, Western Asia and China, between 1980 and 2004. Southeast Asia included the same ASEAN-8 countries as in Suhariyanto and Thirtle (2001). Southeast Asia's agricultural TFP growth rate over the study period is found to be 1.132% annually while that of East Asia was –2.031% and that of East Asia including China was 1.69%. In fact, the authors concluded that China was the driving force behind the healthy Asian agricultural TFP growth of 1.857% over the full study period. Within ASEAN, Myanmar registered the highest TFP growth rate of 2.862% over the full study period. Vietnam experienced a positive growth rate of 0.672% while Laos posted a negative –0.091%. The study, however, did not provide estimates for the remaining ASEAN member states.

Avila and Evenson (2010) applied the accounting relationship derivation to calculate agricultural TFP growth rates for 78 developing economies in Latin America and the Caribbean, Asia and Africa for two periods, namely 1961–1980 and 1981–2011. They found that Southeast Asia, which consists of Cambodia, Indonesia, Laos, Malaysia, the Philippines, Thailand and Vietnam, experienced a decline in its TFP growth rate from 2.37% between 1961–1980 to 0.61% between 1981–2001. Malaysia registered the highest average TFP growth at 2.51% over both periods, followed by Indonesia and Laos at 2.02 and 1.86%, respectively. Cambodia was the only country in ASEAN that posted a negative average TFP growth rate of –1.89%. Nonetheless, Cambodia, Laos and Vietnam were the only three countries in ASEAN that saw an increase in their TFP growth rates between the two periods. The gap between ASEAN and East Asia was again highlighted in this study as East Asia exhibited the highest average TFP growth rate of 3.22% among all regions covered, driven primarily by the stellar performance of China.

In another extensive global study, Fuglie (2012) employed the growth accounting methodology to estimate TFP growth rates for 87 developed and developing countries in five periods: 1961–1970, 1971–1980, 1981–1990, 1991–2000 and 2001–2009. Southeast Asia, which includes ASEAN-8 and Timor Leste, achieved respectable TFP growth from 0.57% between

1961–1970 to 3.29% between 2001–2009, notwithstanding the dip to 0.54% during 1981–1990. Myanmar and Cambodia, in particular, registered the most impressive jumps in TFP growth among the eight ASEAN economies from −1.68% and −0.93% between 1961–1970 to 5.97% and 5.85%, respectively, between 2001–2009. In contrast, signs of slowdown can be seen in Laos, Thailand and Vietnam as these countries experienced lower TFP growth between 2001–2009 compared to 1991–2000.

Within the strand of country-specific literature on the topic, Mundlak, Larson and Butzer (2002) provide a relatively comprehensive analysis of TFP growth in three ASEAN countries, namely the Philippines over the period 1961–1998, Indonesia between 1971–1998 and Thailand between 1971–1995, using the growth accounting methodology. Indonesia posted the highest TFP growth rate of 1.49% over the entire study period as compared to 1.08% for Thailand and 0.25% for the Philippines over their respective study period. Nonetheless, all three countries experienced a decline in TFP growth in the second sub-period as compared to the first one. In particular, the Philippines's TFP growth rate dropped from 0.98% between 1961–1980 to 0.13% between 1980–1998, while that of Thailand fell from 1.27% over the period 1971–1981 to 0.87% in the subsequent years between 1981–1995. Indonesia's decline was of a much lesser extent, from 1.58% in the period 1971–1981 to 1.49% between 1981–1998.

Applying the Malmquist productivity indices, Umetsu, Lekprichakul and Chakravorty (2003) derived TFP growth rates of the rice sector at the national level as well as for 12 regions in the Philippines for the period 1971–1990. They found that the country's rice sector achieved an average TFP growth rate of 0.7% over the full study period. Breaking down the estimation into four sub-periods, they showed that productivity growth was positive and increased from 2.4% to 3.6% between the late 1970s and the early 1980s, but turned negative from the late 1980s. At the sub-national level, only five out of 12 regions registered positive annual TFP growth over the study period.

Son Nghiem and Coelli (2010) provide an estimation of TFP growth for the rice sector in Vietnam over the period 1976–1997. Using two extended DEA-based Malmquist methods, they showed that the average TFP growth rate was between 3.3% and 3.5% per annum over the study period. Nonetheless, variations were observed in the sub-periods before, during and after the incentive reform in 1981. In particular, in the period after the introduction of the reform from 1981 to 1987, the sector's TFP growth surged to 6.5% from 2.7% during the pre-reform period. In the third sub-period between 1987–1997, however, TFP growth declined to 2.4%.

3.4 METHODOLOGY AND DATA[32]

TFP growth can be measured using two main approaches, namely parametric and non-parametric techniques. Each approach has its pros and cons. While more widely used, the accuracy of the parametric approach depends heavily on the specification of the production function. Parametric

[32]This section draws on Tan *et al.* (2016a, forthcoming).

approaches also make strict assumptions on the functional form, which are often difficult to justify in practice. In contrast, the Malmquist index methodology is sensitive to outliers and data quality, but is not dependent on function specification. We thus employ growth accounting methodology as the parametric approach and Malmquist index as the non-parametric approach to gain a comprehensive understanding of the agricultural TFP growth in ASEAN. Given the lack of observations to conduct panel analysis at the country level, growth accounting methodology is only adopted at the regional level and sub-regional level, the details of which will be elaborated in Section 3.4.1 below. The Malmquist index method allows us to decompose TFP growth into technological change and technical efficiency change, which enhances our understanding of the drivers of agricultural TFP change at the country level.

3.4.1 Growth Accounting Methodology

Under the growth accounting methodology, we divide ASEAN-8 into two sub-regional groups according to development status: one consists of Indonesia, Malaysia, Philippines and Thailand, and the other comprises CLMV. The production function for ASEAN's agricultural sector is defined in the Cobb–Douglas form as follows:

$$Y_{it} = A_{it} X_{1it}^{\beta_1} X_{2it}^{\beta_2} X_{3it}^{\beta_3} \tag{1}$$

where i and t represent economy and time respectively. Y_{it} represents the agricultural output per labour for economy i at time t. X_{1it}, X_{2it} and X_{3it} represent machinery per labour, fertiliser per labour and land per labour for economy i at time t, respectively. A_{it} is usually defined as the measure for TFP, which includes technology factors and efficiency factors.

Taking natural logarithms on both sides of Equation (1), one can obtain Equation (2) which determines the contribution of inputs to output and is used to calculate Total Factor Productivity Change (TFPC), or TFP growth:

$$\ln Y_{it} = \ln A_{it} + \beta_1 \ln X_{1it} + \beta_2 \ln X_{2it} + \beta_3 \ln X_{3it} + \varepsilon_{it} \tag{2}$$

TFP growth is the difference between real agricultural output growth and the weighted sum of various inputs, or the residual change in output. Therefore, TFP growth, or TFPC, can be derived by Equation (3) as follows:

$$TFPC_{it} = \Delta y_{it} - \Delta\hat{\beta}_1 X_{1it} - \Delta\hat{\beta}_2 X_{2it} - \Delta\hat{\beta}_3 X_{3it} \tag{3}$$

3.4.2 Malmquist Index Data Envelopment Analysis (DEA) Methodology

In view of the various limitations of parametric approaches in estimating TFP and given the limited time series data, we also apply the non-parametric approach to measure agricultural TFP growth for the individual ASEAN-8 economies using the Malmquist index, which is defined in a distance function by Färe *et al.* (1994).[33] Also according to Färe *et al.* (1994), TFP growth can be decomposed into Technological Change (TC) and Technical Efficiency Change (TEC). Every time point has its specific technology level. Therefore, if a selected time point is represented by t, its corresponding technology level can be labelled as S^t. In this context, x^t can be used to represent all inputs of the production activity, and y^t can be used to represent all outputs of the production activity. Under these circumstances, S^t can be defined in the form of equation (4) which captures all possible combinations of inputs x^t and outputs y^t under the available technology at time t.

$$S^t = \{(x^t, y^t), x^t \text{ produce } y^t\} \tag{4}$$

Since the agricultural production activity in the real world can hardly fully achieve and implement all cutting-edge technologies in a specific time point t, a differentiation between the actual output and the ideal output could be spotted. According to Färe *et al.* (1994), the differentiation can be defined as Equation (5).

$$D_y^t = \inf\left\{\vartheta : \left(x^t, y^t/\vartheta\right) \in S^t\right\} = \left\{\sup\left[\vartheta : \left(x^t, \vartheta y^t\right) \in S^t\right]\right\}^{-1} \tag{5}$$

Equation (5) is an output-oriented distance function, in which D_y^t represents the reciprocal of the maximum change in output y through fully exploiting the input x. Given the input level x^t at time t, Equation (5) can be used to measure the maximum expansion ϑ of the output y^t. Meanwhile, D_y^t can also be interpreted as the reciprocal of the maximum efficiency change of the output y^t by fully exploiting input x^t.

Similarly, as shown in Equation (6), the input-oriented distance function can measure the change of inputs given the output level.

$$D_x^t = \sup\left\{\lambda : \left(x^t/\lambda, y^t\right) \in S^t\right\} \tag{6}$$

$$D_y^t = \inf\left\{\vartheta : \left(x^{t+1}, y^{t+1}/\vartheta\right) \in S^t\right\} = \left\{\sup\left[\vartheta : \left(x^{t+1}, \vartheta y^{t+1}\right) \in S^t\right]\right\}^{-1} \tag{7}$$

Equation (7) represents the reciprocal of maximum change of output, given input x^{t+1}, to make y^{t+1} feasible at time t. Hence, Malmquist index can be written in the form of Equation (8) and Equation (9).

$$M_y^t = \frac{D_y^t\left(x^{t+1}, y^{t+1}\right)}{D_y^t\left(x^t, y^t\right)} \tag{8}$$

---

[33]The other commonly used non-parametric approach is the Törnqvist Index. However, this approach has rather strict data requirements, including prices and quantitative data of both inputs and output. In addition, it assumes, among others, a perfectly competitive market and profit maximisation, which are not suitable for agricultural studies.

$$M_y^{t+1} = \frac{D_y^{t+1}\left(x^{t+1}, y^{t+1}\right)}{D_y^{t+1}\left(x^t, y^t\right)} \tag{9}$$

If a specific time t is selected, the corresponding technology level will become the reference technology level. Under these circumstances, Equation (8) can be used to measure the necessary change to generate the output y^{t+1} through utilising the input x^{t+1} with the feasible technology S^t. In order to avoid choosing an arbitrage benchmark, the geometric mean of two Malmquist indices can be calculated to form a benchmark.

$$\begin{aligned} M_y\left(x^t, x^{t+1}, y^t, y^{t+1}\right) &= \sqrt{M_y^t * M_y^{t+1}} \\ &= \left[\left(\frac{D_y^t\left(x^{t+1}, y^{t+1}\right)}{D_y^t\left(x^t, y^t\right)}\right)\left(\frac{D_y^{t+1}\left(x^{t+1}, y^{t+1}\right)}{D_y^{t+1}\left(x^t, y^t\right)}\right)\right]^{1/2} \end{aligned} \tag{10}$$

By mathematical transformation, Equation (10) can be written in the form of Equation (11), which allows us to decompose the TFP change into technological change (TC) and technical efficiency change (TEC).

$$M_y\left(x^t, x^{t+1}, y^t, y^{t+1}\right) = \frac{D_y^{t+1}\left(x^{t+1}, y^{t+1}\right)}{D_y^t\left(x^t, y^t\right)}\left[\left(\frac{D_y^t\left(x^{t+1}, y^{t+1}\right)}{D_y^{t+1}\left(x^{t+1}, y^{t+1}\right)}\right)\left(\frac{D_y^t\left(x^t, y^t\right)}{D_y^{t+1}\left(x^t, y^t\right)}\right)\right]^{1/2} \tag{11}$$

As specified by Färe *et al.* (1994), Equation (11) can be written in a more illustrative form as Equation (12). The first component represents the TEC between time t and time t+1, and the second component represents the TC between time t and time t+1. TEC refers to activities which enhance the efficiency of production, including better resource utilisation and management. TC refers to a set of developments that can move the productivity frontier forward. We employ the Linear Programming approach to calculate the Malmquist index.

$$TFPC = TEC \times TC \tag{12}$$

In short, Malmquist index methodology not only allows us to decompose TFPC into TEC and TC, but also provides relative rankings of economies based on their distance to the efficiency frontier.

3.4.3 Data

Gross agricultural output for individual ASEAN-8 economies is obtained from FAO and defined as the sum of the production values of 189 crops, fruits, vegetables and livestock commodities, valued at constant, global average prices between 2004–2006 in Standard Local

Currency. Four inputs, namely labour, land, machinery and fertiliser, are retrieved from the USDA.[34] In particular, the number of economically active adults in agriculture from USDA is used as labour input, recorded in terms of number of people. This data is widely used in the strand of literature on agriculture in ASEAN. Area harvested (hectares) is used as land input, which refers to the area from which a crop is gathered. Area sown or planted but without harvest is excluded. Fertiliser input is defined as metric tonnes of fertiliser consumption, weighted by their relative prices. Machinery input is measured by the total metric horse-power (CV) of major farm equipment in use, which is the aggregation of the number of 4-wheel riding tractors, 2-wheel pedestrian tractors, power harvester-threshers, and milking machines, expressed in "40-CV tractor-equivalents."

Our study period is from 2000 to 2011. Since there are concerns about Myanmar's exchange rate data, we employ the implied exchange rates which are derived from GDP in current prices in national currency and US$ for all ASEAN-8 economies. A comparison made between the implied exchange rates and the World Bank official exchange rates indicates there are no significant differences between these two exchange rate data, except for Myanmar.

3.5 EMPIRICAL FINDINGS

Table 3.5 and 3.6 present annual TFP growth in ASEAN-8 estimated using the growth accounting methodology and the Malmquist index, respectively. Overall, ASEAN-8 experienced annual TFP growth between 1.4%, according to the Malmquist index, and 1.6%, according to the Cobb-Douglas production function. The Malmquist index results further indicate that TEC and TC contributed equally to regional average TFP growth between 2000 and 2011 (see Table 3.7). Nonetheless, marked heterogeneity can be observed with some countries experiencing a decline in TFP, as indicated by a score below 1.0, and others registering a positive TFP growth, as shown in their scores which were above 1.0.

In fact, the results of both Cobb–Douglas production function and Malmquist index point to the stark variations in agricultural TFP growth within the Southeast Asian bloc. In particular, the more mature group of economies, which comprises Indonesia, Malaysia, Philippines and Thailand, posted a TFP growth between 2.7% and 3.17% between 2000 and 2011, while the less developed CLMV remained stagnant with a TFP growth of between 0.37% and 0.4% over the same period (see Table 3.5 and 3.6).

The decomposition of TFP growth according to the Malmquist index method also signifies notable diversity in the region (see Table 3.7). For instance, TFP growth in Malaysia was primarily driven by TC, which is in line with the Malaysian government's emphasis on agricultural R&D as a way to circumvent limited land and shrinking agricultural labour. In contrast, TFP growth in Indonesia appears to have been broadly determined by TEC, implying that investment in agricultural R&D and incentives for capitalising on agricultural innovation are imperative for Indonesia to enhance agricultural productivity and maintain food security. Among the top-4 countries with

[34]For detailed definitions of the data, please refer to the USDA International Agricultural Productivity Dataset: http://www.ers.usda.gov/data-products/international-agricultural-productivity.aspx.

Table 3.5: Total Factor Productivity (TFP) Growth in ASEAN-8 Using Growth Accounting Methodology, 2000–2011

	β_1	β_2	β_3	TFPC
ASEAN average	−0.021 (−0.12)	0.077 (0.60)	0.735* (1.97)	1.60%
CLMV	−0.043 (−0.17)	0.085 (0.53)	0.698 (1.46)	0.40%
Indonesia, Malaysia, Philippines and Thailand	0.068 (0.52)	−0.003 (2.53)	0.913 (1.09)	2.70%

Note: t-statistics are reported in parentheses. *, **, and *** indicate statistical significance at the 10%, 5% and 1% level, respectively.
Source: Asia Competitiveness Institute.

Table 3.6: Total Factor Productivity (TFP) Growth in ASEAN-8 Using Malmquist Index DEA Methodology, 2000–2011

	TFPC
ASEAN average	1.40%
Cambodia, Laos, Myanmar and Vietnam	−0.37%
Indonesia, Malaysia, Philippines and Thailand	3.17%

Source: Asia Competitiveness Institute.

Table 3.7: Average Total Factor Productivity Change (TFPC) and Its Decomposition in ASEAN-8, 2000–2011

Rank	Country	TEC	TC	TFPC
1	Malaysia	1	1.041	1.041
2	Philippines	1.012	1.019	1.03
3	Indonesia	1.022	1.007	1.028
4	Thailand	1.016	1.011	1.028
5	Cambodia	1.03	0.986	1.015
6	Vietnam	0.977	1.029	1.005
7	Laos	1	0.998	0.998
8	Myanmar	1	0.968	0.968
	Mean	1.007	1.007	1.014

Source: Asia Competitiveness Institute.

highest TFP growth over the study period, Philippines and Thailand show more balanced contributions of TEC and TC to agricultural TFP growth as compared to Malaysia and Indonesia.

CLMV made up the bottom four in the ranking of average TFP growth over the period of 2000–2011 (see Table 3.7). Nonetheless, while Cambodia and Vietnam attained positive TFP growth, Laos and Myanmar saw a deterioration in TFP over the study period, which can be

Table 3.8: Malmquist Index Robustness Check for ASEAN-7 (Myanmar Excluded), 2000–2011

Rank	Country	TEC	TC	TFPC
1	Malaysia	1	1.041	1.041
2	Philippines	1	1.031	1.031
3	Indonesia	1.006	1.023	1.029
4	Thailand	1.006	1.021	1.027
5	Vietnam	0.976	1.03	1.006
6	Laos	1	0.998	0.998
7	Cambodia	1.03	0.964	0.992
	Mean	1.002	1.015	1.017

Source: Asia Competitiveness Institute.

attributed to a remarkable decline in TC growth. Likewise, Cambodia performed better in terms of TEC compared to TC, implying that more should be done on the technological front. In contrast, Vietnam experienced robust growth in TC but a decline in TEC, which indicates a need for a more efficient and effective allocation of agricultural resources for achieving higher agricultural productivity. The case of Vietnam implies that investment in agricultural R&D may not bring about significant growth in TFP without improved resource allocation and utilisation. While R&D efforts may shift a country's production frontier forward, better resource management can actually increase productivity for a substantially lower cost, which is an important policy implication for low-income and lower middle-income countries.

It is also worth noting that, among the countries with the highest growth rates of labour productivity between 2000 and 2011 (see Table 3.2), only Malaysia remained in the list of countries posting highest TFP growth. There are other notable variations as well between the ranking of labour productivity growth and TFP growth. The Philippines, for example, registered a below average labour productivity growth but ranked second in terms of TFP growth. Cambodia, which ranked second in labour productivity growth, dropped to fifth in terms of TFP growth.

Given concerns about Myanmar's exchange rate data, we excluded Myanmar and re-estimated the results. As shown in Table 3.8, our results are robust with the top-4 performers remaining the same, namely Malaysia, Philippines, Indonesia and Thailand. Vietnam, Laos and Cambodia trailed behind in the fifth, sixth and seventh place, respectively, posting similar results as when Myanmar was included. Overall, ASEAN-8 as a whole registered a positive TFP growth of 1.7%, which is slightly higher than the previous estimate of 1.4%.

3.6 CASE STUDY: AGRICULTURAL DEVELOPMENT IN MALAYSIA

Topping the region in both labour productivity growth and TFP growth, Malaysia provides an instructive example of agricultural development which may hold important lessons to other

ASEAN member states. As such, this section details the key characteristics of Malaysia's agricultural sector, notable agricultural strategies and policies as well as principal challenges and prospects facing the sector. Particular emphasis is on the country's successful development of a competitive agribusiness industry.

3.6.1 Key Features of Malaysia's Agricultural Sector

Malaysia's agricultural sector is most saliently characterised by its distinct duality with the country's manufacturing industries, with the presence of the more developed agro-industrial commodity subsector that includes oil palm, rubber, cocoa and timber which account for about 60% of agricultural GDP, and the less efficient food subsector consisting of rice, vegetables, fruits and others which constitute the remaining 40% of the agricultural GDP.[35] This is in line with the sector's key strategic thrust which is to establish and develop agro-based industries as identified since the first National Agricultural Policy enacted in 1984.[36] Agricultural development efforts have thus been geared towards modernisation and commercialisation of agriculture through increased private sector participation.

Duality is also prevalent in the country's agricultural production modality which comprises a dominant capital-intensive estate sector and a less productive smallholder sector. Nonetheless, smallholders have increasingly been resettled and organised under various public and private land development schemes aiming at exploiting economies of scale, which is particularly crucial for growing agro-industrial crops such as oil palm and rubber. The common characteristic of such schemes is to gather a number of small-scale farmers into an estate-like form of organisation where management, processing and marketing operations are carried out by a public agency.[37] One such notable public scheme is the Federal Land Development Authority (FELDA) established in 1956, which subsequently evolved into a major state-owned agribusiness corporation in the palm oil subsector.[38]

Oil palm not only accounts for the majority of Malaysia's agricultural production but also grows in significance over time (see Figure 3.7). As of 2011, oil palm constituted 67.6% of the country's gross agricultural production in US$, as compared to 26.6% in 1980. In contrast, rubber has seen a notable decline in its share of total agricultural production from 42.1% in 1980 to 9.5% in 2011. Traditionally important crops including cocoa and coconuts also saw a drop in their contribution to total agricultural production over time, whilst vegetables, which is a food crop, became Malaysia's fifth most major crop in 2011, followed by pepper.

Malaysia's agricultural trade showed steady growth despite a sharp plunge in exports in 2009 following the Global Financial Crisis (GFC) (see Figure 3.8). Total agricultural trade grew at an average rate of 9.1% per annum during the two decades between 1981 and 2011. Imports grew at an average annual rate of 9.6% over the same period, slightly higher than the average annual

---

[35] See Ahmad and Suntharalingam (2009, 2).

[36] See Government of Malaysia (1986; 1990; 1996; 2001; 2006; 2011).

[37] See Fold (1994, 75).

[38] See Sutton (2001) for an in-depth case study of FELDA.

Figure 3.7: Share of Major Crops in Malaysia's Gross Agricultural Production by Value (US$), 1980–2011

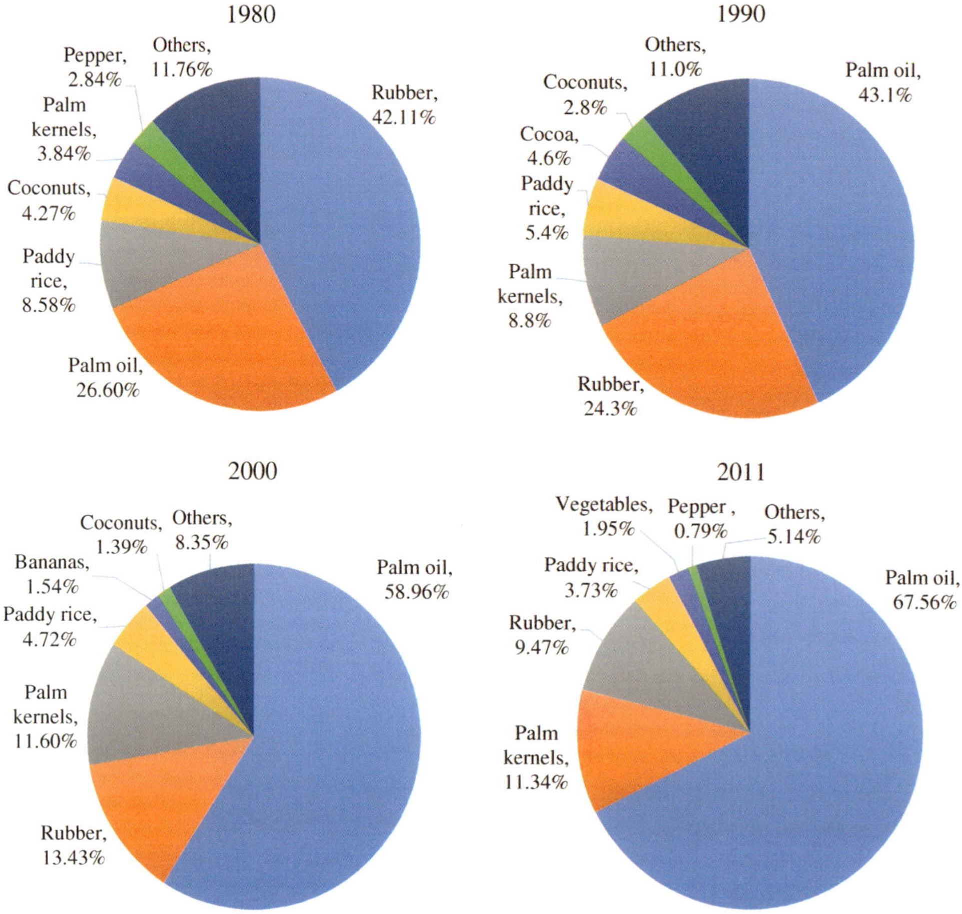

Source: FAO.

Figure 3.8: Malaysia's Agricultural Exports and Imports, 1980–2011

Source: FAO.

Table 3.9: Average Annual Growth Rates of Malaysia's Agricultural Trade, 1981–2011

	Imports	Exports	Total Trade
1981–1990	5.21%	2.66%	3.18%
1991–2000	6.49%	3.90%	4.62%
2001–2011	16.48%	20.12%	18.53%
1981–2011	9.62%	9.26%	9.09%

Source: Calculated from FAO's statistics.

Table 3.10: Trade Performance of Major Subsectors of Malaysian Agriculture (in Thousands US$), 1980–2011

Subsector		1980	1990	2000	2011
	Export	1,155,920	1,626,551	2,558,723	17,452,177
Palm oil	Import	575	6,078	15,719	1,938,254
	Balance of trade	1,155,345	1,620,473	2,543,004	15,513,923
	Export	—	206,312	87,085	203,600
Rubber	Import	—	48	214,934	1,113,342
	Balance of trade	—	206,264	−127,849	−909,742
	Export	23	43	60	423
Rice	Import	59,537	99,739	181,585	606,132
	Balance of trade	−59,514	−99,696	−181,525	−605,709
	Export	6,919	5,327	10,927	22,280
Vegetables	Import	5,053	3,262	12,494	25,722
	Balance of trade	1,866	2,065	−1,567	−3,442
	Export	20	58	129	17,218
Fruits	Import	408	1,330	2,051	14,679
	Balance of trade	−388	−1,272	−1,922	2,539

Source: FAO.

growth rate of 9.3% of exports (see Table 3.9). Over the period from 1981 to 2000, the average annual growth of imports exceeded that of exports by a relatively large margin. However, the recent decade between 2001 and 2011 saw a period of stellar growth of agricultural trade for Malaysia. Exports of Malaysian agricultural products grew at an average rate of 20.1% per annum, surpassing imports which grew at 16.5% annually, in spite of the GFC, which signifies the resilience of Malaysia's agricultural sector.

As shown in Table 3.10, palm oil is evidently the main driver of Malaysia's stellar agricultural trade performance, registering a remarkable increase in trade surplus from US$1.2 billion in 1980 to US$15.5 billion in 2011. Fruits also showed a positive balance of trade in 2011, albeit to a much more limited extent as compared to palm oil. Vegetables' exports rose more than two-fold between 1990 and 2000 although its trade balance deteriorated over the same period. Trade deficits of both rubber and rice widened between 1980 and 2011.

Owing to the success of the palm oil industry, Malaysia is one of the few middle-income developing countries with an internationally competitive agribusiness sector. Among the top-20 largest agro-based companies in the world by assets and revenues in 2007, Malaysia had six companies, surpassing the United States (five companies) and United Kingdom (three companies).[39] Sime Darby, the country's leader in palm oil, took the top spot in the list, followed by KL Kepong (seventh), Kulim (ninth), PPB Group (15th), TSH Resources (17th), and Multi Vest Resources (18th).

3.6.2 Malaysia's Agricultural Strategies and Policies

Malaysia's agricultural policies in the early stages of development in the 1960s and 1970s primarily aimed to reduce poverty in rural areas and narrow the income disparity between the agricultural and non-agricultural sectors through raising incomes of agricultural producers. As land was still abundant during this period, expansion of agricultural land resulted in employment creation in rural areas and increased production of food products for domestic consumption. Agriculture was also promoted as a means to earn foreign exchange. Hence, cash crops such as rubber, palm oil, cocoa were actively developed while food subsectors were protected to insulate producers from competition and save foreign exchange in line with Malaysia's import substitution strategy during this period. It is through these agro-industrial commodities that Malaysia accumulated its early wealth to fund industrialisation.

The enactment of the First National Agricultural Policy (NAP1) in 1984 marked the beginning of agricultural liberalisation, which paved the way for the rapid development of the agro-based industry. Agricultural development strategy in this period centred upon modernisation and commercialisation with an emphasis on improving productivity, efficiency and competitiveness through greater involvement of the private sector.[40] Large-scale plantation-like farming, mechanisation and exploration of agricultural niches were actively encouraged. Enhanced linkages between agriculture and manufacturing were strongly emphasised with the introduction of highly attractive incentives including tax breaks and export rebates for locally processed commodities. Tariffs on agricultural inputs were either abolished or reduced to make imports of raw materials cheaper, thus lowering input costs for downstream industries.

Malaysia has since made concerted efforts to promote agribusiness as the primary agricultural strategy. The Sixth Malaysian Plan 1991–1995 called for the development of an internationally competitive agro-based industry.[41] This was followed by the Second National Agricultural Policy (NAP2) enacted in 1992 which was essentially an extension of the NAP1, aiming at meeting the local and global demand for agro-industrial commodities. As such, in conjunction with the emphasis on enhancing productivity and efficiency of agro-industrial crop production,

[39] See Briones and Rakotoarisoa (2013, 11).
[40] See Government of Malaysia (1986, 296–297).
[41] See Government of Malaysia (1990).

downstream industrial development was emphasised to both increase the value addition in export crops and generate more off-farm employment opportunities for smallholder farmers.[42] Agricultural R&D was intensified during this period, particularly in developing high-yielding varieties, improving processing technologies and diversifying end products.[43]

Agricultural development received renewed attention in the aftermath of the Asian Financial Crisis (AFC) during which domestic food prices went up as a result of higher import costs as the Malaysian ringgit significantly depreciated against the currencies of its trading partners. Social discontent was on the rise, which led the Malaysian government to formulate the Third National Agricultural Policy (NAP3) in 1998 with a view to revitalising the food subsector and pre-empting socio-political instability. In light of resource constraints including land and labour, agricultural strategies under the NAP3 were geared towards increased mechanisation and adoption of technologically-intensive production systems to enhance productivity. Concurrently, the development of export-oriented agro-industrial commodities continued to be stressed, particularly palm oil. Strengthening the agro-based value chains, expanding large-scale commercial farming and developing biotechnology as well as information and communication technology were some notable policies aimed at enhancing productivity in the high-value agricultural subsector.

The Tenth Malaysia Plan 2011–2015 identifies agriculture as one of the 12 National Key Economic Areas. With an aim to enhance food security, a target of 70% rice self-sufficiency has been set in conjunction with "entering long-term contract agreements to import rice with matching agreements to export palm oil or oil."[44] Diversification into new high-value agricultural products is also emphasised including swiftlet farming, aquaculture, organic fruits and vegetables and floriculture. Contract farming undertaken by large-scale food suppliers such as the Malaysian Agriculture Food Corporation (MAFC) is identified as the key policy for poverty alleviation and ensuring sustainable livelihoods for smallholder food producers. In the same year, the Malaysian government enacted the ten-year National Agro-food Policy 2011–2020 (NAP4) which aims to modernise the agro-food industry through augmenting production of food crops, improving productivity, strengthening food value chains and enhancing food safety and quality along the value chains, thereby turning it from being associated with low income and unskilled jobs to a career of choice.[45]

The success of Malaysia's agricultural sector, as evident in the country's strong export performance and remarkable TFP growth being the highest in ASEAN-8 according to our estimates, has been widely attributed to the establishment of a competitive agro-based sector, notably palm oil. While private estates have traditionally played a dominant role in the sector, the government's astute strategies and policies have also effectively contributed to the expansion and upgrading of the sector while ensuring that smallholder farmers are not left behind, thereby achieving both economic and socio-political objectives. Notable policies include: (i) fostering a

---

[42] See Dardak (2015).

[43] See Government of Malaysia (1996, 240).

[44] See Government of Malaysia (2011, 134).

[45] See Government of Malaysia (2015, 1).

competitive agribusiness industry, (ii) close partnership between government agencies, agribusiness firms and research institutes, (iii) integrating smallholders into agro-based value chains, and (iv) provision of infrastructure, credit and extension services.

Fostering a Competitive Agribusiness Industry

While rubber was the major crop throughout the colonial period, rubber experienced a prolonged decline in price throughout the early 1960.[46] As a result, large-scale diversification of former rubber plantations into palm oil, which proved to be more profitable in the global market in the ensuing decades, was promoted and subsidised. Furthermore, as oil palm cultivation requires lower labour inputs, it has since become the preferred crop in the estate sector after independence.

Early incentives to promote downstream palm oil refining started with the promulgation of the Investment Incentives Act of 1968 which conferred a 40% abatement of corporate income tax for two years, with a possibility of extension, and of excess profit for eight years to palm oil companies. Identified as one of the "pioneer" subsectors up until 1974, palm oil refineries were also allowed tax exemptions on capital spending on top of a seven-year tax holiday. A tax allowance of 50% was also extended on qualifying R&D expenditures over a period of ten years.[47] Such generous incentives are generally regarded as the first major impetus to the development of the country's palm oil industry.

Over the period from 1968 to 1984, a system of differential export duties was introduced to stimulate local downstream processing of crude palm oil. While crude palm oil was subject to 5–10% export duties, exemption of processed palm oil from export duties encouraged palm oil manufacturers to shift to exports of processed products.[48] Although the downstream processing industry was initially dominated by Japanese and Indian firms, by the mid-1980s, local and joint-venture plantation companies entered the scene, forming the majority of the budding palm oil subsector in subsequent years. Originated as a public resettlement programme, FELDA also ventured into palm oil processing and exports during this period with the establishment of Felda Refinery Corporation and Felda Marketing Corporation.[49]

As a result of such concerted effort to shift into high value added exports, by 1990 processed products formed the bulk of palm oil exports, thus becoming the largest contributor to agricultural value added. Furthermore, virtually all local refineries in operation had built refining and fractionation plants, and both processes had become integral parts of the palm oil industries. Consolidation of palm oil refineries also took place throughout the 1980s and 1990s with the establishment of new refineries and revamp of existing ones with state-of-the-art refining technologies.[50] Competition pressures hence intensified, leading to various forms of vertical integration of small, inefficient producers into large palm oil estates.

---

[46] See Fold (2000, 475).
[47] See Rasiah and Shahrin (2006, 9).
[48] See Rasiah and Sharin (2006, 11).
[49] See Fold (2000, 476).
[50] See Fold (1994, 77).

Owing to the rapid development of the palm oil industry, the 1990s saw the proliferation of numerous multinational and local suppliers of capital and immediate goods for the palm oil subsector operating in the country. As part of the Industrial Master Plan of 1996, localisation of production of inputs and complementary goods for the palm oil subsector was promoted, including machinery and equipment, packaging and related support services. Major palm oil companies established their own subsidiaries to capture a share of the growing upstream and complementary industries, resulting in a more seamless value chain and lower production costs. Palm oil clusters which comprise plantations and downstream and upstream processing facilities started to be formed during this period.

Similar agro-industrial linkages are also observed in the rubber subsector, which has resulted in the increasing use of natural rubber in local production of final products. While the planting area and production of natural rubber has declined in recent decades in response to price volatility, by 2000, more than 80% of natural rubber output was used by the local gloves, tyre and tube industries.[51]

Close Partnership between Government Agencies, Agribusiness Firms and Research Institutes

The cohesive partnership between government agencies, firms and research institutes has generally been credited for the country's effective agricultural R&D and resultantly stellar export performance in palm oil. One notable example is the establishment of the Palm Oil Research Institute of Malaysia (PORIM) in 1979 which was a result of lobbying from palm oil companies and support from the state-funded Malaysian Agricultural Research and Development Institute (MARDI). Consequently, palm oil firms do not just benefit from R&D incentives such as tax exemption on R&D expenditures but also access to state-led R&D implemented by MARDI, PORIM and universities. In total there are more than 40 agencies that are involved in agricultural research in Malaysia including research universities and private sector laboratories.[52]

The palm oil subsector was further strengthened with the creation of two other institutions, namely the Palm Oil Regulatory and Licensing Association (PORLA) and the Malaysia Palm Oil Promotion Council (MPOPC). PORLA was set up to oversee and promote the development of the palm oil industry. It is tasked with regulating entry of all agents operating in palm oil including suppliers of intermediate and complementary goods, buyers, storage and milling, exporters and importers. PORLA's other focusses are on contractual systems that apply to domestic and international trade and fulfilment of quality standards. The MPOPC was set up as a privately registered but government controlled agency to spearhead marketing and promotion programmes for palm oil with the aim to turn Malaysia into an international leader in palm oil and related products.

PORLA and PORIM eventually merged in the 1990s to form the Malaysian Palm Oil Board (MPOB). Both MPOPC and MPOB have been credited for developing and deepening the palm

[51] See Government of Malaysia (2001, 216).

[52] See Rahman (2012, 31).

oil value chains. These institutions work closely with universities and research institutes, MARDI, the Ministry of International Trade and Industry and industry associations. Such institutional setting has facilitated seamless flows of information and discussion as well as effective coordination between the public and private sector, resulting in flexible formulation of supply strategies in response to prices. As Malaysia accounts for more than half of global exports of many palm oil products, efforts to regulate supply has successfully kept prices, and correspondingly export earnings, stable.[53]

Integrating Smallholders Into Agro-Based Value Chains

As palm oil cultivation is capital intensive in nature, it has been traditionally undertaken by large private estates since independence in Malaysia. Nonetheless, the government has strived to integrate smallholder farmers into the sector with the establishment of FELDA in 1956. The agency was tasked with opening new land, establishing oil palm plantations and resettling impoverished peasants mainly from Malay ethnicity into plantation-like organisations including building houses for the settlers. Individual smallholders are typically allocated a planted oil palm plot of four hectares in a vertically integrated structure of production and processing.[54] Besides proving agricultural infrastructure, FELDA is also responsible for management, providing credit, marketing, processing and refining through its two commercial arms, *viz.* Felda Refinery Corporation and Felda Marketing Corporation.

A large class of organised smallholders thus emerged and by 2011 accounted for about 25% of cultivated area of oil palm in the country.[55] By most measures, FELDA's schemes have been successful: dropout rates are low, loan repayment rates are high, smallholders are able to earn a decent income and the World Bank lauded FELDA as one of the most successful resettlement efforts in the world.[56] FELDA is thus widely regarded as a state-led attempt to spread the benefits of palm oil production among smallholders who would otherwise not be able to grow this capital intensive crop. In this respect, FELDA has been successful in distributing the gains of palm oil more equitably than would be the case if private estates solely dominated the subsector.

In parallel with FELDA's resettlement schemes, private estates also engaged in massive new land development up until the early 1990s in collaboration with local governments under the Regional Development Authorities (RDAs).[57] This led to a substantial extension of basic infrastructure as well as infrastructure required for oil palm cultivation and processing including roads, mills and input supply, which benefits not only the estates themselves but also independent smallholders who typically hold smaller plots of land than organised smallholders. In addition, the Rubber Industry Smallholder Development Authority (RISDA), another government agency coordinating and supporting the rubber smallholder subsector, have been encouraging independent rubber smallholders to integrate oil palm into their production to

---

[53] See Rasiah and Shahrin (2006).
[54] See Bissonnette and De Koninck (2015).
[55] See Barlow (2012).
[56] See Pletcher (1991, 628) and Bissonnette and De Koninck (2015, 8).
[57] See Government of Malaysia (1990, 101).

diversify livelihoods and increase incomes, given the volatility of natural rubber price in the global market.[58] By 2011, independent smallholders were cultivating about 14% of all oil palm land.[59] Malaysia's palm oil boom hence does not just benefit private estates but has also contributed to poverty reduction through successful state-led schemes to integrate smallholder farmers into the palm oil value chains as well as public and private provision of infrastructure.

Provision of Infrastructure, Credit and Extension Services

The Malaysian government's efforts to support smallholders go beyond resettlement schemes under FELDA to also include providing agricultural infrastructure, credit and extension services. While FELDA played a large part in building new infrastructure as it opened new land, the government also implemented various Integrated Agricultural Development Projects (IADPs) throughout the country to provide a package of infrastructural and support facilities. These include the provision of drainage and irrigation systems, flood control, rural roads and machineries. In addition, programmes for consolidating and rehabilitating unproductive farms were carried out by the Federal Land Consolidation and Rehabilitation Authority (FELCRA) which aimed to support smallholders to switch to or incorporate profitable crops into their production to increase productivity and improve livelihoods and incomes. Similar to FELDA schemes, these agricultural development programmes emphasise commercialisation through establishing mini-estates and group-farming projects.

Furthermore, the provision of training and extension services by various public agencies including FELDA, FELCRA, and RISDA has enhanced smallholders' capability to adopt new technologies and sustainable production practices and attain higher productivity. Credit is extended to the smallholder sector mainly through the Centralised Agricultural Credit Scheme under Bank Pertanian Malaysia (BPM). BPM also provides interest-free loans to impoverished farmers to venture into more viable agricultural projects.

The government is also actively involved in marketing of agricultural products in close collaboration with private companies and farmers' organisations. While palm oil promotion is carried out by the MPOPC and subsequently MPOB as noted, the Federal Agricultural Marketing Authority (FAMA) is tasked with marketing other agricultural subsectors in international markets through its participation in international exhibitions and trade missions. For instance, FAMA has successfully promoted Malaysian fruits such as papaya and carambola in East Asia, Europe and Middle East.[60]

3.6.3 Challenges and Prospects

The principal challenges facing Malaysia's agricultural sector are primarily structural and supply-side in nature, notably the increasing shortages of arable land and agricultural labour force. Climate change poses further threats to the country's agricultural production and food

[58] See Mustapha (2011, 156).

[59] See Bissonnette and De Koninck (2015, 8).

[60] See Government of Malaysia (1990, 102).

security. Furthermore, as Malaysia's agriculture is highly dependent on exports of agro-industrial crops, particularly palm oil, trade liberalisation has exposed the sector to intense competition from countries with lower costs of labour and other agricultural inputs. Maintaining the sector's competitiveness is therefore critical in today's globalised world.

Climate Change

As with many countries in the world, Malaysia's agricultural sector is facing formidable challenges arising from climate change. As of 2010, Malaysia ranked as the 30th largest greenhouse gas emitter and as a result, the temperature in the country is projected to rise by 0.3–4.5°C and rainfall is forecast to fluctuate from –30% to +30%. These climate change phenomena threaten to bring about droughts in many areas, which renders the cultivation of some key crops such as oil palm, rubber and cocoa impossible as well as reduces other crops' yields.[61] It has been estimated that a 1% increase in temperature and a 1% increase in rainfall will lead to a 3.4% and 0.12% decrease in rice yield, respectively.[62] While agriculture is often featured prominently in all Malaysia's national economic plans as discussed in Section 3.6.2, climate change adaptation and mitigation were only given due attention in the Tenth Malaysia Plan 2011–2015, which outlines key strategies for tackling climate change including, among others, reducing carbon footprint, promoting green technology and investments in renewable energy and conserving forests.[63]

Labour Force

Notwithstanding the Malaysian government's renewed focus on agricultural development following the 2007–2008 global food crisis, the country's agricultural workforce has continued to decline steadily and reached 1.6 million workers or 11.6% of the total workforce in 2014. More strikingly, 98.3% of the agricultural labour force are unskilled and semi-skilled and 29.3% are above 50 years old as compared to 27.3% in 2005.[64] Shortages in agricultural labour are hampering the development of the sector as even in subsectors where agricultural technologies and mechanisation are heavily utilised like oil palm, massive labour is still required. The lack of skilled labour further aggravates the prospect of Malaysia's agriculture as it hinders the application of R&D and sustainable production practices, which are critical for sustaining agricultural output and productivity.

Trade Liberalisation

Malaysia's agricultural sector has been subject to a competitive trading environment under the frameworks of the World Trade Organisation (WTO), the AFTA as well as multiple ASEAN+1

---

[61] See Alam *et al.* (2012).

[62] See Alam *et al.* (2010).

[63] See Government of Malaysia (2011, 297–306).

[64] See Government of Malaysia (2015, 10).

free trade agreements. Malaysia's accession to the Trans-Pacific Partnership (TPP) and the establishment of the ASEAN Economic Community in 2015 have intensified the competition facing the sector as tariffs as well as non-tariff barriers are further reduced. Participating in such trade liberalisation initiatives means that Malaysia's agricultural sector must be well-positioned to compete with countries with abundant low-cost labour as well as other cheap agricultural inputs. As Malaysia is heavily reliant on export revenues of commercial agro-industrial crops, notably palm oil, for importing rice and foodstuffs, maintaining export competitiveness is vital for not only sustainable agricultural growth but also food security.

Nonetheless, demand-side factors present immense opportunities for Malaysia to further develop and strengthen its agricultural sector. Notably, as income grows in many parts of the world, demand for food has increasingly become more diverse and geared towards healthy and nutritious products. Horticulture and floriculture will also be highly sought after. Such developments open doors to many agricultural niches that Malaysia could explore and diversify into. It is thus clear that Malaysia's success in sustaining a competitive agricultural sector is dependent upon its ability to position itself in the global market. Similar to what Malaysia has achieved with palm oil, a pro-active agribusiness sector is what the country needs to successfully venture into new global agricultural markets.

3.7 CONCLUSION AND POLICY IMPLICATIONS

Even as ASEAN is accelerating its transition into a manufacturing hub of Asia and the world, agriculture remains a key driver of growth and competitiveness in the majority of the member states. Following the 2007–2008 global food crisis, the issue of food security has been thrust to the forefront of regional policy-making and cooperation agenda. Nonetheless, regional efforts for enhancing food security have focused on food accessibility, with the establishment of initiatives such as AIFS Framework in 2009 and APTERR in 2012, while the supply side of the issue has not been given due attention. In this context, ACI conducted a comprehensive quantitative assessment of agricultural productivity in ASEAN-8 (Brunei and Singapore are excluded) with a view to offer insightful policy lessons for improving agricultural production, ensuring adequate food supply as well as enhancing the competitiveness of the agricultural sector in the region. We thus employ both the growth accounting and the Malmquist index DEA methodology on an extensive set of input and output data obtained from FAO and USDA to estimate annual growth of TFP for the eight ASEAN economies between 2000 and 2011.

Our findings show that overall ASEAN-8 experienced an annual growth of agricultural TFP of 1.4% to 1.6% over the period 2000–2011. In view of concerns about Myanmar's exchange rate data, we excluded Myanmar and the results of the Malmquist index remain robust at 1.7% which is only slightly higher than the previous estimates of 1.4%. In light of the development gap within the regional bloc, we divided the region into two sub-groups according to their respective development status. Unsurprisingly, the more mature group comprising Indonesia, Malaysia, Philippines and Thailand posted a much higher TFP growth of between 2.7% and 3.3% while CLMV lagged behind, with a TFP growth of between –0.4% and 0.4% over the study period. A ranking of TFP growth as enabled by the Malmquist index analysis shows that

Malaysia, Philippines, Indonesia and Thailand were the top performers while Cambodia, Vietnam, Laos and Myanmar made up the bottom four. Notably, Laos and Myanmar were the only two countries in the region that registered negative growth rates of TFP between 2000 and 2011.

The decomposition of TFP growth according to the Malmquist index method further sheds light on the drivers of TFP in the region, as well as the remarkable heterogeneity within the regional bloc. In particular, among the top-4 performers, TFP growth in Malaysia was mainly driven by TC as the Malaysian government has been very active in promoting agricultural R&D, particularly in palm oil and rubber. In contrast, TFP growth in Indonesia seems to have generally been determined by TEC, which indicates that efficient allocation and utilisation of agricultural resources have been the key driving force in TFP changes in the country. Philippines and Thailand show more balanced contributions of TEC and TC to agricultural TFP growth as compared to Malaysia and Indonesia.

Diversity in the decomposition of TFP growth is also observed in the bottom four. While Cambodia performed better in terms of TEC compared to TC, Vietnam experienced robust growth in TC but a negative growth rate for TC. This implies that Vietnam needs to improve its allocation and utilisation of resources in order to achieve higher agricultural productivity; meanwhile for Cambodia, investment in agricultural R&D is required. The case of Vietnam is particularly instructive, as it shows that while R&D spending may shift a country's production frontier forward, resulting in an overall improvement of TFP, better resource management can also increase productivity at a much lower cost, which may be a viable policy choice for low-income and lower middle-income countries. Both Laos and Myanmar were stagnant in terms of TEC and deteriorating in terms of TC, which contributed to negative TFP growth over the study period.

Topping the region in both labour productivity growth and TFP growth according to our estimates, Malaysia's agricultural development provides an instructive case in point which other ASEAN members can adopt and adapt to foster their agricultural sectors. Apart from being a global leading exporter of agro-industrial crops, notably palm oil, the country's food subsector has also posted positive growth in recent years following the government's renewed focus on food agriculture following the 2007–2008 global food crisis. Furthermore, Malaysia is one of the handful middle-income developing countries with a relatively developed agribusiness sector with six companies listed in the top-20 agro-based firms in the world. While private estates have played a prominent role in the country's agricultural sector, Malaysia has also been successful in integrating smallholder farmers into agro-based value chains, thereby increasing their incomes and alleviating rural poverty. Malaysia's success in agriculture can be attributed to four key policies: (i) fostering a competitive agribusiness industry, (ii) close partnership between government agencies, agribusiness firms and research institutes, (iii) integrating smallholders into agro-based value chains, and (iv) provision of infrastructure, credit and extension services.

In light of Malaysia's experience, some notable policy implications with respect to agricultural development can be drawn as follows:

i. Active state intervention is prerequisite in modernising agriculture and fostering agribusiness and agro-based industries. While the private sector may take the lead in the agricultural sector, state-led efforts are crucial in identifying product niches, based on each country's natural endowment and core competencies, establishing agro-industrial linkages, developing value chains and coordinating relevant stakeholders.

ii. Structural transformation towards agro-based industries is imperative for a competitive agricultural sector, as evident in Malaysia's distinct duality between agriculture and manufacturing with the country's primary crops geared toward industrial use such as palm oil, rubber and cocoa. Such dual structure would not only unlock ASEAN's potential in both agriculture and manufacturing but also catalyse the region's industrialisation efforts.

iii. Opportunities for vertical fragmentation of the agro-processing industry need to be identified in order to integrate smallholder farmers, who are most vulnerable to volatility in the agricultural market and hence poverty, to production processes where appropriate.

iv. Effective and efficient allocation and utilisation of resources are the key to enhance agricultural TFP, as shown in our decomposition of Malmquist index for ASEAN-8. While investment in R&D would shift a country's production frontier forward, thereby increasing its overall TFP, improvement in resource management would also help to stimulate TFP growth at a substantially lower cost, which may be a viable policy option for less developed economies in the region.

v. Provision and upgrading of infrastructure, particularly roads, drainage and irrigation systems, are of paramount importance for developing agribusiness and improving farmers' livelihoods. Rural infrastructural development has to be environmentally conscious with a view to mitigating and preventing soil erosion and other adverse impacts of climate change.

vi. Close public-private partnership between the government and agri-business firms is vital for establishing sustainable value chains and ensuring stable markets for the farmers' produce. Such partnership would help to augment public resources, particularly in agricultural R&D, and facilitate faster adoption of agricultural best practices and innovations. The remarkable performance of Malaysia's palm oil sector has been widely attributed to effective coordination between government agencies, palm oil companies and research institutes.

vii. Both public and private agricultural R&D need to be incentivised and promoted so as to develop productivity-enhancing, environmentally friendly and sustainable technologies and practices. For instance, developing new varieties of crops with increased resistance to pests, bacteria and viruses would reduce the need for harmful pesticides and herbicides.

viii. Preferential credit schemes and extension services need to be made widely available to farmers to improve their access to capital, new crop varieties and technologies, which will enable them to increase productivity and thus incomes. Such policies are particularly important to help farmers cope with climate change as well.

Beyond the empirics, substantively the following steps could be taken to strengthen regional cooperation in agriculture and food security in ASEAN:

i. Regional cooperation efforts in agriculture and food security should not solely focus on accessibility to food. Although food accessibility is critical, particularly in times of calamities and emergencies, emphasis should also be on enhancing regional supply capacity.
ii. Cooperation agenda should include measures to strengthen supply capacity, particularly: harmonisation of phytosanitary measures for crop products and maximum residue limits of commonly used pesticides, capacity building in pest risk prevention and transfer and adoption of new production and post-harvest technologies. Intra-regional collaboration in agricultural R&D could be helpful to catalyse productivity catch-up and thus narrow the productivity gap within ASEAN as well as between ASEAN and China and India.
iii. Knowledge sharing on new agricultural technologies, sustainable production techniques, resource management, disease controls, food safety, etc. should be promoted to help countries lagging in terms of productivity to catch up with the leading ones. Intra-regional learning is important.
iv. Cross-border contract farming could be beneficial for countries with a less developed agribusiness industry. Such production modality involves pooling smallholder farmers into an estate-like organisation where the contracting company provides inputs, marketing services and possibly credit. This would help to increase farmers' incomes as well as facilitate the adoption of new technologies and sustainable agricultural practices.
v. Regional capacity building in human resource and agro-preneur development would facilitate knowledge sharing, address the shortage of skilled labour in agriculture in many ASEAN countries, and promote entrepreneurship in agro-based industries.
vi. Private sector can play a prominent role in enhancing regional supply capacity given its expertise and R&D capability. An institutional framework for technical collaboration between the ASEAN Secretariat, APTERR and the private sector should be explored. Potential areas of cooperation may include: technology transfer; value chain development; provision of high-yielding varieties, agro-chemicals and other inputs and marketing and trade.

REFERENCES

Ahmad, Tengku Mohd Ariff and Chubashini Suntharalingam. 2009. "Transformation and economic growth of the Malaysian agricultural sector." *Economic and Technology Management Review* 4: 1–10.

Alam, Md. Mahmudul, Basri Talib, Chamhuri Siwar and Mohd Ekhwan Toriman. 2010. *The Impacts of Climate Change on Paddy Production in Malaysia: Case of Paddy Farming in North-West Selangor.* Proceedings of the 4th International Malaysia–Thailand Conference on Southeast Asian Studies, National University of Malaysia, Selangor, March 25–26.

Alam, Md. Mahmudul, Chamhuri Siwar, Basri Talib, Mazlin Mokhtar and Mohd Ekhwan Toriman. 2012. "Climate change adaptation policy in Malaysia: Issues for agricultural sector." *African Journal of Agricultural Research* 7(9): 1368–1373.

ASEAN Secretariat. 2009. *2009 ASEAN Integrated Food security (AIFS) Framework and Strategic Plan of Action on Food Security in the ASEAN Region (SPA-FS) 2009–2013.* http://cil.nus.edu.sg/rp/

pdf/2009%20AIFS%20Framework%20and%20Strategic%20Plan%20of%20Action%20on%20Food%20Security%20in%20the%20ASEAN%20Region-pdf.pdf.

Asian Development Bank. 2008. *Food Prices and Inflation in Developing Asia: Is Poverty Reduction Coming to an End?* Manila: Asian Development Bank.

Avila, Antonio Flavio Dias and Robert E. Evenson. 2010. "Total Factor Productivity Growth in Agriculture: The Role of Technological Capital". In *Handbook of Agricultural Economics: Volume 4*, edited by Prahbu L. Pingali and Robert E. Evenson, 3769–3822. Burlington: Academic Press.

Baldwin, Katherine, Nathan Childs, John Dyck and Jim Hansen. 2012. *Southeast Asia's Rice Surplus*. Washington DC: United States Department of Agriculture Economic Research Service.

Barlow, Colin. 2012. "Agriculture in Malaysia's economic and social transformation." New Mandala, May 9. http://asiapacific.anu.edu.au/newmandala/2012/05/09/agriculture-in-malaysias-economic-and-social-transformation/.

Bissonnette, Jean-François and Rodolphe De Koninck. 2015. "Large Plantations versus Smallholdings in Southeast Asia: Historical and Contemporary Trends". Paper presented at conference on Land Grabbing, Conflict and Agrarian-Environmental Transformations: Perspectives from East and Southeast Asia at Chiang Mai University, Thailand, 5–6 June.

Briones, Roehlano M. 2011. "Regional Cooperation for Food Security: The Case of Emergency Rice Reserves in the ASEAN Plus Three." ADB Sustainable Development Working Paper No. 18, Asian Development Bank, Manila.

Briones, Roehlano M., Alvaro Durand-Morat, Eric J. Wailes and Eddie C. Chavez. 2012. "Climate Change and Price Volatility: Can We Count on the ASEAN Plus Three Emergency Rice Reserve?" ADB Sustainable Development Working Paper No. 24, Asian Development Bank, Manila.

Briones, Roehlano M. and Manitra A. Rakotoarisoa. 2013. "Investigating the Structures of Agricultural Trade Industry in Developing Countries". FAO Commodity and Trade Policy Research Working Paper No. 38, Food and Agriculture Organisation of the United Nations, Rome.

Cahyadi, Eko Ruddy and Hermann Waibel. 2013. "Is Contract Farming in the Indonesian Oil Palm Industry Pro-poor?" *Journal of Southeast Asian Economies* 30(1): 62–76.

Capistrano, Ana Doris and Gerald G. Marten. 1986. "Agriculture in Southeast Asia." In *Traditional Agriculture in Southeast Asia: A Human Ecology Perspective*, edited by Gerald G. Marten. Boulder: Westview Press.

Chandra, Alexander C. and Lucky A. Lontoh. 2010. *Regional Food Security and Trade Policy in Southeast Asia*. Winnipeg: International Institute for Sustainable Development.

Coelli, Tim J., and D. S. P. Rao. 2005. "Total Factor Productivity Growth in Agriculture: A Malmquist Index Analysis of 93 Countries, 1980–2000." *Agricultural Economics* 32(1): 115–134.

Coxhead, Ian, Kim N.B. Ninh, Vu Thi Thao and Nguyen Thi Phuong Hoa. 2010. *A Robust Harvest: Strategic Choices for Agricultural and Rural Development in Vietnam*. San Francisco: The Asia Foundation.

Dardak, Rozhan Abu. 2015. "Transformation of Agricultural Sector in Malaysia Through Agricultural Policy." FFTC Agricultural Policy Platform, Food and Fertilizer Technology Center for the Asian Pacific Region, February 4. http://ap.fftc.agnet.org /ap_db.php?id=386

FAO. 2015. FAOSTATS. http://faostat.fao.org/site/291/ default.aspx.

Färe, Rolf, Shawna Grosskopf, Mary Norris and Zhongyan Zhang. 1994. "Productivity growth, technical progress, and efficiency change in industrialized countries." *The American Economic Review* 48(1): 66–83.

Flaherty, Kathleen, Gert-Jan Stads and Attaluri Srinivasacharyulu. 2013. *Benchmarking Agricultural Research Indicators Across Asia-Pacific*. Washington DC: International Food Policy Research Institute.

Fold, Niels. 1994. "Industrial Organization and Sectoral Linkages: A Study of the Malaysia Palm Oil Industry." *Danish Journal of Geography* 94: 68–81.

Fold, Niels. 2000. "Oiling the Palms: Restructuring of Settlement Schemes in Malaysia and the New International Trade Regulations." *World Development* 28(3): 473–486.

Fuglie, Keith O. 2012. "Productivity Growth and Technology Capital in the Global Agricultural Economy." In *Productivity Growth in Agriculture: An International Perspective*, edited by Keith O. Fuglie, Sun Ling Wang, V. Eldon Ball, 335–372. Oxfordshire: CAB International.

Fuglie, Keith O. and David Schimmelpfennig. 2010. "Introduction to the Special Issue on Agricultural Productivity Growth: A Closer Look at Large, Developing Countries." *Journal of Productivity Analysis* 33: 169–172.

Golan, Amnon and Harinder Kohli. 2013. "ASEAN Dynamism: Agricultural Transformation and Food Security." *Global Journal of Emerging Market Economies* 5(1): 3–21.

Government of Malaysia. 1986. *Fifth Malaysia Plan 1986–1990*. Putrajaya: Economic Planning Unit. http://www.epu.gov.my/en/fifth-malaysia-plan-1986-1990.

Government of Malaysia. 1990. *Sixth Malaysia Plan 1990–1995*. Putrajaya: Economic Planning Unit. http://www.epu.gov.my/en/sixth-malaysia-plan-1990-1995.

Government of Malaysia. 1996. *Seventh Malaysia Plan 1996–2000*. Putrajaya: Economic Planning Unit. http://www.epu.gov.my/en/seventh-malaysia-plan-1996-2000.

Government of Malaysia. 2001. *Eighth Malaysia Plan 2001–2005*. Putrajaya: Economic Planning Unit. http://www.epu.gov.my/en/eighth-malaysia-plan-2001-2005.

Government of Malaysia. 2006. *Ninth Malaysia Plan 2006–2010*. Putrajaya: Economic Planning Unit. http://www.epu.gov.my/en/ninth-malaysia-plan-2006-2010.

Government of Malaysia. 2011. *Tenth Malaysia Plan 2011–2015*. Putrajaya: Economic Planning Unit. http://onlineapps.epu. gov.my/rmke10/rmke10_english.html.

Government of Malaysia. 2015. "Strategy Paper 20: Driving Modernisation in Agro-food." Economic Planning Unit, Prime Minister's Department, Putrajaya. http://rmk11.epu. gov.my /pdf/strategy-paper/ Strategy%20Paper%2020.pdf.

Haggblade, Steven, Duncan Boughton, Khin Mar Cho, Glenn Denning, Renate Kloeppinger-Todd, Zaw Oo, Tun Min Sandar, Tin Maung Than, Naw Eh Mwee Aye Wai, Shannon Wilson, Ngu Wah Win and Larry C.Y. Wong. 2014. "Strategic Choices Shaping Agricultural Performance and Food Security in Myanmar." *Journal of International Affairs* 67(2): 55–71.

Intal, Ponciano S., Jr., Sothea Oum and Mercy J.O. Simorangkir, eds. 2011. *Agricultural Development, Trade & Regional Cooperation in Developing East Asia*. Jakarta: Economic Research Institute for ASEAN and East Asia (ERIA).

International Monetary Fund. 2014. World Economic Outlook Database. http://www.imf.org/external/ pubs/ft/weo/2014/02/weodata/weoselgr.aspx.

Manorom, Kanokwan, David Hall, Xing Lu, Suchat Katima, Maria Theresa Medialdia, Singkhon Siharath and Pinwadee Srisuphan. 2011. *Cross-Border Contract Farming Arrangement: Variations and Implications in the Lao People's Democratic Republic.* Manila: Asian Development Bank.

Mundlak Yair, Donald F. Larson and Rita Butzer. 2002. "Determinants of Agricultural Growth in Indonesia, the Philippines and Thailand." World Bank Policy Research Working Paper No. 2803, World Bank, Washington DC.

Mustapha, Nik Hashim Nik. 2011. "Technical Efficiency for Rubber Smallholders under RISDA's Supervisory System using Stochastic Frontier Analysis." *Journal of Sustainability Science and Management* 6(1): 156–168.

Ong, Keng Ho. 2014. *Strategic Plan of Action on ASEAN Cooperation in Crops (2016–2020): A Study Report.* http://www.fao.org/fileadmin/templates/rap/files/meetings/2014 /140718-report.crops.pdf.

Pletcher, James. 1991. "Regulation with Growth: The Political Economy of Palm Oil in Malaysia." *World Development* 19(6): 623–636.

Poapongsakorn, Nipon. 2011. "R&D and Performance of the Thai Agriculture and Food Processing Industry: The Role of Government, Agribusiness Firms and Farmers." In *Agricultural Development, Trade & Regional Cooperation in Developing East Asia*, eds. Ponciano S. Intal, Jr., Sothea Oum and Mercy J.O. Simorangkir. Jakarta: Economic Research Institute for ASEAN and East Asia (ERIA).

Rahman, Zaharah A. 2012. "Agricultural Research and Development in Malaysia." *Journal of International Society for Southeast Asian Agricultural Sciences (ISSAAS)* 18(2): 22–33.

Rasiah, Rajah and Azmi Shahrin. 2006. "Development of Palm Oil and Related Products in Malaysia and Indonesia." Paper, University of Malaya, Kuala Lumpur.

Reardon, Thomas, C. Peter Timmer and Bart Minten. 2012. "Supermarket revolution in Asia and emerging development strategies to include small farmers." *Proceedings of the National Academy of Sciences* 109(31): 12332–12337.

Rungsuriyawiboon, Supawat and Xiaobing Wang. 2009. "Agricultural Productivity Growth in Traditional and Transitional Economies in Asia." *Asian-Pacific Economic Literature* 23(2): 52–72.

Son Nghiem, H. and T. Coelli. 2002. "The Effect of Incentive Reforms Upon Productivity: Evidence from the Vietnamese Rice Industry." *The Journal of Development Studies* 39(1): 74–93.

Squires, Victor. 2009. "Land Degradation and the Food Crisis in the ASEAN Region." Discussion paper prepared for the ASEAN-United Nations Summit, Hua Hin, February 28–March 1.

Sriboonchitta, Songsak and Aree Wiboonpoongse. 2008. "Overview of Contract Farming in Thailand: Lessons Learned." ADB Institute Discussion Paper No. 112, Asian Development Bank Institute, Tokyo.

Suhariyanto, Kecuk and Colin Thirtle. 2001. "Asian Agricultural Productivity and Convergence." *Journal of Agricultural Economics* 52(3): 96–110.

Sutton, Keith. 2001. "Agribusiness on a Grand Scale — FELDA's Sahabat Complex in East Malaysia." *Singapore Journal of Tropical Geography* 22(1): 90–105.

Tan, Khee Giap, Kartik Rao and Ramkishen S. Rajan. 2015. "How Productive is the Agricultural Sector across Indian States?" *International Journal of Development Issues* 14(3): 231–248.

Tan, Khee Giap, Haoran Suo and Ramkishen S. Rajan. 2016a (forthcoming). "How Productive is the Agricultural Sector in Association of South East Asian Economies?" *International Journal of Technological Learning, Innovation and Development*.

Tan, Khee Giap, Haoran Suo and Ramkishen S. Rajan. 2016b (forthcoming). "Estimating Provincial Agricultural Total Productivity in Mainland China." *International Journal of Business Performance Management*.

Tan, Khee Giap, Nurina Merdikawati and Ramkishen S. Rajan. 2016c (forthcoming). "Agricultural Productivity in Indonesian Provinces." *Journal of Agribusiness in Developing and Emerging Economies*.

Thapa, Ganesh and Raghav Gaiha. 2011. "Smallholder Farming in Asia and the Pacific: Challenges and Opportunities." Paper presented at the International Fund for Agricultural Development (IFAD) Conference on New Directions for Smallholder Agriculture, Rome, January 24–25.

The Ministry of Agriculture and Rural Development of Vietnam and FAO. 2013. *Vietnam Country Programming Framework 2012–2016*. http://www.fao.org/fileadmin/user_upload/faoweb/vietnam/docs/CPF2012_2016.pdf.

Tiongco, Marites M. 2011. "Philippines: Food Security versus Agricultural Exports." In *Agricultural Development, Trade & Regional Cooperation in Developing East Asia*, edited by Ponciano S. Intal, Jr., Sothea Oum and Mercy J.O. Simorangkir. Jakarta: Economic Research Institute for ASEAN and East Asia (ERIA).

Trethewie, Sally. 2013. "Is the ASEAN Plus Three Emergency Rice Reserve (APTERR) the Answer to Southeast Asia's Food Security Challenges?" NTS Alert No. AL13-02, RSIS Centre for Non-Traditional Security (NTS) Studies, S. Rajaratnam School of International Studies (RSIS), Singapore.

Umetsu, Chieko, Thamana Lekprichakul and Ujjayant Chakravorty. 2003. "Efficiency and technical change in the Philippine rice sector: a Malmquist total factor productivity analysis." *American Journal of Agricultural Economics* 85(4): 943–963.

United States Department of Agriculture Economic Research Service. 2014. *Agricultural Total Factor Productivity Growth Indices for Individual Countries, 1961–2011*. http://www.ers.usda.gov/data-products/international-agricultural-productivity.aspx.

Warr, Peter. 2011. *The effect of research on agricultural productivity in Indonesia*. Canberra: Australian Centre for International Agricultural Research.

Woods, Kevin. 2015. "CP maize contract farming in Shan State, Myanmar: a regional case of a place-based corporate agro-feed system." Paper presented at conference on Land Grabbing, Conflict and Agrarian-Environmental Transformations: Perspectives from East and Southeast Asia at Chiang Mai University, Thailand, 5–6 June.

World Bank. 2015. World Development Indicators. http://data.worldbank.org/indicator/ NV.AGR.TOTL.ZS.

Discussant Notes 3.1

Dr Roehlano Briones
Senior Research Fellow,
Philippine Institute for Development Studies
Philippines

So far we have heard researchers say, we do so-and-so, steps 1, 2, 3... but when it's the discussant's turn, we hear: "let's step back for a while." So — let's step back for a while. First some preliminaries. Actually, I had a number of technical considerations I wanted to take up based on previous versions, but these seem to be mostly addressed in the presentation today. A few remain:

- Regarding the summation of output, it does not really make sense to add up sugar cane, rice etc. in quantities. What should be used is value of production and fixed at some base year prices.
- Regarding summing up of trade at the ASEAN level: it is more useful to look at net trade, so ASEAN as a whole. I suspect the data on the slide is a simple sum of ASEAN imports which is why palm oil is reported as so large.

Moving on to more conceptual issues. Consider labour productivity: this indicator is a function not just of the numerator (agricultural output), but also the denominator (number of workers). Where labour productivity is stagnant, it may mean agricultural workers are stuck in agriculture for whatever reason. However, if their moving out can be facilitated, e.g. with easy availability of jobs outside agriculture, then that could be a way to accelerate growth of labour productivity of agriculture.

That said, the focus is more on the numerator, which is output. Here we may consider the issue of productivity. The way to understand the usefulness of productivity is that it illuminates the pace and components of growth, i.e. it is a kind of growth accounting. I understand the study's particular methodology does not adopt growth accounting framework, but I adopt the framework as it is generally applicable and useful for deriving policy implications.

Accordingly: output growth can arise from growth of inputs, or from growth in productivity. If growth is mostly or entirely due to growth of inputs, then it is not likely to be sustainable in some sense. However, if growth is based on productivity, that might be sustainable, as it essentially depends on knowledge which can in principle expand without limits. That is, humans are becoming cleverer and cleverer in growing our food and agriculture products.

Considering the level of growth: acceleration of growth can be readily done in the short run expanding the land area, adding more workers here, adding machines, fertilisers — in short, by boosting inputs and factors of production. However, in the long run, sustained growth, especially in a world which (it has been mentioned before) will require 70% more food by 2050, must ultimately rely on growth of productivity.

Such growth of productivity is ideally independent of inputs. This rules out common measures which are partial productivity measures, i.e. yield — for expansion in yield may be due to fertilizer usage. The better measure is therefore TFP. The indicator may be understood in relation to the production frontier of a farm, where actual production may fall below the frontier. Then TFP change may be decomposed into: a movement towards the frontier or an improvement in technical efficiency; and a shift in the frontier itself or an improvement in technology, or technical progress. Explaining such movements is difficult — for as stated in the presentation, TFP is that component of growth not explained by changes in inputs — so in principle it could be anything other than those changes. Still it seems worthwhile to make the attempt.

As for technical progress: some of study's discussion on policy implication, regarding R&D and so on, are well taken. Technical efficiency is trickier: what prevents farmers from reaching the production frontier? This is a very interesting question. Are they nuts? Are they ignorant? Are they constrained in some sense? It seems that for highly aggregative exercise as is done in the study, i.e. the entire agriculture sector summed up to a single output, and inputs likewise aggregated as in the FAO data set, findings about technical inefficiency, the problem of omitted variables is surely critical. For instance, there is no variable for soil quality. I suspect a lot of inefficiency is due to unaccounted constraints, rather than say irrational or ignorant behaviour of farmers. However this objection is not specific to this study, but all such studies using similar methodology and data set.

Moving on to more policy-oriented comments: a previous discussant has highlighted the R&D. I would like to highlight another matter of perhaps equal importance, which is somewhat less emphasised even in the presentation. This is the role of the extension system. Extension is a service that raises technical efficiency, in case farmers are already deploying a new technology (e.g. the most recent line of hybrid rice released by the national breeding centre) but are not applying the right inputs; it can also raise the production frontier, in case farmers are yet to be updated with new technologies.

In Vietnam, according to Bo (2012), the extension systems were established fairly recently, so basically before farmers were on their own to find out new technologies. However, even with that, it is still a national governance system. It is quite understaffed. The FAO standard is 7.5 workers per 10,000 farming population, but in fact there are only four per 10,000 households. So, it is seriously underfunded. Moreover the extension budget is only two dollars per household per year. Based on quality indicators, the extension system in Vietnam, according to that author, is still very top-down in terms of approach. There is little feedback that goes back to the extension system, to check whether the introduced technologies are really useful.

Moving to the Philippines, as described in Cardenas *et al.* (2012), there was a big decentralisation of the agricultural extension system in 1991. A near-universal criticism has been that such a move has led to a weakening of the system. Of course this is not the case for every local government unit (LGU). There are some LGUs, especially in the city level, where the extension system actually improved, with better funding and staffing, and active projects. Such LGUs have much more into their agriculture sector post-decentralisation, compared to pre-decentralisation resources from the national government. However these are selected cases; by and large, decentralization has not worked. Consider the perennial problem of understaffing: for the smaller municipalities, there is just one extension worker; following the FAO benchmark, there should be at least three. Given the usual size of even a small but rural-based municipality. Especially for coastal communities, there should be an agronomist and a fisheries/aquaculture graduate; there should be a horticulturist as well, especially for LGUs with uplands within their jurisdictions. However, in many cases, the agronomist is also needed to serve the fishing communities, which is not really suitable. The reality is that many LGUs do not prioritise agriculture. Agriculture extension workers are often assigned with some other duties, such as on detail to the planning office or office of LGU chief executive.

The Philippine agricultural extension system can be made more effective by providing more resources to local governments, leveraging technologies, facilities, and staff in state university and colleges, and by increasing reliance on information technologies to deliver services. Again, similar to the comments made for Vietnam, a more feedback oriented, community based, participative type of extension service is needed; this enables farmers to quickly learn and master new technologies that they really need, and is most suitable for their own agro-climatic conditions.

REFERENCES

Bo, Nguyen Van. 2012. "Agricultural extension in Vietnam: its roles, problems, and opportunities." Background Paper, Beijing Roundtable 2012, Syngenta Foundation for Sustainable Agriculture.

Cardenas, V., M. T. Velasco, M. Magsino, and E. Sulabo. 2012. "Institutional partnership for extension, education, training, and communication." In Rola, A., J. Hernandez, J. Medina, M. Paunlagui, and L. Velaco, eds. *Partnership for Food Security.* Department of Agriculture, Bureau of Agricultural Research, University of the Philippines Los Baños, Los Baños, Laguna, 91–108.

Discussant Notes 3.2

Dr Bustanul Arifin

Professor, University of Lampung and
Board of Founder/Senior Economist,
Institute for Development of Economics and Finance
Indonesia

What we are trying to contribute here is that our comments are to enrich the arguments in the papers. Ladies and gentlemen, I would like to congratulate the team, everybody actually. Firstly, there is no surprise, as Professor Ramkishen said earlier about the issues. Interestingly, your TFP results are comparable to the previous TFP studies, even though the data sets are quite different. That's very important. Now, how can you contribute to the literature and the policy process on this kind of thing? I think the explanation matters. Maybe you could explain more on the composing of the index, but I think there are literatures about that.

The second point is on the quality of the data. As we have mentioned earlier, the data set you have is quite generalised, I think. For example, if you have labour, it is just simply labour. There is no quality, decomposition of education and so on. We want to have some explanation on the labour productivity. That is what we called limitation of the data set you use. Limitation does not necessarily become a weakness. It is only meant to warn the future researchers who want to do that, to accommodate or to improve this kind of data.

For example, we talked about Malaysia. Malaysia did quite well, because Malaysia did very well on human investment during the 1980s and 1990s. Now the results are announced. I hope I am correct. They did very well and reformed the whole education system. The results can be seen 20 years later. So their labour productivity stays very high, which is not surprising. The second, the TFP in Malaysia explains more about the technological changes, instead of technological efficiency. Why is that? Again, it is about the similar explanation. Malaysia did very well in investing in R&D. I have some figures at home, where the percentage of R&D fund of GDP is 0.08% in Indonesia. It is not even 0.1%. It is very small. I myself and some people were appointed as members of the National Innovation Council. We try to improve that, but we are not very successful. I could say that, because the whole arrangement of budget allocation is a totally different story after the regional autonomy. That is a big challenge. I think if you go to East Java or Maluku, the whole arrangement about the local budgets are really different, which affects the Indonesian case. That is why our technological improvements are quite different. I did a study on decomposing the different paths of historical agriculture development. If you have data on that and try to decompose, you will see the story could be different. Maybe we can find somebody else to do that, to decompose the paths into two periods with the point of the year 2000 or with the point of the economic crisis. The story could be different.

One last point about institution. We are really fighting very hard to put R&D as a kind of elite group of the country. Again, this can be linked to the competitiveness study we have discussed earlier, it is about innovation. There are two entry points. One is food security. Of course there is no question about that, with the whole dimension. The second point is about the competitiveness of the agriculture business, agriculture products in general. Malaysia did very well, as we feel, Malaysia is consistent in connecting the up-stream and down-stream. That's very important, but we in Indonesia, are not connecting very well. Probably, administratively, Malaysia has a specific minister of primary industry. However, we do not have. What we have is still under the traditional name of agriculture. It looks like that the new government are trying to reshuffling this kind of thing which could influence the whole setting of R&D. It is not decided until 20th of October. There will be a talk — Indonesia will divide the ministry of education. Basic education and middle education will form one part, higher education will be separated and connected with the ministry of research and technology. There could be differences in terms of R&D, I hope. Let's see what will happen. We are confident that technological change will contribute more to the TFP.

Discussant Notes 3.3

Dr Waleerat Suphannachart

Assistant Professor,
Department of Agricultural and Resource Economics
Kasetsart University, Thailand

This is a review of the study entitled "Agricultural Productivity for Cambodia, Laos, Myanmar, Vietnam, Malaysia, Indonesia, Thailand and Philippines," presented at the ACI's Review Seminar on August 26th, 2014. The study aims to measure and examine productivity growth for the agriculture sector in ASEAN-8 countries during 2000–2011. Investigating agricultural productivity, particularly TFP, has long been a central focus of many studies as it has important implications on sustaining agricultural growth and competitiveness, alleviating poverty and improving standard of living. Nonetheless, the calculation and interpretation of TFP has long been a debating issue and it is really a matter of empirical study. So this is an important study that provides empirical evidence on the TFP growth for ASEAN using relatively new data set. The results can be beneficial to policy makers in developing policy tools that could help stimulating production potentials of the agricultural sectors and strengthening ASEAN's attractiveness as a promising food basket.

Methodology

The study employs DEA-based Malmquist productivity index which is a non-parametric method that can be decomposed into two components: change in efficiency (TEC) and technological change (TC). This method serves the purpose of the study and the available data. However, it should be noted that agriculture involves lots of noises or uncontrollable factors and the nonparametric method does not take into account these noises. As TFP growth is conceptually represented by a residual of output growth that cannot be explained by the conventional input growth, the results are subject to the method of the study and empirical data. Careful measurement and interpretation should be undertaken. If possible, the same data set shall be applied to other measurement methods that are commonly used in existing studies, for instance the stochastic frontier and the growth accounting approaches, so that the results can be compared among different methods. In addition, the DEA-based Malmquist approach does not provide explanation on the TEC and TC and so what factors drive these two components of the TFP are still in doubt. Sources of TEC and TC should be investigated empirically so that more targeted policy can be developed. Alternatively, results from existing studies that investigated sources of TFP growth could be used to back up policy implications of this study.

Results on Labour Productivity

Although the study focuses on the TFP, it also examines the labour productivity. The study shows the average annual growth rate of labour productivity in Thailand during 2000–2011 to be an estimated 5.52%, ranked third among ASEAN-8. The finding is consistent with Suphannachart (2009), though using different data set and time period. Labour productivity in Thailand has been rising with positive average annual growth rate suggesting that farm income per person has increased over time. Evidence from the rice sector, which is the largest sector in Thai agriculture, shows that labour productivity is determined by farm machinery, farm size and rice research budget (Suphannachart, 2013). This is consistent with the theory that growth of labour productivity is determined by growth in stock of capital (such as farm machinery) and technological improvement which results from R&D investment. Nonetheless, it should be noted that the measures of labour productivity, both in this study and the previous study, do not take into account the role of foreign migrant workers. Migrant workers from neighbouring countries notably Myanmar, Cambodia and Laos have played an increasingly important role in Thai agriculture since 1996 (Chalamwong, 1996). The number of migrant workers should be accounted for when examining the agricultural productivity in Thailand.

Results on Total Factor Productivity

The results show that TFP of Thai agriculture during 2000–2011 is modest, ranked fourth among ASEAN-8. Technical efficiency change and technological change has contributed roughly evenly to the TFP growth. Regarding the sources of technical efficiency, Krasachat (2000) found that increasing farm size can increase efficiency levels and the availabilities of new land, the diversity of climate and natural resources could have had an influence on technical efficiency in Thai agricultural production. Regarding the sources of technological change, Suphannachart and Warr (2011, 2012) found that the major factors affecting the TFP in Thai agriculture are the public investment in agricultural research, international research spillovers, rural roads and case-specific factors. Factors positively affecting the technological change are the public investment in agricultural research and the spillovers of foreign research. The private sector and university research also play an increasingly important role. In addition, the previous studies confirm that it requires certain period of time for research investment to have a full impact on productivity and research lags depend on types of product. For the case of crop sector, it takes approximately 8 years for crop research to have a full and stable impact on the TFP (Suphannachart, 2009). As a result, the current slowdown in R&D investment could be translated into a decelerated growth of the TFP in the near future. Table 1 shows that Thailand is the only country experiencing a continuous decline in the public agricultural R&D intensity ratios over the period of 1996–2008, which could pose a threat to the long-term agricultural growth.

Policy Implications

Policy implications should be focusing on the factors driving the TFP growth. Regarding the role of R&D, it has been almost a census among existing studies that R&D plays an important role in driving agricultural TFP growth. In the case of Thai agriculture, Suphannachart and Warr (2011) have shown that the rate of returns to public investment in agricultural research is estimated at around 30%, implying underinvestment in research. Given the declining trend in the

Table 1: Public Agricultural R&D Intensity Ratios in ASEAN

	Public Agricultural R&D Spending as a Share of Agricultural GDP		
	1996	**2002**	**2008**
Cambodia	n.a.	n.a.	0.16
Myanmar	0.06	0.03	n.a.
Laos	n.a.	0.30	n.a.
Vietnam	0.09	0.17	0.17
Malaysia	1.15	1.92	1.05
Indonesia	0.37	0.28	0.31
Philippines	0.34	0.48	0.33
Thailand	0.69	0.51	0.32
Asia-Pacific average	0.62	0.70	0.63

Source: Beintema *et al.* (2012).

public agricultural research spending as a share of agricultural GDP (Table 1), Thailand should now invest more heavily in its own agricultural research capacity, especially in the area that adds value to its agricultural commodities. Improving research collaboration among the public sector, the private sector and foreign research institutes can also help strengthen R&D as a source of productivity growth. To sustain growth of the agricultural sector of Thailand as one of ASEAN food basket policy commitment to provide continued support on R&D investment should be emphasized. This commitment is also needed for the rest of the ASEAN countries.

REFERENCES

Beintema, Nienka, Gert-Jan Stads, Keith O. Fuglie and Paul Heisey. 2012. *ASTI Global Assessment of Agricultural R&D Spending: Developing Countries Accelerate Investment*. Washington DC: International Food Policy Research Institute.

Chalamwong, Yongyuth. 1996. An Estimate of Undocumented Migrant Workers in Thailand (in Thai). Thailand Development Research Institute, Bangkok.

Krasachat, Wirat. 2000. "Technical Efficiency in Thai Agriculture". The Chao Phraya Delta: Historical Development, Dynamics and Challenges of Thailand's Rice Bowl: Proceedings of the International Conference, Kasetsart University, Bangkok, December 12–15.

Suphannachart, Waleerat. 2013. *Labour productivity of Main and Second Rice Production in Thailand.* A study prepared for TDRI research project (in Thai).

Suphannachart, Waleerat and Peter Warr. 2012. "Total Factor Productivity in Thai Agriculture: Measurement and Determinants." In *Productivity Growth in Agriculture: An International Perspective*, eds. Keith O. Fuglie, Sun Ling Wang and V. Eldon Ball. Oxfordshire, UK: CAB International.

Suphannachart, Waleerat and Peter Warr. 2011. "Research and Productivity in Thai Agriculture." *Australian Journal of Agricultural and Resource Economics* 55(1): 35–52.

Suphannachart, Waleerat. 2009. *Research and Productivity in Thai Agriculture*. Ph.D. dissertation in Economics, Australian National University, Canberra.

Chapter 4

Drivers of Growth in the Travel and Tourism Industry in Malaysia, Thailand and Singapore: A Geweke Causality Analysis

4.1 INTRODUCTION

The significance of the travel and tourism industry as a driver of economic growth and development has long been established in the economic literature. As countries increasingly open up their borders to join the wave of global integration, several countries globally have seen their domestic travel and tourism industries flourish. The development of the travel and tourism industry has become a source of substantial foreign exchange revenue for several economies.

Further, countries have increasingly witnessed the significant contribution of travel and tourism industry to their overall socio-economic development through generating employment opportunities and denting poverty rates. Boosting the local economy through the development of domestic travel and tourism industry generates both direct and indirect employment as well as facilitates backward and forward linkages with other allied industries. This not only helps in terms of achieving greater economic growth, but also enhances a country to scale up its competitiveness in the global markets. Thus, investment in the development and promotion of tourism and travel industry remains a policy priority globally.[1]

Inbound tourism as a service traded across borders has emerged as one of the world's important trading categories, with tourism exports accounting for as much as 30% of the total exports of commercial services in the world. Visitor exports, which refer to expenditure by foreign visitors, stood at about 9% as a share of global exports in 2014.

The growing significance of tourism in the world can also be tracked by various factors including the growth in international tourist arrivals, the consequent growth in tourism revenues, its total contribution to foreign exchange earnings as well as gross output in a particular country. For instance, the travel and tourism industry globally generated 9% of Gross Domestic Product (GDP) and one in 11 jobs via direct, indirect and induced impacts for the global economy in 2014.[2] Further, the total number of international tourist arrivals globally was over one billion and tourists from emerging economies represented almost half of these international arrivals. International tourism receipts crossed US$1 trillion worldwide in 2014, registering an increase of over 3.7% in real terms from 2013.

---

[1] See Gopalan (2013).

[2] See United Nations World Tourism Organisation (UNWTO) (2015).

Among the various regions in the world, Europe and Asia–Pacific have consistently stood out as the world's most preferred destinations, with both regions constituting on average 60% of global tourist arrivals during the last decade or so. Within Asia–Pacific, the Southeast Asian region spearheads the growth in international tourist arrivals with an annual average growth of about 10% over the last decade, although it has slowed down after 2013, mainly driven by the political crisis in Thailand. In terms of tourism receipts from international visitor expenditure on accommodation, food and drink, shopping and entertainment, Europe accounted for over 40% of worldwide international tourism receipts and also witnessed the largest increase from 2013. Consistent with the trends on international tourist arrivals, Asia–Pacific cornered a 30% share in global tourism receipts which crossed US$1 trillion in 2014. Southeast Asian countries represented a third of this share within Asia–Pacific in 2014 in terms of international tourism receipts.

While the tourism and travel industry on the whole matters for several countries within the Southeast Asian region, three countries, namely Malaysia, Thailand and Singapore, stand out. The travel and tourism industry in Malaysia has undergone tremendous growth and development since 2000. In 2014, it represented about 6% of total direct employment in Malaysia and about 14% of indirect employment, underlining its importance as a crucial engine and source of growth for the country.[3] In terms of its contribution to the output of the economy, the travel and tourism industry directly contributed about 6% and indirectly represented over 15% of the Malaysia's GDP.

In terms of the share of international tourism receipts in the region, Thailand and Singapore represent over half of the international tourism receipts generated by the Southeast Asian region on average for the last decade. Specifically, in the last five years after the global financial crisis, the Southeast Asian region broadly has seen a resurgence in international tourism receipts, growing from about US$68 billion to nearly US$110 billion in 2014, which translates into a jump from about 25–30% of the international tourism receipts in less than five years. This growth was spearheaded by Thailand and Singapore, with the sum of international tourism receipts rising from US$34 billion to about US$58 billion between 2010 and 2014, which represented an average consistent share of over 50% of receipts generated by Southeast Asia.[4]

While, in general, Thailand's tourism industry has been recognised for its resilience even during difficult times like the global financial crisis, the 2014 political turmoil in Thailand appeared to have caused a severe setback to the tourism industry.[5] This is likely to pose some macroeconomic challenges to Thailand's economy, considering that the economy is heavily reliant on the tourism industry for both as a source of exports and economic growth.

Further, it is also interesting and important to note that relative tourism growth in Singapore tends to exhibit a highly positive correlation with that of Thailand (see Figure 4.1). This implies that the growth of tourism industry, proxied by international tourist arrivals in Thailand, tends to co-move with that of Singapore, underlining the need to examine the drivers of tourism in both countries together.[6] Further, we believe that this strong correlation, exhibited by both the countries for a sustained period of time, is another reason to believe that the causal factors

[3]See World Travel & Tourism Council (2014).

[4]See United Nations World Tourism organisation (UNWTO) (2015).

[5]See Credit Suisse (2014).

[6]See Credit Suisse (2014, Figure 4.1).

Figure 4.1: Growth of Visitor Arrivals in Singapore and Thailand, 2007–2013

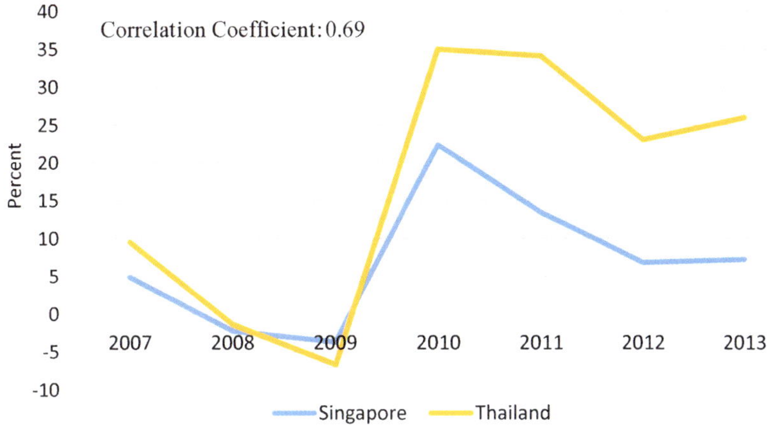

Source: World Bank.

determining the growth of tourism industry in both Thailand and Singapore could perhaps be similar. Finally, it is also worth emphasising that given the significant share of tourism revenues that these two countries together represent in the Southeast Asian region, understanding the causal determinants of tourism growth in these economies will prove vital for policy makers in this region to ensure that the industry continues to remain robust.

The chapter is interested in empirically understanding the causal factors that drive the growth of the tourism industry. Specifically, the chapter attempts to provide an empirical explanation for the sources of international tourist arrivals in Malaysia, Thailand and Singapore by identifying the key causal determinants of the growth of the travel and tourism industry. To do this, the chapter employs a novel empirical approach, the Geweke causality analysis, which helps us understand and measure the linear dependence and feedback between multiple time series. To that effect, we performed both a bi-variate as well as multi-variate causality analysis. The unidirectional causality analysis that is popularly employed in the literature does not investigate the degree of dependence or the extent of various kinds of feedback between different time series variables.[7] However, the method suggested by Geweke (1982) overcomes this problem and helps measure the linear dependence and feedback between multiple time series variables. In essence, the linear dependence is defined as "the sum of the measure of linear feedback from the first series to the second, linear feedback from the second to the first, and instantaneous linear feedback."[8] The direction of causality and the interplay of the variables can be examined in a more detailed manner with a reduced form quantitative framework under the Geweke causality analysis.[9]

The remainder of the chapter is organised as follows. Section 4.2 provides a brief background of the significance of the travel and tourism industry in Malaysia, Thailand and Singapore.

[7]See Calderón and Liu (2003).

[8]See Geweke (1982).

[9]For a more detailed treatment and discussion of causality analysis, see Geweke (1982) and Granger (1988).

Section 4.3 provides an overview of the related literature which forms the basis for developing empirically testable hypotheses. Section 4.4 presents the empirical methodology and data, followed by the discussion of the empirical findings in Section 4.5. Finally, Section 4.6 concludes with policy implications for Malaysia, Thailand and Singapore.

4.2 BACKGROUND OF THE TRAVEL AND TOURISM INDUSTRY IN MALAYSIA, THAILAND AND SINGAPORE

Tourism is one of the world's largest industries and one of its fastest growing economic sectors in Southeast Asian countries, especially Malaysia, Thailand and Singapore. In many countries, tourism and travel has been promoted as an integral part of the national development strategies for decades. Given the high degree of regional integration among the Southeast Asian countries in terms of trade and investment, the region on the whole could also benefit from the various inter-linkages and spillover effects from the national strategies that are taken by various countries to boost their domestic tourism and travel industry. With the establishment of the ASEAN Economic Community (AEC) in December 2015,[10] it is imperative that countries in the region continue to focus and develop plans to bolster their travel and tourism industry that will distribute the benefits of economic growth to the region.

4.2.1 Malaysia

Malaysia, as per the statistics produced by the World Travel and Tourism Council (WTTC) in 2014, was ranked 26th out of the 184 countries in the world in terms of the relative importance of the contribution of travel and tourism to the country's output 2014.[11] Malaysia has a long history of promoting the travel and tourism industry and the government has played a very active role in the process by adopting and formulating several laws, providing the appropriate institutional and legal framework to ensure sustainable tourism.

Though Malaysia's tourism efforts go back to the 1970s, given that the Malaysian economy was largely an exporter of primary commodities in the 1970s, the tourism industry was rather underdeveloped at the time. With the establishment of the Tourism Development Corporation of Malaysia (TDC) in 1972, the country took a major step in the right direction to make the TDC the official development authority to promote tourism and travel. However, the relatively modest performance of the TDC to exploit the potential of the tourism sector left Malaysia a largely unexplored and unknown destination for tourism in the Southeast Asian region relative to other touristy places such as Thailand or Indonesia.[12]

The real tipping point in Malaysia's history, specific to the tourism industry came about in the early 1980s when there was a global tourism boom which coincided with the collapse of primary commodity prices during the same time. This paved the way for the Malaysian policy makers to

---

[10] See Chapter 1 of this book for an in-depth analysis of the AEC.

[11] See WTTC (2015).

[12] See Mosbah and Al Khuja (2014).

recognise the immense potential of the tourism sector as a tool to foster economic and social development, and in promoting greater national as well as regional integration.[13]

The country started taking an active interest in the development of the tourism sector since then through the drafting and enacting of the five-year plans that placed the promotion of the tourism and travel industry as a policy priority. These five-year plans were also concomitantly supplemented with a series of policy initiatives such as the National Tourism Policy (NTP) that was enacted in 1992, the National Ecotourism Plan (NEP) in 1996 and the most recent being Malaysian Tourism Transformation Programme (MTTP) in 2010. All these policy initiatives were meant to tap into the unexploited and unexplored endowments as well as the potential that the home grown tourism and allied industries possess, which will provide a fillip to greater and sustained economic growth.

In light of all the policy discussions in Malaysia about the need for the country to develop a robust and strong services sector, which would in turn facilitate the country's transition from the middle-income to the high-income status, the Malaysian government launched the Economic Transformation Programme (ETP) in 2010. One of the integral parts of this initiative pertains to the tourism sector.

The travel and tourism industry have been explicitly recognised as both a National Key Economic Area (NKEA) and a Strategic Reform Initiative (SRI) contributing toward this economic transformation, in the government's New Economic Model (NEM) of 2010 and the Tenth Malaysia Plan.[14] Under these overarching national development plans, which have identified the industry as an "economic sweet spot," the government has put forward aspirational plans that include: bolstering the importance of eco-tourism; developing urban tourism, tourism clusters and new tourism products; and investing in the industry and related infrastructure. These efforts to develop the domestic tourism infrastructure are expected to enhance the country's competitiveness relative to other neighbouring countries in the region, which could make Malaysia a more attractive tourist destination in the region.

Concurrently, the Malaysia Tourism Transformation Programme (MTTP) was formulated to tap the growth potential of tourism sector. The MTTP aims to attract 36 million international tourists with MYR168 billion in revenue by 2020, which is three times the amount of its present foreign exchange earnings.[15] The MTTP also aims to engage in capacity building domestically so as to enhance business tourism.

Furthermore, a report published by the ETP on tourism in 2012 reveals that there will be renewed focus on alleviating domestic capacity constraints in infrastructure by channelling more government expenditure to enhance the quality of tourism related infrastructure. In addition, the report points out that the priority of the country moving forward would be to project Malaysia as a global biodiversity hub by developing eco-nature resorts, as well as creating dedicated entertainment zones and improving the quantity and quality of hotels, which will establish Malaysia as a leading destination for business tourism.

---

[13] See Mosbah and Al Khuja (2014) and Nanthakumar *et al.* (2007).

[14] See National Economic Advisory Council (2010) and Economic Planning Unit (2010).

[15] See Ministry of Tourism and Culture of Malaysia (2012).

In a speech dated 27th August 2015, Datuk Seri Najib Razak, the Prime Minister of Malaysia, was quoted saying:

> "Our renowned Malaysian hospitality and culture of service, our rich national heritage, our cultural diversity and of course, now more than ever, the value for money that foreign tourists will experience in Malaysia compared with many other destinations because of the exchange rates. All of these are huge selling points that need to be amplified and communicated across all channels to drive up tourism numbers" [16]

4.2.2 Thailand

To Thailand, the importance of tourism is extensively recognised. As per the rankings produced by the WTTC, Thailand was ranked 17th out of the 184 countries in the world in terms of the absolute size and 35th in terms of relative size measured by contribution of the tourism and travel industry to the country's output in 2013.[17]

Thailand has also experienced impressive growth in terms of the direct contribution of its travel and tourism industry to the output. As Figure 4.2 shows, Thailand's direct contribution of its travel and tourism industry to the country's output is significant, with the average contribution around 7.5% or so between 2005 and 2013, with the share peaking at around 9% in 2013, on par with the world average.

As for the number of international tourist arrivals, shown in Figure 4.3, Thailand has experienced constant increase in international tourist arrivals, which consequently lead to the growing awareness of the importance of the tourism industry. This has also consequently led to a

Figure 4.2: Direct Contribution of Tourism and Travel as a Percentage of GDP in Malaysia, Singapore and Thailand, 2005–2013

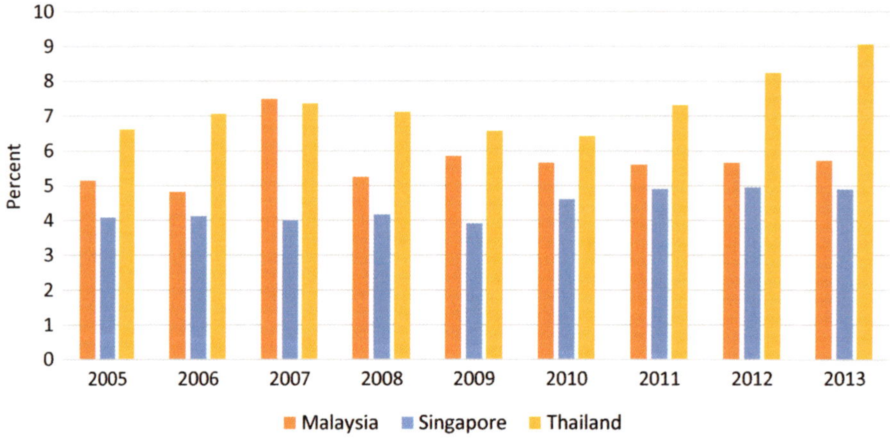

Source: WTTC Reports, various years.

[16] See Dhillon (2015).

[17] See World Travel and Tourism Council (WTTC) (2015).

Figure 4.3: Number of International Tourist Arrivals (in Millions) to East Asia and Pacific, Malaysia, Singapore and Thailand, 2000–2013

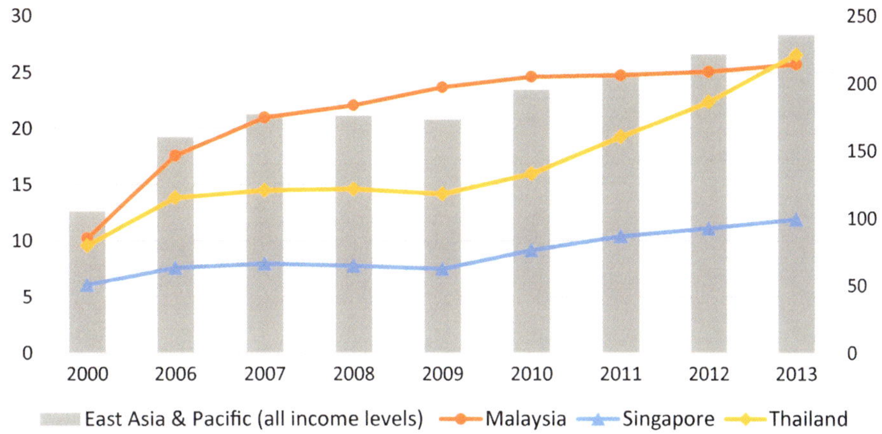

Source: World Bank.

Figure 4.4: International Tourism Receipts as a Percentage of Total Exports in East Asia and Pacific, Malaysia, Singapore and Thailand, 2006–2013

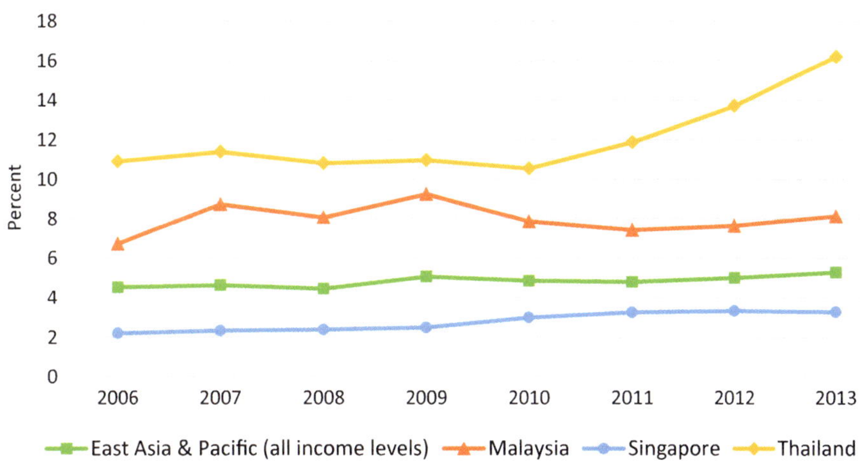

Source: World Bank.

significant rise in tourism receipts both as a share of total exports (see Figure 4.4) and in absolute terms (see Figure 4.5). Thus, inbound tourism as a service traded across borders has emerged as one of the world's important trading categories, with tourism receipts accounting for as much as 16% of the total exports for Thailand and averaged about 12% between 2006 and 2013.

Several initiatives have been taken to promote the development of the domestic tourism and travel industry in Thailand. The most recent and significant policy initiatives taken by Thailand for instance were the Tenth Economic and Social Development Plan, covering the period of 2007–2011, which included a master plan to jumpstart the economy through the active promotion of domestic and international tourism, more specifically to enhance the attractiveness of the

Figure 4.5: International Tourism Receipts (in Current US$ Billion) in East Asia and Pacific, Malaysia, Singapore and Thailand, 2000–2013

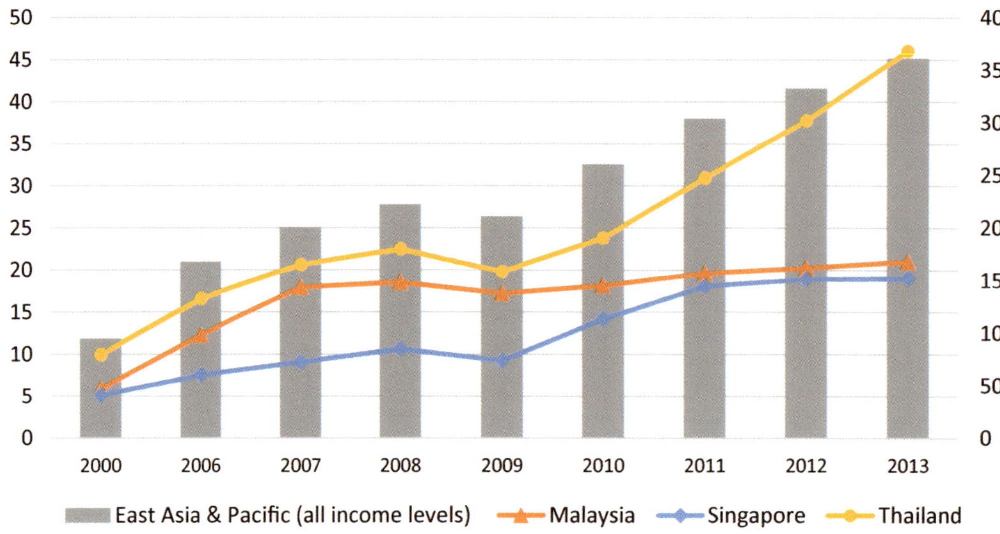

Source: World Bank.

country as a preferred destination for international tourists. Yet another policy initiative comes from the national tourism development plan covering the subsequent period of 2012–2016, which has an explicit focus of turning Thailand into one of the top five destinations in Asia.[18] In addition, several tourism promotional campaigns have been launched with the most recent one being the plan by the Tourism Authority of Thailand (TAT) to promote the country as a "Quality Leisure Destination through Thainess," which marks a new era for Thai tourism. As noted by the acting Governor of TAT:

> "This plan marks the opening of a new chapter for Thai tourism. It is based on the fact that there is more than adequate accessibility to Thailand via excellent air, road and sea connections, as well as smooth facilitation that allows visa-free or visa-on-arrival access for citizens from 68 countries and territories .... we can confidently say that the era of promoting "quantity" is over. The era of promoting "quality" has begun."[19]

4.2.3 Singapore

According to the WTTC, Singapore was ranked 35th out of the 184 countries in the world in terms of the absolute size and 83rd in terms of relative size measured by contribution of the tourism and travel industry to the country's output in 2013.[20] Singapore has also registered impressive growth rates in terms of the direct contribution of the travel and tourism industry to the

[18] See Board of Investment of Thailand (2011).
[19] See Tourism Authority of Thailand Newsroom (2015).
[20] See WTTC (2015).

output of the country. As Figure 4.2 shows, Singapore's direct contribution of its travel and tourism industry to the country's output has been consistently hovering around 5% on average between 2005 and 2013.

The importance of the tourism industry to Singapore has been a direct consequence of the flurry of international tourist arrivals which has been seen over the last decade, as shown in Figure 4.3. This has consequently led to a significant rise in tourism receipts both as a share of total exports (see Figure 4.4) and in absolute terms (see Figure 4.5). Thus, tourism receipts account for as much as 3.3% of the total exports for Singapore and averaged about 3% or so between 2006 and 2013 (see Figure 4.5).[21]

The development of the tourism and travel industry has always been a prominent feature of the policy agenda of Singapore, all the way from the 1970s, in order to make it a top tourist destination in the region. Considering especially the fact that Singapore is a small city-state with relatively few natural resources compared to other countries in the region, the Singapore government duly recognised the potential that the travel and tourism industry can offer to the Singapore economy. The Singapore Tourism Board (STB) has been playing an instrumental role in promoting tourism through a combination of developing "garden attractions and modern hotels" and marketing the city state as "Instant Asia."[22]

Since the 1990s, the STB has elevated its tourism strategies to the next level by coming up with a master plan which accords primacy to tourism in the 21st century. The objectives are threefold: to multiply the tourism receipts, to double visitor arrivals and to create significant employment opportunities in the industry. Marking Singapore's 50th anniversary after independence, the Singapore government announced a host of policy initiatives to be undertaken by the STB, with the most prominent one being to ramp up global marketing efforts to boost Singapore's image. At the annual Tourism Industry Conference in 2015, the Singapore Minister for Trade and Industry Mr S. Iswaran was quoted saying:

> "Outbound travel to the Asia–Pacific is expected to continue to grow, so too intra-Asia travel, as Asian economies and disposable incomes rise. We must be ready to seize our share of this growth and, I would like highlight three ways we can do so: deepening innovation, boosting promotion and marketing and enhancing capabilities."[23]

It is clear from the foregoing discussion that all of the three countries, Malaysia, Thailand and Singapore, having well recognised the significance of the contribution of the tourism and travel industry to their respective economies, are putting in place a host of policy initiatives to boost the industry. This chapter thus seeks to empirically examine the causal factors that drive the

[21] Visitor exports refer to expenditures by foreign visitors which remains an important and growing share of direct contribution of the travel and tourism industry. The ability of the tourism industry to address current account imbalances of a country is also widely recognised in the literature and there are a number of studies that examine the importance of tourism through its implications for the balance of payments, especially in developing countries. See Malik *et al.* (2010) and Thano (2015).

[22] See Meng *et al.* (2013).

[23] See Kotwani and Ariffin (2015).

growth of the travel and tourism industry in these three countries as a contribution to the literature as well as policy discourse in the three countries of interest.

4.3 LITERATURE REVIEW AND HYPOTHESES

Most of the related literature explaining the economic significance of the travel and tourism industry examines its potential impact on poverty alleviation, especially in emerging and developing countries. For instance, a comprehensive survey study by the Overseas Development Institute (ODI) (2007) offers a useful conceptual framework to illustrate this link between tourism and poverty alleviation. As noted by ODI (2007), there are three major "pathways": namely "direct" "indirect," and "dynamic", through which tourism can impact poverty reduction.

The direct effects pertain to the incomes generated from tourism jobs. The indirect channel relates to the "tourism value chain," where the earnings are generated from allied sectors such as food or transportation industries. Studies suggest that inter-sectoral impacts are substantial and make up about 60–70% of earnings in the industry in developing economies, in addition to the direct effects of tourism.[24] Finally, the dynamic channel spans a wide spectrum as the development of the tourism industry in an economy can enhance the livelihoods of local households, improve the business environment for small enterprises and also ease the infrastructural bottlenecks in an economy, all of which can contribute to poverty reduction through growth.[25]

There is also a related and extensive literature dating back to the 1990s which has investigated the determinants of tourism demand. The literature points out that the specific variables that have been identified as plausible determinants of tourism demand in any country are real per capita income of the host countries, the (effective) exchange rate and relative prices prevailing in the host country as well as the quality of tourism infrastructure.[26]

These variables are generally thought to be significant in explaining tourism demand for various countries. For the case of Turkey for instance, one of the early studies on tourism demand by Uysal and Crompton (1984) shows that real per capita income in the host countries, relative exchange rates, transportation costs and promotional expenses drive international tourist arrivals. In similar vein, Onder *et al.* (2009) also find that the most significant variables explaining international tourist arrivals in Turkey were real exchange rates, per capita GDP of host as well as home countries and quality of transportation.

It is interesting how there is a clear convergence of the choice of explanatory variables and the results in various country studies trying to investigate the drivers of tourism demand are quite similar. As an illustration, a study by Song *et al.* (2003) for Hong Kong finds that the costs of tourism in the destination country contribute significantly to the demand for tourism in Hong Kong.

In a more recent paper, Culiuc (2014) estimated the impact of both supply and demand side determinants of tourism to find that tourism flows respond strongly to changes in the destination

---

[24] See Ashley *et al.* (2007).

[25] See Gopalan (2013).

[26] See Sheldon (1993), Nanthakumar *et al.* (2007), Tan *et al.* (2016a, forthcoming), (2016b, forthcoming).

country's real exchange rate. The paper also finds that tourism in smaller island countries is less sensitive to changes in the country's real exchange rate, although more sensitive to the quality of infrastructure and connectivity, specifically the introduction and removal of direct flights.

Beyond economics, tourism policy that is designed to market the distinctiveness of local traditions and culture can pave the way for better cultural exchanges between countries which can foster bilateral relations between countries. The literature also points out that, since tourism by its very nature involves and requires inter-country cooperation on many fronts, tourism policy assumes significance in order to shaping a country's engagement with the rest of the world, for instance Prideaux (2005).

The foregoing discussion leaves us with a broad idea about the variables highlighted in the literature as determinants of tourism demand. However, we go beyond just the related literature and also complement it with the thrust of the various national policy initiatives the governments of Malaysia, Thailand and Singapore have taken in order to identify the growth drivers of the travel and tourism industry.

To that end, we empirically try to understand the casual factors that affect the growth of travel and tourism industry in Thailand and Singapore. We do this by employing a novel empirical methodology, the Geweke causality analysis, to specifically analyse what factors, ranging from economic to infrastructure to environmental indicators, have a significant causal impact on the growth of the industry.

This chapter sets International Tourism Arrivals (ITA) as the dependent variable, which has been identified by the relevant literature as one of the most appropriate for analysing the industry.[27] The following hypotheses will be tested: *Government policy, increased purchasing power for international tourists, improvements in the environment and infrastructure, and safety and security measures are the causal factors driving the growth of international tourist arrivals into Malaysia, Thailand and Singapore.*

The independent and control variables that we propose to include in the model can be divided into several categories. First, government policy, which includes government expenditure as well as capital investment on the travel and tourism industry. Second, purchasing power indicators specifically refer to the prevailing cost dynamics in the countries by focusing on the consumer price index (CPI). Third, environment and infrastructure indicators, which include airport facilities, air quality and internet users. Fourth, safety and security measures, which include public security and safety and religious unrest indicators. In the next section we discuss the methodology employed and the data sources for the variables used in the chapter.

4.4 METHODOLOGY AND DATA

4.4.1 The Geweke Causality Analysis

The methodology presented here begins by describing the two methods used in the selection of explanatory variables, namely the utilisation of stepwise regression in the selection of the "best"

---

[27] See Lim (1997).

set of explanatory variables, and the use of factor analysis in the construction of differing policy environment. Finally, the Geweke causality analysis is discussed in its application to the research.

We utilised a stepwise regression in the selection of the "best" set of explanatory variables before we identify the variables that need to be tested for causal relationships. As the first step, the stepwise regression (forward selection) helped us choose the "best" set of explanatory variables by introducing one explanatory variable one at a time. The decision to keep or drop a variable was based on their contribution to the variance due to regression, based on the F-test[28].

Further to this, we also employed a factor analysis in the construction of a policy environment comprising several plausible independent variables that could possess significant causal power in explaining the growth of tourist arrivals. Factor analysis consists of an "array of structure-analysing procedures used to identify the interrelationships among a large set of observed variables and then, through data reduction, to group a smaller set of these variables into dimensions or factors that have common characteristics."[29] In this chapter we adopted a principle component analysis method for factor analysis.

Once the variables were identified, we employed a novel empirical framework, the Geweke causality analysis, to assess the causal factors determining the growth of the tourism industry in Malaysia, Thailand and Singapore. Geweke causality analysis is used as a tool to identify causal relationships in the field of economics and neuroscience. Some recent and prominent applications of Geweke causality to macroeconomic issues include Tan and Cheng (1995), Calderón and Liu (2003) and Aizenman and Noy (2006). While Tan and Cheng (1995) employed Geweke's approach to examine the causal nexus of money, output and prices in Malaysia, Calderón and Liu (2003) explored the direction of causality between financial development and ecnomic growth. Aizenman and Noy (2006) investigated the two-way linkages between foreign direct investment (FDI) and trade.

The causality and linear feedback between two linear systems were defined in Granger (1963) and Granger (1969), and Sims (1972) provides complementary tests for the existence of unidirectional causality. Although the determinants of single economic variable are likely to be multidimensional, most applications found in the literature are focused on the bivariate cases. Geweke (1982) developed the concept further by including the instantaneous (two-way) linear feedback between multiple time series, i.e. $F_{X,Y} = F_{X\rightarrow Y} + F_{Y\rightarrow X} + F_{X\cdot Y}$, which indicates that the measure of linear dependence, $F_{X,Y}$ is the sum of the measure of linear feedback from the first series to the second, $F_{X\rightarrow Y}$, linear feedback from the second to the first, $F_{Y\rightarrow X}$ and instantaneous linear feedback, $F_{X\cdot Y}$. The measures are non-negative and zero only when feedback (causality) of the relevant type is absent.

The multivariate causality test proposed by Geweke (1982) is a test between two vectors of the variables. The equivalence of linear dependence measures as proved in one of the theorems in Geweke (1982) provides us with an elegant way to conduct the multivariate test, which is as convenient as bivariate tests. Like Granger (1969), Sims (1972) and Geweke (1982), we focus

---

[28] See Gujarati (2004).

[29] See Pett, Lackey and Sullivan (2003, 4).

on wide-sense stationary and purely non-deterministic time series $X = \{x_t, t \text{ real}\}$. By wide-sense stationary, we infer that the mean of x_t exists and is independent of t, and for all t and s cov (x_t, x_{t+s}) exists and depends on but not on t. By purely non-deterministic, it presumes that the correlation of x_{t+p} and x_t vanishes as increases.

The idea of causality between multiple time series X and Y can be summarized as follows:

$$F_{X,Y} = F_{X \to Y} + F_{Y \to X} + F_{X \bullet Y} . \tag{1}$$

A non-deterministic and stationary multiple time series can be considered as follows:

$$Z_t = \sum_{s=1}^{\infty} B_S Z_{t-s} + \varepsilon_t , \tag{2}$$

Where ε_t is white noise and $\mathbf{z}_t$ can be partitioned into $k \times 1$ and $l \times 1$ subvectors X_t and Y_t .

The partition of vector can be shown diagrammatically as follows:

$$\mathbf{Z}_t = \begin{bmatrix} z_{1,1} & z_{1,2} & \cdots & z_{1,N} \\ z_{2,1} & z_{2,2} & \cdots & z_{2,N} \\ \vdots & \vdots & \ddots & \vdots \\ z_{k,1} & z_{k,2} & \cdots & z_{k,N} \\ & & & \\ z_{k+1,1} & z_{k+1,2} & \cdots & z_{k+1,N} \\ \vdots & \vdots & \ddots & \vdots \\ z_{k+l,1} & z_{k+l,2} & \cdots & z_{k+l,N} \end{bmatrix} = \begin{pmatrix} \mathbf{X}_t \\ \mathbf{Y}_t \end{pmatrix}.$$

Geweke (1982) showed a canonical form for the wide sense stationary time series $Z_t = (X_t, Y_t)$ is of the form:

$$x_t = \sum_{s=1}^{\infty} E_{1s} x_{t-s} + u_{1t} \quad \text{var}(u_{1t}) = \Sigma_1, \tag{3}$$

$$x_t = \sum_{s=1}^{\infty} E_{2s} x_{t-s} + \sum_{s=1}^{\infty} F_{2s} y_{t-s} + u_{2t} \quad \text{var}(u_{2t}) = \Sigma_2, \tag{4}$$

$$x_t = \sum_{s=1}^{\infty} E_{3s} x_{t-s} + \sum_{s=0}^{\infty} F_{3s} y_{t-s} + u_{3t} \quad \text{var}(u_{3t}) = \Sigma_3, \tag{5}$$

$$x_t = \sum_{s=1}^{\infty} E_{4s} x_{t-s} + \sum_{s=-\infty}^{\infty} F_{4s} y_{t-s} + u_{4t} \quad \text{var}(u_{4t}) = \Sigma_4, \tag{6}$$

$$y_t = \sum_{s=1}^{\infty} G_{1s} y_{t-s} + v_{1t} \quad \text{var}(v_{1t}) = T_1, \tag{7}$$

$$y_t = \sum_{s=1}^{\infty} G_{2s} y_{t-s} + \sum_{s=1}^{\infty} H_{2s} x_{t-s} + v_{2t} \quad \text{var}(v_{2t}) = T_2, \tag{8}$$

$$y_t = \sum_{s=1}^{\infty} G_{3s} y_{t-s} + \sum_{s=0}^{\infty} H_{3s} x_{t-s} + v_{3t} \quad \text{var}(v_{3t}) = T_3, \tag{9}$$

$$y_t = \sum_{s=1}^{\infty} G_{4s} y_{t-s} + \sum_{s=-\infty}^{\infty} H_{4s} x_{t-s} + v_{4t} \quad \text{var}(v_{4t}) = T_4. \tag{10}$$

Based on equation system (3), (4), (7) and (8), Geweke (1982) defined the measure of linear feedback from y to x as:

$$F_{Y \to X} = \ln(|\Sigma_1| / |\Sigma_2|). \tag{11}$$

The measure $F_{Y \to X}$ is always non-negative and takes the value of zero only if the linear feedback running from y to x is absent. Symmetrically, the measure of linear feedback from x to y is:

$$F_{X \to Y} = \ln(|T_1| / |T_2|) \tag{12}$$

and the measure of instantaneous feedback is:

$$F_{X \cdot Y} = \ln(|T_1| * |\Sigma_2|) / |Y|), \tag{13}$$

where

$$Y = \text{var}\begin{pmatrix} u_{2t} \\ v_{2t} \end{pmatrix} = \begin{bmatrix} \Sigma_2 & C \\ C' & T_2 \end{bmatrix}.$$

We can decompose the measure of linear dependence between any two groups of time series X and Y, $F_{X,Y}$, as the sum of the measure of linear feedback from the X to Y, $F_{X \to Y}$, linear feedback from the Y to X, $F_{X \to Y}$, and instantaneous linear feedback between the two series, $F_{X \cdot Y}$, as shown in equation (1). In the notion of equations (3)–(10), Geweke (1982) also proved that the following set of equations are equivalent:

$$F_{X,Y} = \ln(|\Sigma_1|/|\Sigma_4|) = \ln(|T_1|/|T_2|), \tag{14}$$

$$F_{X\to Y} = \ln(|\Sigma_3|/|\Sigma_4|) = \ln(|T_1|/|T_2|), \tag{15}$$

$$F_{Y\to X} = \ln(|\Sigma_1|/|\Sigma_2|) = \ln(|T_3|/|T_4|)F_{X\bullet Y} = \ln(|\Sigma_2|/|\Sigma_3|) = \ln(|T_2|/|T_3|). \tag{16}$$

Based on this equivalence, we can simplify the estimations to equations (3) to (6), and avoid the estimations (7) to (10) which require multivariate vector y as the dependent variables. Note that the error terms in nested estimation equations (3)–(6) should become smaller as we include more dependent variables. Therefore, we shall expect the ordering of variance as $\Sigma_1>\Sigma_2>\Sigma_3>\Sigma_4>$, which further implies that all the measures are non-negative and zero only when feedback (causality) of the relevant type is absent.

To ensure the white-noise property, we adopted the approach by Tan and Cheng (1995) and conducted an informal test by running the first-order autoregressive model on the residuals of each equation. Under the null hypothesis that there is no unidirectional causality running from y to x, i.e. if $\hat{F}_{Y\to X} = 0$, then $n\hat{F}_{Y\to X} \sim \chi^2(klp)$.

Similarly, it follows that if $\hat{F}_{X\to Y} = 0$, $n\hat{F}_{X\to Y} \sim \chi^2(klp)$ and if $\hat{F}_{X.Y} = 0$, $n\hat{F}_{X.Y} \sim \chi^2(kl)$. As these are nested hypotheses, all three restrictions could be tested simultaneously since if $\hat{F}_{X,Y} = 0$, then $n\hat{F}_{X,Y} \sim \chi^2[kl(2p+1)]$.

The 95% confidence interval (CI) could be calculated approximately as follows:

For $\hat{F}_{Y\to X} = 0$,

$$\left\{\left[\left(\hat{F}_{Y\to X} - \frac{klp-1}{3n}\right)^{1/2} \pm \frac{1.96}{\sqrt{n}}\right]^2 - \frac{2klp+1}{3n}\right\}. \tag{17}$$

For $\hat{F}_{X\to Y} = 0$,

$$\left\{\left[\left(\hat{F}_{X\to Y} - \frac{klp-1}{3n}\right)^{1/2} \pm \frac{1.96}{\sqrt{n}}\right]^2 - \frac{2klp+1}{3n}\right\}. \tag{18}$$

For , $\hat{F}_{X.Y} = 0$,

$$\left\{\left[\left(\hat{F}_{X\bullet Y} - \frac{klp-1}{3n}\right)^{1/2} \pm \frac{1.96}{\sqrt{n}}\right]^2 - \frac{2kl+1}{3n}\right\}. \tag{19}$$

For $\hat{F}_{X,Y} = 0$,

$$\left\{\left[\left(\hat{F}_{X\bullet Y} - \frac{kl(2p+1)-1}{3n}\right)^{1/2} \pm \frac{1.96}{\sqrt{n}}\right]^2 - \frac{2kl(2p+1)+1}{3n}\right\}. \tag{20}$$

4.4.2 Data

The data used in this research was collected from several key sources and is a combination of both hard and survey data. This sub-section proceeds by first describing the dependent variable and then providing a description of the independent variables adopted in the analysis.

The dependent variable is the number of ITA which have been recognised as the best proxy capturing the growth of travel and tourism industry in a country. The data was obtained from the World Bank and captures the total amount of international inbound visitors into Malaysia, Thailand and Singapore between 2000 and 2012.

As far as the independent variables are concerned, we proposed four major environments and respective indicators, which represent the various dimensions of the economy that could affect the growth of the travel and tourism industry. We assessed their relevance in causally explaining the travel and tourism industry's growth in Malaysia, Thailand and Singapore, as summarised in Table 4.1.

In what follows, we will explain the master set of independent variables that we have considered for our study, although the final choices of the variables used for Geweke causality analysis will be a function of the results from the stepwise regressions.

The independent variables under the Government Policy Environment (GPE) include Government Expenditure on Tourism as a percentage of GDP and Capital Investment on Tourism as a percentage of GDP, which were both collected from the WTTC. The Government Expenditure on Tourism (GET) indicator describes the government's spending as a percentage

Table 4.1: Potential Drivers of Growth in the Travel and Tourism Industry — Four Environments and Nine Indicators

Environment	Indicator
Environment and Infrastructure	Airport Facilities Air Quality Internet Users
Purchasing Power	Exchange Rate Consumer Price Index
Government Policy	Government Expenditure on Tourism (% of GDP) Capital Investment on Tourism (% of GDP)
Safety and Security	Public Security and Safety Religious Unrest

Source: Asia Competitiveness Institute.

of total GDP on travel and tourism services directly linked to visitors and Capital Investment on Tourism (CI) includes spending by all sectors involved in the travel and tourism industry.

The variables under the Environment and Infrastructure Environment (EIE) include Airport Facilities, Air Quality and Internet Users. Airport Facilities (AF) was obtained from the World Economic Forum's (WEF) Global Competitiveness Index's survey data which asked the respondents to rate the quality of the country's air transportation infrastructure with 1 being "extremely underdeveloped" and 7 being "extensive and efficient by international standards" and was utilised to examine the quality of travel infrastructure. The data on Air Quality (AQ) was obtained from the Global Market Information Database (GMID) and examines the amount of fine particulates, micrograms per cubic metre, that are in the air which are capable of causing health damage. The final indicator in this environment, Internet Users (IU) indicates the number of individuals with access to the Internet per 100 people and was collected from the World Bank.

The independent variables considered under the Purchasing Power Environment (PPE) include both Exchange Rate and CPI. The Exchange Rate (ER) data was collected from the International Monetary Fund (IMF) Economic Outlook Database. The CPI data was collected from the World Bank and the indicator describes the changes in the cost to the average consumer of purchasing a specific basket of goods and services on a yearly basis.

Finally, we also emphasise the importance of a volatile and unstable security environment in understanding the growth in tourist arrivals. As discussed earlier, this issue has been a major concern for Thailand and given that the fortunes of Thailand and Singapore are tied together when it comes to political and security related instability affecting their respective tourism industry, we construct the Safety and Security Environment (SSE), which consists of two independent variables, Public Security and Safety and Religious Unrest. Both the Public Security and Safety (PSS) and Religious Unrest (RU) indicators were collected from the WEF's Global Competitiveness Index and are survey data to represent the security related explanatory variables. The PSS indicator is a composite of two survey questions combined to access the state of security and safety in the country; the first question asked to what extent does organised crime impose costs on businesses in the country with 1 being "to a great extent" and 7 being "not at all," and the second question asked to what extent can police services be relied upon to enforce law and order with 1 being "cannot be relied upon" and 7 being "can be completely relied upon." The RU indicator asked respondents to what extent does the threat of terrorism impose costs on businesses in the country with 1 being "to a great extent" and 7 being "not at all."[30]

All variables used in empirical estimations were subjected to stationary tests and those that were found to have unit roots were transformed to a stationary series before being used in the causality analysis. The sample of our study spans from 2000 to 2012 and due to limited observations for the period of 2000–2012, we converted the annual data into quarterly data using quadratic-sum/average method.

[30]See World Economic Forum (2014).

4.5 EMPIRICAL FINDINGS

4.5.1 Geweke Causality Analysis for Malaysia

As discussed, we employed the Geweke causality analysis to identify the causal determinants of the growth of travel and tourism industry. Before we undertook the causality analysis, we ran a stepwise regression in order to narrow down the "best" set of explanatory variables among the list of nine variables we identified earlier.

Table 4.2 presents the results of the stepwise regression for Malaysia. Amongst the nine independent variables, GET, CI, AF and RU appear to be significant for Malaysia, which were used for the Geweke analysis.

As can be inferred from the bi-directional causality results shown in Table 4.3,[31] there exists a significant bi-directional association between ITA and GET at the 1% level of significance and the total causality is primarily made up of the instantaneous association. In examining the relationship between ITA and AF, there is evidence of not only a strong bi-directional association between the dependent and explanatory variable at the 1% level of significance, but also strong feedback from AF to ITA at the 1% level of significance. While there is a significant bi-directional association between the ITA and AF, the feedback from AF to ITA underlines the important role that AF can have in contributing to increased ITA. In the case of ITA and RU, the results highlight that there is a bi-directional association between ITA and RU at the 1% level of significance and a weak feedback from RU to ITA at the 10% level of significance. The bi-directional association between ITA and CI is not significant.

Analysing the results in Table 4.4 of the multivariate causality between ITA on the x vector and GET, CI, AF and RU on the y vector highlights that there is a strong association between x

Table 4.2: Summary of Linear Stepwise Regression Analysis on Tourism Development in Malaysia

Dependent Variable: International Tourism Arrivals	$y_i = \partial_i + \sum_{k=1}^{K} \beta_k x_{ik} + \epsilon_i$			**Number of obs = 51** **F = 6.84** **Prob> F = 0.0002** **Adjusted R^2 = 0.3186**
Independent Variables	β	**Standard Error**	***t* Statistics**	***p*-Value**
Airport Facilities (AF)	−757593.2***	211494.7	−3.58	0.001
Capital Investment on Tourism (% of GDP) (CI)	75256.91***	19438.2	3.87	0.000
Government Expenditure on Tourism (% of GDP) (GET)	1295848**	490886.8	2.64	0.011
Religious Unrest (RU)	283919.4***	99380.75	2.86	0.006
Constant	215294.6	1419340	0.15	0.880

Note: *, ** and *** denote 10%, 5% and 1% level of significance, respectively.
Source: Asia Competitiveness Institute.

[31] For the subsequent Geweke causality tables, figures of $F_{x,y}$ may not be equal to the sum of figures of $F_{x\to y}$, $F_{y\to x}$ and $F_{x\cdot y}$ due to round-up.

Table 4.3: Estimated Measures of Bi-Directional Feedback between International Tourism Arrivals (ITA), Capital Investment on Tourism (% of GDP) (CI), Government Expenditure on Tourism (% of GDP) (GET), Airport Facilities (AF) and Religious Unrest (RU) for Malaysia, 2000–2012[a,b]

Economic Aggregates		$H_0\ (F_{x,y} = F_{x\to y} + F_{y\to x} + F_{x\cdot y})$			
x	y	$F_{x,y}$	$F_{x\to y}$	$F_{y\to x}$	$F_{x\cdot y}$
ITA	CI	0.0407 (0.8611)	0.0095 (0.8009)	0.0284 (0.5131)	0.0028 (0.7146)
CI	ITA	0.0167 (0.9779)	0.0053 (0.8827)	0.0086 (0.8173)	0.0028 (0.7146)
ITA	GET	0.5846*** (0.0000)	0.0407 (0.1578)	0.0069 (0.5616)	0.5370*** (0.0000)
GET	ITA	0.5938*** (0.0000)	0.0379 (0.1730)	0.0190 (0.3350)	0.5370*** (0.0000)
ITA	AF	1.4886*** (0.0000)	0.0010 (0.9765)	0.3063*** (0.0007)	1.1813*** (0.0000)
AF	ITA	1.3966*** (0.0000)	0.1245* (0.0536)	0.0908 (0.1184)	1.1813*** (0.0000)
ITA	RU	0.7741*** (0.0000)	0.0031 (0.9301)	0.1300* (0.0472)	0.6410*** (0.0000)
RU	ITA	0.6983*** (0.0000)	0.0081 (0.8272)	0.0492 (0.3148)	0.6410*** (0.0000)

[a]*, ** and *** denote 10%, 5% and 1% level of significance, respectively.
[b]Confidence interval can be provided upon request.
Source: Asia Competitiveness Institute.

Table 4.4: Estimated Measures of Multi-Directional Feedback between International Tourism Arrivals (ITA), Capital Investment on Tourism (% of GDP) (CI), Government Expenditure on Tourism (% of GDP) (GET), Airport Facilities (AF) and Religious Unrest (RU) for Malaysia, 2000–2012[a,b]

Economic Aggregates		$H_0\ (F_{x,y} = F_{x\to y} + F_{y\to x} + F_{x\cdot y})$			
x	y	$F_{x,y}$	$F_{x\to y}$	$F_{y\to x}$	$F_{x\cdot y}$
ITA	CI GET AF RU	2.2707*** (0.0000)	0.2604 (0.1409)	0.4086** (0.0138)	1.6017*** (0.0000)

[a]*, ** and *** denote 10%, 5% and 1% level of significance, respectively.
[b]Confidence interval would be provided upon request.
Source: Asia Competitiveness Institute.

and *y* at the 1% level of significance, and linear feedback from *y* to *x* at the 5% level of significance and instantaneous association at 1% of significance. The results indicate that the linear feedback from the independent variables to the dependent variable highlights the role that the explanatory variables can jointly and simultaneously have in contributing to increased ITA.

As noted, we constructed four environments, namely (i) Environment and Infrastructure (EIE), (ii) Purchasing Power (PPE), (iii) Government Policy (GPE), and (iv) Safety and Security (SSE), to bolster the previous results in order to analyse the relationship between the specific policy environments and the dependent variable of interest. As the results in Table 4.5 highlight, there is a strong bi-directional association between ITA and GPE at the 1% level of significance, as well as a strong bi-directional association between ITA and SE at the 1% level of significance. Interestingly, in examining the relationship between ITA and PPE, the relationship between the two is not significant. Additionally, EIE was dropped from the bivariate and multivariate tests, as the environment bi-directional causality did not pass the white noise test.

Proceeding to the Geweke multi-directional causality in Table 4.6, in which GPE, PPE and SSE were collated on the *y* vector and ITA on the *x* vector, we find that there exists a strong association between the *x* and *y* at the 1% level of significance. Furthermore, the instantaneous association between *x* and *y* are significant at 1% and contribute the majority of the total causality.

Table 4.5: Estimated Measures of Bi-Directional Feedback between International Tourism Arrivals (ITA), Purchasing Power Environment (PPE) and Government Policy Environment (GPE) and Safety and Security Environment (SSE) for Malaysia, 2000–2012[a,b][32]

Economic Aggregates		$H_0\ (F_{x,y} = F_{x\to y} + F_{y\to x} + F_{x\cdot y})$			
x	y	$F_{x,y}$	$F_{x\to y}$	$F_{y\to x}$	$F_{x\cdot y}$
ITA	PPE	0.0208 (0.7972)	0.0036 (0.6747)	0.0001 (0.9421)	0.0171 (0.3607)
PPE	ITA	0.0254 (0.7421)	0.0005 (0.8703)	0.0078 (0.5357)	0.0171 (0.3607)
ITA	GPE	0.5846*** (0.0000)	0.0407 (0.1578)	0.0069 (0.5616)	0.5370*** (0.0000)
GPE	ITA	0.5938*** (0.0007)	0.0379 (0.1730)	0.0190 (0.3350)	0.5370*** (0.0000)
ITA	SSE	0.8878*** (0.0000)	0.0082 (0.8244)	0.0893 (0.1227)	0.7904*** (0.0000)
SSE	ITA	0.8394*** (0.0000)	0.0039 (0.9117)	0.0451 (0.3466)	0.7904*** (0.0000)

[a]*, ** and *** denote 10%, 5% and 1% level of significance, respectively.
[b]Confidence interval can be provided upon request.
Source: Asia Competitiveness Institute.

[32]The Environment and Infrastructure environment was not included as its bi-directional causality with international tourism arrivals could not pass white-noise test.

Table 4.6: Estimated Measures of Multi-Directional Feedback between International Tourism Arrivals (ITA), Purchasing Power Environment (PPE), Government Policy Environment (GPE) and Safety and Security Environment (SSE) for Malaysia, 2000–2012[a,b]

Economic Aggregates		$H_0\ (F_{x,y} = F_{x\to y} + F_{y\to x} + F_{x\cdot y})$			
x	y	$F_{x,y}$	$F_{x\to y}$	$F_{y\to x}$	$F_{x\cdot y}$
ITA	GPE PPE SSE	1.0921*** (0.0000)	0.0588 (0.8378)	0.2092 (0.1319)	0.8241*** (0.0000)

[a]*, ** and *** denote 10%, 5% and 1% level of significance, respectively.
[b]Confidence interval can be provided upon request.
Source: Asia Competitiveness Institute.

To summarise, the Geweke causality results suggest that Malaysia's government expenditure on tourism promotions as well as infrastructure investment such as on enhancing airports facilities emerge as causal and significant determinants of the growth in the travel and tourism industry.

4.5.2 Geweke Causality Analysis for Thailand

As noted, we adopted the stepwise forward selection to select the "best" set of explanatory variables. As Table 4.7 shows, only CPI and RU are the two variables that turn out to be the most significant variables used for Geweke analysis for Thailand.

Table 4.8 provides the first set of results for bi-directional causality for Thailand. For the subsequent Geweke causality tables, figures of $F_{X,Y}$ may not be equal to the sum of figures of $F_{X\to Y}$, $F_{Y\to X}$, and $F_{X\cdot Y}$ due to round-up. As can be inferred from the bi-directional causality results we do not find any significant unidirectional association between ITA and CPI on either direction. However, we find that there is an instantaneous causal relationship between the two variables that is significant at the 5% level, which cannot be ignored. Further, it is useful to note that

Table 4.7: Summary of Linear Stepwise Regression Analysis on Tourism Development in Thailand

Dependent Variable: International Tourism Arrivals	$y_i = \partial_i + \sum_{k=1}^{K} \beta_k x_{ik} + \epsilon_i$			**Number of obs = 51** **F = 19.71** **Prob> F = 0.0000** **Adjusted R^2 = 0.4280**
Independent variables	β	**Standard Error**	***t*-Statistics**	***p*-value**
Consumer Price Index (CPI)	81656.53***	19811	4.12	0.000
Religious Unrest (RU)	–702575.7***	131989.1	–5.32	0.000
Constant	11890.11	16561.69	0.72	0.476

Note: *, ** and *** denote 10%, 5% and 1% level of significance, respectively.
Source: Asia Competitiveness Institute.

Table 4.8: Estimated Measures of Bi-Directional Feedback between International Tourism Arrivals (ITA), Consumer Price Index (CPI) and Religious Unrest (RU) for Thailand, 2000–2012[a,b]

Economic Aggregates		$H_0\ (F_{x,y} = F_{x\to y} + F_{y\to x} + F_{x\cdot y})$			
x	y	$F_{x,y}$	$F_{x\to y}$	$F_{y\to x}$	$F_{x\cdot y}$
ITA	CPI	0.0886 (0.2268)	0.0019 (0.7619)	0.0008 (0.8450)	0.0860** (0.0401)
CPI	ITA	0.0936 (0.2047)	0.0002 (0.9242)	0.0075 (0.5457)	0.0860** (0.0401)
ITA	RU	0.8041*** (0.0000)	0.0008 (0.8413)	0.0505 (0.1156)	0.7528*** (0.0000)
RU	ITA	0.7774*** (0.0000)	0.0155 (0.3830)	0.0091 (0.5049)	0.7528*** (0.0000)

[a]*, ** and *** denote 10%, 5% and 1% level of significance, respectively.
[b]Confidence interval can be provided upon request.
Source: Asia Competitiveness Institute.

Table 4.9: Estimated Measures of Multi-Directional Feedback between International Tourism Arrivals (ITA), Consumer Price Index (CPI) and Religious Unrest (RU) for Thailand, 2000–2012[a,b]

Economic Aggregates		$H_0\ (F_{x,y} = F_{x\to y} + F_{y\to x} + F_{x\cdot y})$			
x	y	$F_{x,y}$	$F_{x\to y}$	$F_{y\to x}$	$F_{x\cdot y}$
ITA	CPI RU	1.6695*** (0.0000)	0.0027 (0.9355)	0.0581 (0.2410)	1.6087*** (0.0000)

[a]*, ** and *** denote 10%, 5% and 1% level of significance, respectively.
[b]Confidence interval can be provided upon request.
Source: Asia Competitiveness Institute.

this instantaneous causation does not translate into total causality between the two variables, in either direction. Considering the bi-directional causality results between ITA and RU, we find that the causality again is bi-directional, although the association between the dependent and explanatory variable at the 1% level of significance is being driven by instantaneous feedback.

Table 4.9 reports the multi-directional causality between ITA on the x vector which comprises CPI and RU. The results are consistent with what we found earlier in the bilateral case, and there is a strong overall causal relationship between the vector of explanatory variables and tourism growth, although the feedback is instantaneous and $(F_{X\cdot Y})$ accounts for almost all the variation in total multi-directional linear dependence. These results indicate that while CPI and social stability in Thailand could simultaneously influence international tourism arrivals, the direction of causality individually is difficult to ascertain from the results.

Table 4.10: Estimated Measures of Bi-Directional Feedback between International Tourism Arrivals (ITA), Purchasing Power Environment (PPE), Government Policy Environment (GPE), Safety and Security Environment (SSE) and Environment and Infrastructure Environment (EIE) for Thailand, 2000–2012[a,b]

Economic Aggregates		$H_0 (F_{x,y} = F_{x\to y} + F_{y\to x} + F_{x\cdot y})$			
x	y	$F_{x,y}$	$F_{x\to y}$	$F_{y\to x}$	$F_{x\cdot y}$
ITA	PPE	0.0962 (0.1940)	0.0000 (0.9704)	0.0013 (0.7997)	0.0949** (0.0311)
PPE	ITA	0.1034 (0.1670)	0.0021 (0.7501)	0.0065 (0.5738)	0.0949** (0.0311)
ITA	GPE	0.1106 (0.1437)	0.0009 (0.8381)	0.0020 (0.7520)	0.1077** (0.0216)
GPE	ITA	0.0594 (0.1036)	0.0002 (0.9131)	0.0004 (0.3471)	0.0587** (0.0216)
ITA	SSE	0.6568*** (0.0000)	0.0013 (0.7985)	0.0466 (0.1308)	0.6089*** (0.0000)
SSE	ITA	0.6358*** (0.0000)	0.0155 (0.3840)	0.0115 (0.4528)	0.6089*** (0.0000)
ITA	EIE	0.1238 (0.1085)	0.0292 (0.2319)	0.0054 (0.6085)	0.0893** (0.0365)
EIE	ITA	0.1304* (0.0942)	0.0061 (0.5837)	0.0350 (0.1904)	0.0893** (0.0365)

[a]*, ** and *** denote 10%, 5% and 1% level of significance, respectively.
[b]Confidence interval can be provided upon request.
Source: Asia Competitiveness Institute.

The final bit of analysis for Thailand is to examine the bi-directional and multi-dimensional causality of ITA with the various explanatory variables lumped into factors through a principle component analysis. We construct different environments which proxy the different determinants of tourism. They can be grouped into: Environment and Infrastructure, Purchasing Power Government Policy, and Safety and Security as mentioned earlier. The intuition is to tease out the causal relationships that could exist between these set of variables and tourism growth.

Table 4.10 furnishes the estimated bi-directional feedbacks between ITA and the four environments mentioned previously. Consistent with the bi-directional and multi-directional results we found earlier, we find only a strong instantaneous association between the different variables. With the exception of the total causality for ITA and the Environment and Infrastructure Environment, almost every other environment has a strong and statistically significant relationship with ITA. However, as noted before, the direction of causality cannot be ascertained because we find only a simultaneous or instantaneous feedback. The multi-dimensional counterpart of

Table 4.11: Estimated Measures of Multi-Directional Feedback between International Tourism Arrivals (ITA), Purchasing Power Environment (PPE), Government Policy Environment (GPE), Safety and Security Environment (SSE) and Environment and Infrastructure Environment (EIE) for Thailand, 2000–2012[a,b]

Economic Aggregates		$H_0\ (F_{x,y} = F_{x \to y} + F_{y \to x} + F_{x \cdot y})$			
x	y	$F_{x,y}$	$F_{x \to y}$	$F_{y \to x}$	$F_{x \cdot y}$
ITA	PPE GPE SSE EIE	1.4171*** (0.0000)	0.0282 (0.8477)	0.0865 (0.3744)	1.3024*** (0.0000)

[a]*, ** and *** denote 10%, 5% and 1% level of significance, respectively.
[b]Confidence interval can be provided upon request.
Source: Asia Competitiveness Institute.

the same set of regressions is shown in Table 4.11 and we find that the results broadly concur with the results shown in Table 4.10. We find evidence only for an instantaneous feedback.

In summary, the empirical results point out a couple of interesting results. One, we do not find an explicit uni-directional feedback in either direction for the variables considered. Two, however, there is a joint or simultaneous causation that can be observed between inflation and tourist arrivals as well as between religious unrest and tourism growth. These results point out the need for the policy-makers to be sensitive to the macroeconomic shocks as well as domestic political conditions which could adversely affect tourism growth that in turn is likely to have a spillover effect on overall economic growth.

4.5.3 Geweke Causality Analysis for Singapore

We followed a similar template for Singapore, as we did for Malaysia and Thailand, and tried and ascertained the direction of causality between the concerned variables. One notable exception from the Thailand results is that, using our stepwise forward selection, we find more relevant variables to be included in our causality analysis. Specifically, we find that GET (% of GDP), CI, CPI, AF as well as the indicator capturing RU appear to become the most significant variables to be used for Geweke causality analysis (see Table 4.12).

Table 4.13 starts with bi-directional causality results. With the exception of a weak unidirectional causality between ITA and AF, there is no other set of variables that exhibit strong causality in either direction. However, there are two interesting results that come out of this analysis. One, consistent with the results for Thailand, we find that there is evidence only for an instantaneous feedback between the variables and no evidence whatsoever for unidirectional causal associations individually. Two, however, unlike Thailand, the variables for which we find a significant instantaneous feedback, are exactly those variables that were not part of the Thailand estimation, namely, we find that inflation and religious unrest to be insignificant whereas all the other variables were significant in the case of Singapore.

Table 4.12: Summary of Linear Stepwise Regression Analysis on Tourism Development in Singapore

Dependent Variable: International Tourism Arrivals	$y_i = \partial_i + \sum_{k=1}^{K} \beta_k x_{ik} + \epsilon_i$			**Number of obs = 51** **F = 44.54** **Prob> F = 0.0000** **Adjusted R^2 = 0.8132**
Independent variables	β	**Standard Error**	***t*-Statistics**	***p*-value**
Government Expenditure on Tourism (% of GDP)	3253069.00***	358715.80	9.07	0.000
Capital Investment on Tourism (% of GDP)	–50283.18***	10181.20	–4.94	0.000
Consumer Price Index (CPI)	20341.80***	7154.60	2.84	0.007
Airport Facilities	–351607.80**	170567.20	–2.06	0.045
Religious Unrest	–46283.62**	22789.90	–2.03	0.048
Constant	8506.50	6361.20	1.34	0.188

Note: *, ** and *** denote 10%, 5% and 1% level of significance, respectively
Source: Asia Competitiveness Institute.

Table 4.13: Estimated Measures of Bi-Directional Feedback between International Tourism Arrivals (ITA), Government Expenditure on Tourism (% of GDP) (GET), Capital Investment on Tourism (% of GDP) (CI), Consumer Price Index (CPI), Airport Facilities (AF) and Religious Unrest (RU) for Singapore, 2000–2012[a,b]

Economic Aggregates		$H_0\ (F_{x,y} = F_{x\to y} + F_{y\to x} + F_{x\cdot y})$			
x	y	$F_{x,y}$	$F_{x\to y}$	$F_{y\to x}$	$F_{x\cdot y}$
ITA	GET	1.2279*** (0.0000)	0.0010 (0.8274)	0.0025 (0.7270)	1.2245*** (0.0000)
GET	ITA	1.2553*** (0.0000)	0.0002 (0.9185)	0.0306 (0.2209)	1.2245*** (0.0000)
ITA	CI	0.4247*** (0.0001)	0.0014 (0.7950)	0.0035 (0.6804)	0.4198*** (0.0000)
CI	ITA	0.4216*** (0.0001)	0.0006 (0.8674)	0.0013 (0.8041)	0.4198*** (0.0000)
ITA	CPI	0.0503 (0.4818)	0.0010 (0.8259)	0.0002 (0.9295)	0.0491 (0.1207)
CPI	ITA	0.0551 (0.4406)	0.0001 (0.9675)	0.0059 (0.5915)	0.0491 (0.1207)
ITA	AF	0.8076*** (0.0000)	0.0038 (0.6663)	0.0662* (0.0718)	0.7377*** (0.0000)
AF	ITA	0.7642*** (0.0000)	0.0212 (0.3078)	0.0053 (0.6088)	0.7377*** (0.0000)
ITA	RU	0.0257 (0.7384)	0.0002 (0.9222)	0.0020 (0.7558)	0.0236 (0.2825)
RU	ITA	0.0252 (0.7447)	0.0005 (0.8812)	0.0012 (0.8103)	0.0236 (0.2825)

[a]*, ** and *** denote 10%, 5% and 1% level of significance, respectively.
[b]Confidence interval can be provided upon request.
Source: Asia Competitiveness Institute.

Table 4.14: Estimated Measures of Multi-Directional Feedback between International Tourism Arrivals (ITA), Government Expenditure on Tourism (% of GDP) (GET), Capital Investment on Tourism (% of GDP) (CI), Consumer Price Index (CPI), Airport Facilities (AF) and Religious Unrest (RU) for Singapore, 2000–2012[a,b]

Economic Aggregates		$H_0\ (F_{x,y} = F_{x\to y} + F_{y\to x} + F_{x\cdot y})$			
x	y	$F_{x,y}$	$F_{x\to y}$	$F_{y\to x}$	$F_{x\cdot y}$
ITA	GET CI CPI AF RU	2.4647*** (0.0000)	0.0063 (0.9975)	0.0923 (0.4768)	2.3661*** (0.0000)

[a]*, ** and *** denote 10%, 5% and 1% level of significance, respectively.
[b]Confidence interval can be provided upon request.
Source: Asia Competitiveness Institute.

Table 4.14 shows the multi-directional causality results between ITA on the x vector and the y vector of variables comprising GET, CPI, CI, AF and RU. We find that that there is strong causality between x and y ($F_{X,Y}$) at the 1% level of significance. Although there is no significant linear feedback running from x to y or y to x ($F_{X\to Y}$; $F_{Y\to X}$) the instantaneous association between x and y ($F_{X\cdot Y}$) is strongly significant at 1% level and contributes 96% of total multi-directional linear dependence. These results indicate that the government policies on tourism, inflation rates, airport construction and social stability in Singapore could effectively and simultaneously affect international tourism arrivals.

Following what we did for Thailand, we next consider the causality results for the different environments and tourism growth. It is important to highlight that the environment capturing environment and infrastructure as well as government policies have a strong instantaneous causal feedback with growth in international tourist arrivals, while there is no significant association between the other two environments comprising of purchasing power and security (see Table 4.15).

The multi-directional counterpart is shown in Table 4.16 and the results appear to be consistent with what we discussed earlier in that the there is a strong and significant linear feedback running from y to x ($F_{Y\to X}$), which perhaps is being driven by the significance of the government policy and environment and infrastructure environment. The instantaneous association ($F_{X\cdot Y}$) is also highly significant at the 1% level, contributing to over 90% of the total multi-directional linear dependence.

To summarise the results, our causality analysis shows two interesting trends: The first pertains to the consistency with that of Thailand, where we find evidence only for an instantaneous feedback between the explanatory variables considered and international tourist arrival. This implies that there is no unidirectional causal association individually that we could establish. The second conclusion to come out of this analysis is that we find the quality of infrastructure as proxied by airport facilities causes tourism growth, which is consistent with the story of Singapore's investment in its airport infrastructure. Further, we also find that the lists of variables which exhibit significant instantaneous feedback are those that were not significant for Thailand.

Table 4.15: Estimated Measures of Bi-Directional Feedback between International Tourism Arrivals (ITA), Purchasing Power Environment (PPE), Government Policy Environment (GPE), Environment and Infrastructure Environment (EIE) and Safety and Security Environment (SSE) for Singapore, 2000–2012[a,b]

Economic Aggregates		$H_0\ (F_{x,y} = F_{x\to y} + F_{y\to x} + F_{x\cdot y})$			
x	y	$F_{x,y}$	$F_{x\to y}$	$F_{y\to x}$	$F_{x\cdot y}$
ITA	PPE	0.0723 (0.3150)	0.0015 (0.7874)	0.0085 (0.5182)	0.0623* (0.0805)
PPE	ITA	0.0761 (0.2925)	0.0001 (0.9486)	0.0136 (0.4138)	0.0623* (0.0805)
ITA	GPE	1.6515*** (0.0000)	0.0053 (0.6115)	0.0102 (0.4786)	1.6360*** (0.0000)
GPE	ITA	1.6679*** (0.0000)	0.0001 (0.9786)	0.0318 (0.2117)	1.6360*** (0.0000)
ITA	EIE	0.5322*** (0.0000)	0.0093 (0.5000)	0.0526 (0.1084)	0.4703*** (0.0000)
EIE	ITA	0.5132*** (0.0000)	0.0200 (0.3220)	0.0229 (0.2895)	0.4703*** (0.0000)
ITA	SSE	0.2283** (0.0108)	0.0130 (0.4251)	0.1985*** (0.0018)	0.0168 (0.3646)
SSE	ITA	0.1155 (0.1294)	0.0543 (0.1029)	0.0444 (0.1400)	0.0168 (0.3646)

[a]*, ** and *** denote 10%, 5% and 1% level of significance, respectively.
[b]Confidence interval can be provided upon request.
Source: Asia Competitiveness Institute.

Table 4.16: Estimated Measures of Multi-Directional Feedback between International Tourism Arrivals (ITA), Purchasing Power Environment (PPE), Government Policy Environment (GPE), Environment and Infrastructure Environment (EIE) and Safety and Security Environment (SSE) for Singapore, 2000–2012[a,b]

Economic Aggregates		$H_0\ (F_{x,y} = F_{x\to y} + F_{y\to x} + F_{x\cdot y})$			
x	y	$F_{x,y}$	$F_{x\to y}$	$F_{y\to x}$	$F_{x\cdot y}$
ITA	PPE GPE EIE SSE	2.6134*** (0.0000)	0.0263 (0.8634)	0.2300** (0.0237)	2.3571*** (0.0000)

[a]*, ** and *** denote 10%, 5% and 1% level of significance, respectively.
[b]Confidence interval can be provided upon request.
Source: Asia Competitiveness Institute.

4.6 CONCLUSION AND POLICY IMPLICATIONS

Tourism is one of the vibrant economic sectors in Southeast Asian countries, being spurred by the growth of this industry in Malaysia, Thailand and Singapore. These countries have also promoted several national development strategies for decades. In this light, this chapter has tried to empirically understand the causal factors driving the growth of international tourist arrivals in Malaysia, Thailand and Singapore, using quarterly data between 2000 and 2012, under a Geweke causality framework. The findings of the chapter leave us with rich policy insights that could serve as a useful guidance to policy-makers to boost the capacity and enhance the quality of the domestic travel and tourism industry in each country.

In summary, the empirical results point out a couple of interesting results. Our results show that, for Malaysia, there exists a strong causal link between airport facilities and international tourist arrivals, with airport facilities significantly driving the growth of international tourist arrivals. This result highlights the prominent role that the quality of air transportation facilities can have on contributing to the growth of the travel and tourism industry into Malaysia. As demonstrated by the empirical results, government policies also can have a substantial effect on international tourist arrivals into Malaysia; both government expenditure on tourism and the government policy environment have a significant bi-directional association with international tourist arrivals. While the empirical results confirm the importance of the state intervention on the development of the travel and tourism industry, additional resources from both the public and private sector should be utilised to bolster travel and tourism services and attractions.

Additionally, there is also a significant association between ITA and the SSE in the case of Malaysia. Further, we also find a bi-directional association between RU and ITA while there is a weak reverse feedback from RU to ITA. These results indicate that a country's security is an important determinant of growth in tourist arrivals, which highlights the need for the government to ensure a safe and secure environment. In order to ensure the safety of the local populace and international visitors, it is important that Malaysia continues to invest in the police force and state mechanisms of law and order.

For Singapore and Thailand, while overall our results are stronger for Singapore than Thailand, we still find that there is a strong linear instantaneous feedback between ITA and the different policy environments we selected for the respective countries. For Thailand specifically, we find that RU, capturing an element of domestic instability and turbulence, seems to affect ITA significantly. This appears to be consistent with what happened to Thailand after 2013, where political turmoil caused a severe setback to the growth of the tourism industry and also halted the country's economic growth. This is also likely suggestive that the policy-makers must pay attention to the economic damage that such instability can bring about to the tourism industry in particular but the economy more generally. In the case of Singapore, we find that ITA are driven by infrastructural variables covering AF as well as government policy variables such as GET industry.

The strong causal relationship we find between these variables is consistent with the national tourism policies of Thailand and Singapore although our causality results are indicative of an instantaneous feedback rather than a strong unidirectional feedback in either direction. The only exception being the weak unidirectional relationship that we find in the case of Singapore where the causality runs from quality of airports to international tourist arrivals, which again stands testimony to the efforts taken by the Singapore government in maintaining a world-class airport like the Changi airport. In fact, Changi airport is an example of an aviation hub that has played an instrumental role in the transformation of economy into an important destination by leveraging the tourism–transport policy complementarities.

REFERENCES

Aizenman, Joshua, and Ilan Noy. 2006. "FDI and trade — Two-Way Linkages?" *The Quarterly Review of Economics and Finance* 46(3): 317–337.

Ashley, Caroline, Peter De Brine, Amy Lehr and Hannah Wilde. 2007. *The role of tourism sector in expanding economic opportunity*. Cambridge: John F. Kennedy School of Government, Harvard University.

Board of Investment of Thailand. 2011. *National Tourism Development*. http://www.boi.go.th/tir/issue/201104_21_4/155.htm.

Calderón, César, and Lin Liu. 2003. "The Direction of Causality between Financial Development and Economic Growth." *Journal of Development Economics* 72(1):321–334.

Credit Suisse. 2014. *Thailand's political crisis: What it means for ASEAN tourism*. www.creditsuisse.com/researchandanalytics.

Culiuc, Alexander. 2014. "Determinants of International Tourism." IMF Working Paper No. WP/14/82, International Monetary Fund, Washington DC.

Dhillon, Rajina. 2015. "Tourism industry should promote Malaysia's selling points, stress on value for money." The Rakyat Post, August 27. www.therakyatpost.com/news/2015/08 /27/tourism-industry-should-promote-malaysias-selling-points-deflect-negativity-says-najib/.

Economic Planning Unit. 2010. *Tenth Malaysia Plan 2011–2015*. Putrajaya: Economic Planning Unit. http://onlineapps.epu.gov.my/rmke10/rmke10_english.html.

Geweke, John. 1982. "Measurement of Linear Dependence and Feedback between Multiple Time Series." *Journal of the American Statistical Association* 77(378): 304–313.

Gopalan, Sasidaran. 2013. "Mapping India-EU tourism flows." CARIM-India Research Report No. 2013/15, Robert Schuman Centre for Advanced Studies, European University Institute, Florence.

Granger, Clive W.J. 1963. "Economic Processes Involving Feedback." *Information and Control*, *6*(1): 28–48.

Granger, Clive W.J. 1969. "Investigating Causal Relations by Econometric Models and Cross-Spectral Methods." *Econometrica 37*(3):424–438.

Granger, Clive W.J. 1988. "Some Recent Developments in a Concept of Causality." *Journal of Econometrics* 21: 161–194.

Gujarati, Damodar N. 2004. *Basic Econometrics* Fourth Edition. New York: McGraw-Hill.

Kotwani, Monica and Nur Afifah Ariffin. 2015. "STB to launch S$20m global campaign to boost tourism numbers." *Channel NewsAsia,* April 7. www.channelnewsasia .com /news/business/singapore/stb-to-launch-s-20m/1769792.html.

Lim, Christine. 1997. "Review of International Tourism Demand Models." *Annals of Tourism Research* 24(4): 835–849.

Malik, Shahnawaz, Imran Sharif Chaudhry, Muhammad Ramzan Sheikh, and Fareed Shareef Farooqi. 2010. "Tourism, Economic Growth and Current Account Deficit in Pakistan: Evidence from Co-integration and Causal Analysis." *European Journal of Economics, Finance and Administrative Sciences* 22: 21–31.

Mazumder, Mohammad Nurul Huda, Mast Afrin Sultana, and Abdullah Al-Mamun. 2013. "Regional Tourism Development in Southeast Asia." *Transnational Corporations Review* 5(2): 60–76.

Meng, Xianming, Mahinda Siriwardana, and Tien Pham. 2013. "A CGE Assessment of Singapore's Tourism Policies." *Tourism Management* 34: 25–36.

Ministry of Tourism and Culture of Malaysia. 2012. "ETP: transforming tourism to the new heights." *Tourism Malaysia*, July 9. http://www.tourism.gov.my/media/view/etp-transforming-tourism-to-the-new-heights.

Mosbah, Aissa and Mohamed Saleh Abd Al Khuja. 2014. "A Review of Tourism Development in Malaysia." *European Journal of Business and Management* 6(5):1–9.

Nanthakumar, Loganathan, Yahaya Ibrahim and Madzli Harun. 2007. "Tourism Development Policy, Strategic Alliances and Impact of Consumer Price Index on Tourist Arrivals: The Case of Malaysia." *TOURISMOS: An International Multidisciplinary Journal of Tourism* 3(1): 83–98.

National Economic Advisory Council. 2010, *New Economic Model for Malaysia.* Putrajaya: National Economic Advisory Council. http://www.epu.gov.my/en/new-economic-model

Önder, A. Özlem, Aykan Candemir, and Neşe Kumral. 2009. "An Empirical Analysis of the Determinants of International Tourism Demand: The Case of Izmir." *European Planning Studies* 17(10): 1525–1533.

Overseas Development Institute. 2007. "Can tourism offer pro-poor pathways to prosperity? Examining evidence on the impact of tourism on poverty". Briefing Paper No. 22, Overseas Development Institute, London.

Pett, Marjorie A., Nancy R. Lackey and John J. Sullivan. 2003. *Making Sense of Factor Analysis: The Use of Factor Analysis for Instrument Development in Health Care Research.* Thousand Oaks: Sage Publications.

Prideaux, Bruce. 2005 Factors affecting bilateral tourism flows. *Annals of Tourism Research* 32(3): 780–801.

Sheldon, Pauline J. 1993. "Forecasting Tourism: Expenditures versus Arrivals." *Journal of Travel Research* 13(1):13–20.

Sims, Christopher A. 1972. "Money, income, and causality." *The American Economic Review* 62(4): 540–552.

Song, Haiyan, Kevin KF Wong, and Kaye K.S. Chon. 2003. "Modelling and Forecasting the Demand for Hong Kong Tourism." *Hospitality Management* 22(4): 435–451.

Tan, Khee Giap and Chee-Seng Cheng. 1995. "The Causal Nexus of Money, Output and Prices in Malaysia." *Applied Economics* 27(12): 1245–1251.

Tan, Khee Giap, Evan Tan Beng Kai and Vincent Kwan Wen Seng. 2016 (forthcoming). "Drivers of Growth in the Travel and Tourism Industry in Thailand and Singapore: A Geweke Causality Analysis." *Public Enterprise Journal* 22(1).

Tan, Khee Giap, Sasidaran Gopalan and Ye Ye Denise. 2016 (forthcoming). "Drivers of Growth in the Travel and Tourism Industry in Malaysia: A Geweke Causality Analysis." *Economies* 4 (1).

Thano, Rakela. 2015. "The Impact of Tourism on the Balance of Payments." *American Journal of Economics, Finance and Management* 1(5): 529–536.

Tourism Authority of Thailand Newsroom. 2015. "Thai tourism to enter new era with total focus on "quality". July 6. http://www.tatnews.org/thai-tourism-to-enter-new-era-with-total-focus-on-%E2%80%9Cquality%E2%80%9D/.

United Nations World Tourism Organisation. 2015. *UNWTO Tourism Highlights*. Madrid: United Nations World Tourism Organisation.

Uysal, Muzaffer, and Joln L. Crompton. 1984. "Determinants of Demand for International Tourist Flows to Turkey." *Tourism Management* 5(4): 288–297.

World Travel & Tourism Council. 2014. *Travel & Tourism Economic Impact 2014 Malaysia.* London: World Travel & Tourism Council.

World Travel & Tourism Council. 2015. *Travel & Tourism Economic Impact 2015 Thailand and Singapore.* London: World Travel & Tourism Council.

Chapter 5

Concluding Remarks and Future Research Agenda

5.1 SUMMARY OF KEY FINDINGS

With the formal launch of the ASEAN Economic Community (AEC) on 31st December 2015, ASEAN is at another critical crossroads, where deepening liberalisation has to be carefully managed in conjunction with earnest domestic reforms for regional integration efforts to bear fruits. As argued by the Asia Competitiveness Institute (ACI) at the Lee Kuan Yew School of Public Policy (LKYSPP), National University of Singapore (NUS), in an editorial opinion for The Straits Times on 25th February 2016, economic integration could be a "make or break" for less competitive economies if they do not step up their reform efforts to systematically address their shortfalls.[1] With this compendium on *2015 Annual Competitiveness Analysis, Agricultural Productivity and Development Vision for ASEAN-10*, we present not only our latest competitiveness ranking and simulation analysis for the ten ASEAN member states but also our rigorous studies of various economic issues that are of paramount importance to the region, with a view to facilitating the full formation of the AEC, as well as contributing insights to policy formulation in the region beyond the AEC milestone.

With such objectives in mind, this book starts with a comprehensive assessment of the progress of the AEC in Chapter 1. While the AEC was officially established in December 2015, only about 80% of the stipulated requirements in the AEC blueprint was completed by then, which raises questions as to whether the formation of AEC could be feasibly completed by the end of 2016. Our in-depth analysis has shed light on the thorny issues where implementation has been problematic, most notable of which are: (i) non-tariff barriers (NTBs) to trade, (ii) trade facilitation through the establishment of the ASEAN Single Window (ASW), (iii) trade in services, (iv) free movement of skilled labour, and (v) the formation of the ASEAN Single Aviation Market (ASAM), or the Open Skies policy.

In the year 2016, the initiative which may see substantial progress is likely to be the ASAM, as domestic pressure has mounted for Indonesia to ratify the Open Skies agreement. That said, it is unlikely for all outstanding deliverables to be fully implemented by the end of 2016. First, for NTBs the current "notify-negotiate-eliminate" approach is plagued with serious shortcomings, in terms of incentive misalignments as well as institutional weaknesses in some of the ASEAN member states. Second, achieving the 70% target of foreign ownership in all services sectors by 2016 also seems unrealistic in view of the lack of progress even in priority integration sectors such as air transport.

[1] See Tan *et al*. (2016). See also Appendix 1 for the related policy brief.

Third, the operationalisation of the ASW is also likely to remain outstanding as the ongoing full-fledged pilot of the ASEAN only involves a handful of ASEAN members and the legal framework for its eventual implementation has not been ratified by all members. The National Single Windows (NSWs) have not even been fully established in Cambodia, Laos and Myanmar. Fourth, in the area of skilled labour mobility, notwithstanding the conclusion of the various Mutual Recognition Arrangements (MRAs) for eight priority professions, restrictive domestic regulations remain, particularly in issuance of work visas, quotas on foreign labour, licensing requirements, impeding the cross-border movement of professionals in the region.

Furthermore, the AEC blueprint itself leaves much to be desired. For instance, its scope in services trade liberalisation is particularly limited, as only 70% of foreign ownership is targeted. Additionally, measures for facilitating skilled labour mobility only affect a small fraction of total workers in ASEAN. At the same time, many flexibilities allowed under the blueprint, such as the "ASEAN minus x" formula, are vaguely worded and thus leave ample room for discretion, as exemplified by the case of Indonesia. Equally important is the fact that the establishment of the AEC has largely been a state-led top-down project with limited inputs and awareness amongst the private sector in the region.

The formal launch of the AEC in December 2015 should thus be seen as a stepping stone, rather than the end, in ASEAN's trajectory toward deeper economic integration. For a start, a more far-reaching agenda beyond what is stipulated in the AEC blueprint is required for ASEAN to meaningfully deepen regional integration. Flexibilities such as the "ASEAN minus x" arrangement, while allowing ASEAN member states to come on board when ready, is an evident testament to the fact that the road to deep integration in ASEAN remains long.

More importantly, the AEC will only be truly formed when each ASEAN member state and their people feel that they are sharing the economic dividends of being in the community, and take a greater ownership and stake in shaping its direction. Without a shared sense of ASEAN identity and community spirit, opportunities arising from an integrated regional grouping will not be fully capitalised.

Beyond the 2015 milestone, enormous work awaits the ten ASEAN member states to realise the common dream of inclusive and sustainable prosperity. For one, narrowing the development divide remains a work in progress, which requires, first and foremost, far-reaching domestic reforms, particularly in the less developed Cambodia, Laos, Myanmar and Vietnam (CLMV) grouping. On the other hand, economic convergence could be expedited if ASEAN steps up to the challenge and takes the lead in trade and investment initiatives in the broader Asia-Pacific region. Of particular emphasis are the Regional Comprehensive Economic Partnership (RCEP) and the Free Trade Area of the Asia-Pacific (FTAAP), which have both grained traction as critical vehicles for enhancing economic growth in Asia Pacific in light of the currently stalled Doha Round. Domestic reforms and global integration must go hand in hand as a two-pronged strategy in the pursuit of inclusive, balanced and sustainable growth.

Equally imperative is the need for the regional bloc to maintain centrality and unity amidst geopolitical uncertainties, especially the ongoing power tussles between the United States (US) and China which is threatening to divide ASEAN if member states begin to take sides between the two powers. Such potential threats of disintegration must be artfully managed and mitigated;

otherwise, community building efforts will be rendered fruitless. In brief, balancing power rivalry while driving global cooperation efforts toward bolstering economic development in the region is the way forward for ASEAN in its post-2015 journey.

Chapter 2 presents our 2015 annual update on ACI's competitiveness ranking and simulations for the ten ASEAN member states conducted through an eclectic and comprehensive framework as well as a rigorous quantitative methodology. Amalgamating various data sources for 121 indicators over the period from 2000 to 2012, our competitiveness analysis indicates wide variations not only in levels of economic development but also paces of catching-up within the bloc. Singapore and Malaysia remain the two best-performing economies in the region although Singapore seems to deteriorate in 2012 as compared to 2011 and 2000, while Malaysia shows signs of improvement in 2012 from 2011 but seems to be losing competitiveness as compared to the year 2000. Thailand at third appears to be recovering from its political tumult as its scores improved steadily over the study period.

While Brunei seems to be stagnant at the fourth spot, Indonesia has been surging firmly at fifth, particularly in sub-environments such as Economic Vibrancy and Financial Deepening and Business Efficiency, as shown in our results of its top –20% strongest indicators and median spider web chart. Nonetheless for Vietnam, the catching-up momentum seems to be running out of steam as its score plunged further from the regional average and as compared to the year 2000. Cambodia and Laos seems quietly rising with marked improvements in many aspects of competitiveness. The Philippines risks being outstripped by the newer members as its standardised scores plunged more negatively over time. Thailand, too, cannot be complacent on the socio-political front if it is to restore trust and confidence in both domestic and foreign investors.

It is thus evident in our annual competitiveness study that ASEAN is still grappling with the issue of narrowing the development divide between the older, more mature ASEAN-6 and the less developed grouping of CLMV. On the one hand, diversity in development stages can be a powerful source of synergy, complementing and supplementing one another for increased overall regional competitiveness as a "single market and production base." On the other hand, the fruits of integration through various initiatives including the AEC, the RCEP and potentially the FTAAP, would only be equitably distributed if individual economies in ASEAN are more earnest in implementing their own domestic economic reforms, addressing their economic malaises while strengthening their capabilities. Our annual competitiveness analysis and simulation offers an instructive roadmap for such pursuits.

In fact, our competitiveness simulation shows strong potential of convergence within ASEAN-10. If each economy improves on its top-20% weaknesses to catch up with regional average, Vietnam would jump from seventh to fifth in the Overall Competitiveness ranking, and Cambodia, Laos and Myanmar would move up one spot from eighth to seventh, ninth to eighth and tenth to ninth, respectively in the same ranking. In terms of each of the four environments, Cambodia would remarkably leap from eighth to fifth for Quality of Life and Infrastructure Development and eighth to sixth for Financial, Businesses and Manpower Conditions and Government and Institutional Setting. Vietnam would climb two places from eighth to sixth and seventh to fifth for Financial, Businesses and Manpower Conditions and Quality of Life and

Infrastructure Development, respectively. Likewise, Myanmar would step up from tenth to eighth for Government and Institutional Setting and Quality of Life and Infrastructure Development. Our simulation analysis thus highlights the critical significance of deepening domestic reforms for harnessing the benefits of integration and ensuring sustainable and inclusive economic development.

Chapter 3 examines the agricultural sector in ASEAN and offers insights as to how agricultural productivity could be improved for greater food security and enhanced agricultural competitiveness. While industrialisation has gathered pace in ASEAN, agriculture remains a critical sector, accounting for a substantial share of Gross Domestic Product (GDP) and total employment in many ASEAN economies. Further, the global food crisis of 2007–2008 accentuates the region's vulnerability to agricultural shocks with the lack of coordinated responses from both the net food importers and the net food exporters. In light of the importance of agriculture to ASEAN, we conducted an empirical estimation of the annual change of agricultural Total Factor Productivity (TFP) — broadly defined as the portion of output not explained by the amount of inputs used in the production — in eight ASEAN countries (Brunei and Singapore excluded) over the period from 2000 to 2011.

Our findings show that overall ASEAN-8 experienced an annual growth of agricultural TFP of 1.4–1.6% over the period 2000–2011, with the more mature group comprising Indonesia, Malaysia, Philippines and Thailand posting a much higher TFP growth of between 2.7% and 3.3% and CLMV lagging behind with a TFP growth of between –0.4% and 0.4% over the study period. Notably, Laos and Myanmar were the only two countries in the region that registered negative growth rates of TFP between 2000 and 2011. The decomposition of TFP growth indicates heterogeneity within the bloc in terms of drivers of agricultural development. For instance, for Malaysia and Vietnam, agricultural growth has been driven by technological change, while for Indonesia and Cambodia, technical efficiency change has played a more dominant role.

The case of Malaysia as the best performer in the region in both agricultural labour productivity growth and TFP growth was also explored in great detail. Malaysia's success in agriculture can be attributed to four key policies: (i) fostering a competitive agribusiness industry, (ii) close partnership between government agencies, agribusiness firms and research institutes, (iii) integrating smallholders into agro-based value chains, and (iv) provision of infrastructure, credit and extension services. The Malaysian experience thus sheds important light on how ASEAN as a whole, and CLMV in particular, could foster their agricultural sectors as ASEAN progresses toward a "single market and production base." Our research also highlights that regional cooperation efforts in agriculture and food security should not solely focus on accessibility to food. Although food accessibility is critical, particularly in times of calamities and emergencies, emphasis should also be on enhancing regional supply capacity, which should be part of the post-2015 ASEAN agenda.

Chapter 4 provides an empirical analysis of the causal factors that drive the growth of the travel and tourism industry in Malaysia, Thailand and Singapore. Applying the novel Geweke causality framework on an extensive dataset across four different environments between 2000 and 2012, we find that for Malaysia, there exists a strong causal link between airport facilities and religious unrest and international tourist arrivals, which is the standard proxy representing the growth of the travel and tourism industry. Additionally, both government expenditures on tourism and the government policy environment have a significant bidirectional association with international tourist arrivals, indicating the importance of state intervention in bolstering the travel and tourism industry.

For Thailand, we find that religious unrest, capturing an element of domestic instability and turbulence, seems to affect international tourist arrivals significantly. This appears to be consistent with what happened to Thailand after 2013, where political turmoil caused a severe setback to the growth of the tourism industry and also halted the country's economic growth. In the case of Singapore, we find that international tourist arrivals are driven by infrastructural variables covering airport facilities as well as government policy variables such as government expenditures on the tourism industry.

Overall, our results point to the fact that infrastructure, socio-political security and government policy are the three most significant drivers of growth in the travel and tourism industry in Malaysia, Singapore and Thailand, albeit to varying degrees. This finding is particularly pertinent as the ten ASEAN member states have been fostering physical and people-to-people connectivity as part of the Master Plan on ASEAN Connectivity (MPAC).[2] In order to boost the various dimensions of intraregional connectivity, regional initiatives must be accompanied by a concrete national policy framework for enhancing security, upgrading infrastructure and promoting the tourism sector in individual countries.

5.2 DEVELOPMENT VISION AND STRATEGIES FOR ASEAN-10

Recognising that community building is an ongoing and dynamic process, even before the formal launch of the AEC in December 2015, ASEAN leaders have committed to formulating and implementing the post-2015 agenda with the signing of the Nay Pyi Taw Declaration on the ASEAN Community's Post-2015 Vision on 12 December 2014, which include, among others, the following key components:[3] (i) promoting Southeast Asia as a region of peace, stability and prosperity; (ii) promoting ASEAN as a people-oriented community; (iii) maintaining an outward-looking ASEAN, enhancing ASEAN centrality and ASEAN's relevance in the global community, and (iv) strengthening ASEAN's institutional capacity.

Building on our studies presented in this book as well as our Asia Economic Connectivity Vision 2030 (AECV2030), which is discussed in great detail in Chapter 1 of our *2014 Annual Competitiveness Ranking and Simulation Study for ASEAN-10 and Development Strategies to*

---

[2]See ASEAN Secretariat (2010).

[3]The Nay Pyi Taw Declaration on the ASEAN Community's Post-2015 Vision is accessible from: www.asean.org/storage/images/pdf/2014_upload/Nay%20Pyi%20Taw%20Declaration%20on%20the%20ASEAN%20Communitys%20Post%202015%20Vision%20w.annex.pdf.

Enhance Asia Economic Connectivity,[4] ACI proposes that the ten ASEAN member states aspire to a bolder and more concrete development vision where:

(i) ASEAN becomes a highly integrated and competitive manufacturing hub and a productive food basket of Asia and the world;
(ii) The prevailing wide development gap within ASEAN is substantially and sustainably narrow;
(iii) The people, companies and various communities within ASEAN share a common ASEAN identity and benefit from the fruits of integration; and
(iv) ASEAN is a key fulcrum of global economic integration for greater regional prosperity.

Our development vision can be crystallised as five integrative and mutually reinforcing elements of ASEAN integration, complementing and supplementing one another, as illustrated in Figure 5.1.

ACI's development vision for ASEAN-10 can thus be seen as an advancement of the post-2015 vision signed by ASEAN leaders in Nay Pyi Taw in November 2014. While prosperity, inclusiveness and outward orientation are already featured in the Declaration on the ASEAN Community's Post-2015 Vision, what ACI espouses is for ASEAN to become the lynchpin in the ever-evolving regional and global architecture. After all, ASEAN has the potential to become the fourth largest economy in the world by 2030.[5] ACI's vision for ASEAN-10 offers a succinct

Figure 5.1: ACI's Development Vision for ASEAN-10

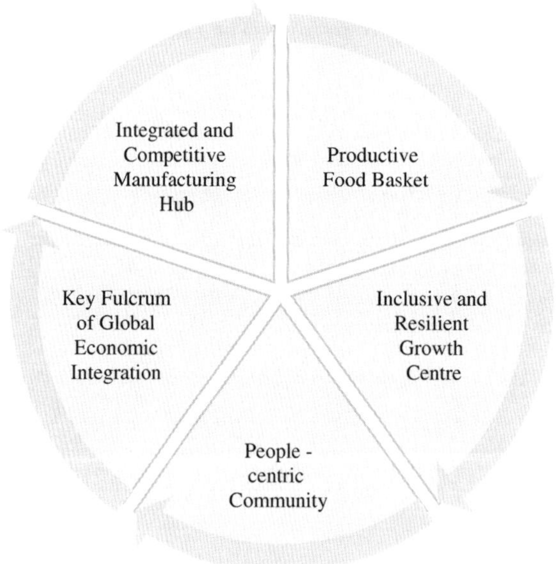

Source: Asia Competitiveness Institute.

[4]See Tan *et al*. (2015a).
[5]See Heng (2015).

conception of how such aspirations can be achieved, together with concrete and astute strategies under each element. These strategies can be further fine-tuned in accordance with the progress of integration initiatives, taken by the participating regional bloc as well as domestic reforms in individual member states.

To be clear, ACI's development vision for ASEAN-10 is anchored in the pursuit of sustainable and inclusive economic development and the integration momentum that has been under way in ASEAN. While ASEAN should evolve into a "single market and production base" as stipulated in the AEC blueprint,[6] the ten-member bloc should not get side-tracked from its growth trajectory by transforming itself into a highly bureaucratic and complex structure akin to the European Union (EU). As also argued elsewhere[7] although ASEAN has to maintain a significant degree of cohesiveness and institutional efficacy, the regional grouping should not lose its pragmatism. In view of the diversity in development levels within the bloc, ASEAN should not aspire toward political or policy integration in the same fashion as the EU. Concentration on growth and economic development should be the way forward.

Element 1: ASEAN as a Highly Integrated and Competitive Manufacturing Hub

ASEAN as one of the most enduring and cohesive regional grouping in the developing world, is well positioned to be a leading manufacturing base in Asia, driven largely by Malaysia, Thailand, Indonesia, and to a lesser extent Vietnam, Cambodia and Laos. Taken as a whole, the regional bloc is home to more than 600 million people and has the third largest labour force in the world, behind China and India.[8] ASEAN's demographic dividend offers an invaluable opportunity which, if harnessed well, will propel the region into another factory of the world.

Underpinning such aspiration should be private sector dynamism and the robust growth of small- and medium-sized enterprises (SMEs) in the region. In this regard, it is important to note that the establishment of the AEC has largely been a top-down and state-led project.[9] Engaging the private sector, particularly SMEs, in the process of community building is therefore paramount for ASEAN to become the manufacturing centre of Asia and beyond. A strong and competitive private sector would entrench ASEAN as a key node in the global value chains. Concomitantly, the growth of private enterprises would facilitate the growth of the middle class in the region, thereby increasing domestic and regional demand, which will in turn further spur the development of the regional manufacturing sector.

The first element in our development vision thus envisages the relocation of manufacturing from Japan, South Korea, Taiwan, Greater China, the US and the EU to ASEAN as the primary way to both broaden and deepen ASEAN's participation in the global value chains. Such process will be predominantly driven by the Trans-Pacific Partnership (TPP), which has been signed, and the outcomes of both the RCEP and FTAAP.

---

[6] See ASEAN Secretariat (2008).

[7] See Asian Development Bank Institute (2014) and Intal *et al.* (2014).

[8] See McKinsey Global Institute (2014).

[9] See Chapter 1 of this book for an in-depth assessment of the AEC.

Apart from efforts to liberalise regionally and globally through the foregoing trade initiatives, the realisation of this element would require individual ASEAN member states to put in place measures including:

(i) Identifying opportunities for upgrading manufacturing capabilities within the global value chains;
(ii) Developing cost-effective manufacturing clusters through partnership with industrial partners in Japan, South Korea, Greater China, the US and the EU;
(iii) Building a conducive investment climate through streamlining rules and regulations and substantially reducing the costs of doing business;
(iv) Strengthening laws and enforcement of intellectual property rights (IPRs), including credible patent systems in order to protect foreign investors' interests and to establish trust in the local business environment, in order to generate stability for a stronger investment climate;
(v) Levelling the playing field between state-owned and local private sector and foreign investors, removing barriers to entry and promoting healthy competition amongst all firms;
(vi) Fostering local SMEs to become reliable suppliers of multinational corporations (MNCs) and able to absorb technologies and adopt global best practices;
(vii) Promoting technology and knowledge transfers from MNCs to local enterprises;
(viii) Encouraging inter-industry linkages for deepening the value chains;
(ix) Developing high-quality human capital capable of undertaking increasingly technology- and knowledge-intensive activities in the value chains;
(x) Improving transportation networks and increasing accessibility to inland production centres through maintenance and expansion of strategic roads, highways, sea ports and airports;
(xi) Particularly for CLMV, designating special industrial and economic zones with preferential tax treatment for MNCs as a springboard for regulatory, corporate and institutional development, thereby paving the way for deepening structural reforms.

A highly integrated ASEAN is crucial to the decisions of relocation and expansion by MNCs. According to the ASEAN Business Outlook Survey 2014 conducted by the US Chambers of Commerce (USCC), about three-fourths of the survey respondents from US companies operating in ASEAN view ASEAN integration as vital to their business in ASEAN. Likewise, the Federation of Japanese Chambers of Commerce and Industry in ASEAN has highlighted to the Secretary General of ASEAN that deepening of production networks by Japanese firms would be better executed if ASEAN becomes more integrated.[10] In light of this, strategies for future ASEAN integration should include:

(xii) Completing the implementation of the Master Plan on ASEAN Connectivity (MPAC) for greater physical, institutional and people-to-people connectivity, facilitating a more seamless flow of trade and investment and thus distribution of production networks across the region;

[10]See Intal *et al.* (2014, 95–96).

(xiii) Fully establishing the ASW for harmonising customs procedures and minimising trade costs;
(xiv) Developing an institutional framework for public-private partnerships (PPPs) in infrastructure development, consolidating existing public and private financial resources and tapping into resources from the newly established Asia Infrastructure Investment Bank (AIIB) in which all ten ASEAN countries are founding members;
(xv) Working with MNCs to coordinate industrial investments amongst the ten ASEAN member states, compounding the net benefits of regional integration in line with individual members' comparative advantage;
(xvi) Promoting intra-regional information sharing and learning, especially on joint industrial projects between ASEAN members and external partners, for instance the Malaysia–China Kuantan Industrial Park and China–Singapore Suzhou Industrial Park, which can serve as inputs for cooperation blueprints between ASEAN and extra-regional partners.

Element 2: ASEAN as a Productive Food Basket of Asia and the World

As noted in Chapter 3 of this book, as a region endowed with vast fertile land, diverse topography and abundant labour, ASEAN has enormous potential not only to achieve self-sufficiency in food but also to become the food basket of Asia and the world. Furthermore, agriculture remains a vital economic sector in the majority of ASEAN member states. Agricultural growth is thus paramount to both regional food security and economic competitiveness. Nonetheless, while food accessibility has been addressed in multiple regional cooperation efforts such as the ASEAN Integrated Food Security (AIFS) Framework launched in 2009 and the ASEAN Plus Three Emergency Rice Reserve (APTERR) enacted in 2012, the supply side of food security, particularly agricultural productivity, has not been given due attention.

The case of Malaysia as discussed in Chapter 3 underscores the fact that agriculture can be integral to the industrialisation process in a dual economic structure, bolstering the overall national competitiveness and economic development. In fact, Malaysia's success in agricultural development has been underpinned by its concerted efforts to foster a competitive agro-based industry. Drawing on our empirical study on agricultural TFP growth in ASEAN-8 and in light of Malaysia's experience in agricultural development, the following important policy lessons can be drawn for ASEAN as a whole to become a key food basket in the world:

(i) Identifying and developing agricultural niches based on climate, natural endowment and core competencies, with a view to becoming internationally competitive in those particular products;
(ii) Fostering strong linkages between agriculture and resource-based manufacturing in a dual economic structure;
(iii) Developing agro-based industries through nurturing the growth of competitive agribusiness companies and PPPs;
(iv) Identifying opportunities for vertical fragmentation of the agro-based industry, integrating smallholder producers into the agribusiness value chains, thereby raising their incomes and alleviating rural poverty;
(v) Upgrading agricultural infrastructure including drainage and irrigation systems as well as roads, sea ports and electricity for improving farmers' access to the market;

(vi) Promoting and incentivising public and private agricultural R&D to develop productivity-enhancing, environment friendly and sustainable technologies and practices, for instance, creating new varieties of crops with increased resistance to pests;
(vii) Improving resource management for stimulating agricultural TFP growth;
(viii) Providing preferential credit schemes and extension services to farmers to enhance their access to capital and new technologies, thereby augmenting their productivity and capability to tackle climate change.

Additionally, the following measures could be incorporated in future regional cooperation initiatives in agricultural development and food security:

(ix) Strengthening regional supply capacity through: harmonisation of phytosanitary measures for crop products and maximum residue limits of commonly used pesticides, capacity building in pest risk prevention and transfer and adoption of new production and post-harvest technologies;
(x) Promoting and facilitating intra-regional collaboration in agricultural R&D for enhancing regional agricultural TFP as well as narrowing the productivity gap within ASEAN;
(xi) Encouraging knowledge sharing on new agricultural technologies, sustainable production techniques, resource management, disease controls, and food safety;
(xii) Implementing capacity building in human resource and agro-preneur development to augment the supply of skilled labour in agriculture and promote entrepreneurship in agro-based industries;
(xiii) An institutional framework for technical collaboration between the ASEAN member states and the private sector should be explored. Areas of collaboration could include: technology transfer, value chain development, provision of high-yielding varieties, agro-chemicals and other inputs and marketing and trade.

Element 3: ASEAN as an Inclusive and Resilient Growth Centre

One of the most salient characteristics of ASEAN relative to other regional groupings is its distinct heterogeneity in levels of development with two high-income countries (Brunei and Singapore), one upper-middle-income country (Malaysia), five lower-middle-income countries (Indonesia, Laos, Philippines, Thailand and Vietnam) and two low-income countries (Cambodia and Myanmar). Narrowing this development gap has been featured in the AEC blueprint as a key measure under the third pillar, namely Equitable Economic Development. While emphasis has been on identifying and providing technical assistance and capacity building programmes for CLMV under the Initiative for ASEAN Integration (IAI) framework, much remains to be done for the gap to be substantially reduced for the CLMV grouping to become the next growth frontier in ASEAN.

One way to achieve this is to leverage on the corresponding wide difference in wages across the ten ASEAN member states for the expansion and distribution of regional production networks. The spread of regional production networks, largely from Japan and Europe, has been the cornerstone of economic development in the East Asian, newly industrialised economies (NIEs), namely Hong Kong, Singapore, South Korea and Taiwan and subsequently China, a phenomenon that has been crystallised in the "flying geese model" pioneered by Kaname Akamatsu in the 1930s.[11] Facilitating the distribution of regional supply chains could play a pivotal role in narrowing the development divide in ASEAN. Strengthening economic integration with East Asia, the US and the EU is thus paramount in this process. In this respect, the conclusion of the RCEP and FTAAP would be monumental.

To be sure, such migration of industries by developed economies does not automatically result in dissemination of technology and knowledge to the host countries. For middle- and low-income economies in ASEAN to move up the industrial ladder, foreign direct investment (FDI) inflows from developed economies must be accompanied by appropriate technological capabilities and human resource for ensuring effective transfer of technology- and knowledge-intensive manufacturing activities from MNCs rather than mere employment creation in low value-added processes. This is where policies for attracting the right FDI with potential for upgrading, strengthening human capital and nurturing best-performing SMEs must be heeded from the outset. Strategies for CLMV could therefore include:

(i) Identifying and attracting industrial activities with potential for technological upgrading and transfer of capital and know-how;
(ii) Facilitating migration of industries from developed economies through developing a pro-business investment climate with competitive tax regimes and sound governance;
(iii) Providing appropriate incentives for sharing of knowledge and technologies;
(iv) Strengthening human capital through more rigorous and relevant technical education, preferably in collaboration with MNCs, industry associations and research institutes;
(v) Deepening domestic reforms with particular emphasis on addressing market failures, substantially reducing the state's direct involvement in the economy, improving bureaucratic capabilities and moving towards a rule-based, transparent and competitive economic environment.

Building on the IAI framework, additional measures can be taken to bolster existing programmes for narrowing the development divide between ASEAN-6 and CLMV as follows:

(vi) Deepening economic integration with extra-regional powers through concluding the RCEP and expediting the negotiation process under the FTAAP, as well as the ASEAN–EU Free Trade Agreement (FTA) which has been suspended since 2009;
(vii) For ASEAN-6, particularly Singapore, Malaysia and Thailand: facilitating relocation of industries from East Asia and potentially the US and EU to CLMV. Models such as the Ota Technopark which serves as a hub for Japanese SMEs operating in Thailand could be applied in CLMV;
(viii) Facilitating intra-regional technology and knowledge transfers, particularly from the top performers such as Singapore and Malaysia to CLMV, and exploiting synergy and network effects across different member states for strengthening firms in the CLMV grouping;

[11] See Tan *et al.* (2015a, Chapter 1, Sec. 1.2.2) for a detailed discussion on this phenomenon.

(ix) Encouraging outward direct investment from Singapore, Malaysia and Thailand to CLMV, for example the Vietnam–Singapore Industrial Park;
(x) Intra-regional learning and technical assistance should be extended to not only trade and investment facilitation but also institutional capability-building and macro-prudential management to strengthen CLMV's capabilities in tackling economic crises;
(xi) Cross-border contract farming could be encouraged for improving agricultural productivity and enhancing the competitiveness of the agribusiness industry in CLMV.

Element 4: ASEAN as a People-centric Community

Regional economic integration would only be meaningful if the benefits of integration are shared by the people and businesses in the region. However, as discussed in Chapter 1 of this book, the process of establishing the AEC has primarily been a top-down state-led project with few inputs as well as limited awareness amongst the private sector in ASEAN.[12] The issue of lack of awareness about the AEC has been reiterated by ASEAN policy-makers in multiple forums. For instance, in a discussion on ASEAN integration in September 2015, former ASEAN Secretary-General Mr Ong Keng Yong noted that, "So far we have not been able to demonstrate, in a simple way, the benefits of ASEAN membership to citizens."[13]

While active engagement with relevant stakeholders has been incorporated as one of the components in the Nay Pyi Taw Declaration on the ASEAN Community's Post-2015 Vision, more measures could be taken for ASEAN to become a truly people-centric community such as:

(i) Organising regular regional forums between ASEAN leaders and the private sector for enhancing awareness of the ASEAN Community, soliciting feedback for the post-2015 ASEAN agenda and facilitating greater utilisation of the rules of origin and preferential tariffs amongst regional enterprises;
(ii) Adopting a consultative mechanism which guarantees meaningful, representative and inclusive exchange of ideas between the business community, civil society and ASEAN leaders, and ensuring the participation of private enterprises, non-governmental organisations as well as think-tanks in the region in shaping the direction of ASEAN;
(iii) Broadening and deepening the involvement of private sector in regional cooperation efforts, starting with PPPs in infrastructure development;
(iv) Promoting inter-state learning, knowledge sharing and cultural exchange;
(v) Facilitating greater interaction and exchange in all sectors of society and encouraging people-led initiatives;
(vi) Cultivating an ASEAN identity amongst all the ten member states;
(vii) The role of the ASEAN Secretariat could be expanded to facilitate effective community and engagement of the business community, civil society as well as the people in the member states.

---

[12] See also Hu (2013).
[13] See Chia (2015).

Element 5: ASEAN as a Key Fulcrum of Global Economic Integration

ASEAN has been relatively more outward-oriented than many regional groupings in the world. For one, ASEAN as a whole, trades more with extra-regional economies. Most of the ten ASEAN member states are also reliant on FDI inflows from outside the region. A globally connected ASEAN as a driver of multilateral economic integration has been the cornerstone of many ASEAN-led initiatives, including the formation of the AEC, as analysed in-depth in Chapter 1 of this book.

While ASEAN has been rather successful in bringing together multiple developed and emerging economic powers under the ASEAN+1 FTAs umbrella as well as in many ASEAN+x forums such as the East Asia Summit, the successful signing of the high-standard US-led TPP, in which four ASEAN economies are members, are as contrast to the sluggish negotiation process of the ASEAN-initiated RCEP, which failed to be concluded by the 2015 deadline, accentuates the risk that ASEAN may lose its centrality in the fast-changing global and regional economic architecture.

Not less paramount is the meteoric rise of China, whose economic and geopolitical clout has been deeply felt not only in Asia but also globally with a series of initiatives announced in 2014 both as counter-recessionary measures as part of its economic strategy in the New Normal[14] while fulfilling the China dream.[15] The Belt & Road (B&R) initiative is a system of multi-modal and multi-track infrastructure development project aiming, in particular, to enhance connectivity across Southeast Asia, South Asia and Africa, which is supported by the AIIB established in December 2015 with 30 founding member-states including ASEAN-10. Whether or not China is pursuing its hegemonic ambition, the question of ASEAN unity becomes even more pressing in light of the "Sino-American rivalry for influence" in the region, with a real possibility of ASEAN "devolving into a house divided against itself" if member states begin to choose sides between the two powers.[16]

A case in point is the failure of ASEAN's foreign ministers to produce a joint communique at the ASEAN Summit chaired by Cambodia in July 2012 — the first time in the bloc's 45-year history — owing to disagreement over the South China Sea dispute, which is widely believed a reflection of China's expanding influence in ASEAN's agenda. As noted in our AECV2030, amidst the geopolitics of the South China Sea dispute concerning many in ASEAN and with Japan taking a harsher military and naval stance over Diaoyu/Senkaku Islands, with China, on the other hand, pursuing the AIIB and B&R, the geo-economics has to be rendered more realistic.[17]

In other words, balancing power rivalry while driving global cooperation efforts toward bolstering economic development in the region is the way forward for ASEAN. After all, geo-economics holds critical significance in ASEAN's path toward prosperity as the fourth largest economy in the world by 2030. ACI's vision for ASEAN-10 thus extends beyond the

---

[14]See Tan *et al*. (2015b) for an analysis of China's economic strategies under the New Normal.

[15]See Tan *et al*. (2015a, Chapter 1).

[16]See Mahbubani (2015, 140).

[17]See Tan *et al*. (2015a, Chapter 1).

"coherent approach towards external economic relations" as stipulated in the AEC blueprint,[18] for ASEAN to rise as a key fulcrum of global economic integration, starting with its leadership in initiatives for widening physical and information connectivity in Asia as proposed in our AECV2030. Beyond connectivity, even more imperative is the need for ASEAN to speak with one voice, with the goal of regional prosperity in mind, in the global drive toward deeper integration.

In this regard, the ongoing negotiation of the RCEP would serve as the litmus test for ASEAN as it takes the lead in achieving a high-standard, comprehensive FTA involving many of its major trading partners including Japan and China. As demonstrated in Chapter 1 of this book, the net welfare effects would be the largest if the FTAAP is concluded, encompassing 21 economies from both sides of the Pacific Ocean, and the outcome of the RCEP would be a key determinant of the process. As the FTAAP involves both traditional and emerging major powers including the US, Japan and China, ASEAN has a crucial role to play in ensuring that power rivalry will not derail integration efforts, and that the FTAAP agreement would help to bridge the development gap between ASEAN and developed economies as well as within ASEAN itself.

For ASEAN to effectively preserve ASEAN centrality and become a key fulcrum in global economic integration, the following measures should be heeded:

(i) Drawing on ASEAN's experience with the AEC as well as multiple bilateral and multilateral FTAs signed by the member states, including the TPP, to lead the negotiation process of the RCEP, with the aim for it to become a high-quality and encompassing trade agreement, thereby bridging the ambition gaps with the TPP;
(ii) Deepening economic relations with APEC economies and driving the formulation and negotiation process of the FTAAP, with special emphasis on developmental components including technical assistance and capacity building for lower-middle-income and low-income members;
(iii) ASEAN member states must speak with one voice in articulating ASEAN's common interests in global forums and initiatives, especially APEC and the China-led B&R, and in shaping the governance structure and operational guidelines for new global institutions such as the AIIB where all ten ASEAN economies are members;
(iv) Pushing for the conclusion of the Code of Conduct on the South China Sea with China while ensuring the implementation of the 2002 Declaration of the Conduct of Parties in the South China Sea, and emphasising multilateralism as well as peaceful means of resolving the dispute;
(v) Insofar as infrastructure development is concerned, ensure the China-led B&R and other existing infrastructural funding mechanisms serve as complementary and supplementary efforts to bolster the MPAC and enhance intra-ASEAN land and sea connectivity;
(vi) Strengthening mutual trust amongst the ten ASEAN member states;
(vii) Identifying fault lines in regional and national agenda which may generate potential threats of disintegration in ASEAN and strategizing accordingly in accordance with the principles of unity and consensus. A pertinent example would be the possibility of

[18] See ASEAN Secretariat (2008, 25).

constructing the Kra Isthmus Canal across southern Thailand mooted by China and Thailand, which would compete directly with ports in Singapore's Strait of Malacca;[19]

(viii) While ASEAN as a whole stands to benefit from the Sino–American rivalry for influence in the region as both have sought to strengthen economic and diplomatic ties with the regional grouping, the ten ASEAN member states must resist the geopolitical pressures to take sides. It is in the collective interest of ASEAN that neither the US or China is alienated and both continue to be engaged in ASEAN-led initiatives.

5.3 FUTURE RESEARCH AGENDA

In view of ACI's five-element development vision for ASEAN-10 discussed in the foregoing section and building on ACI's strengths in evidence-based quantitative policy analysis, our research agenda in the next publications on ASEAN may include at least three policy areas including: (i) an empirical analysis of the impacts of exchange rates on FDI and trade, (ii) a multi-country global vector autoregressive model on economic interdependencies, and (iii) tracking firm-level productivity, efficiency and competitiveness for the 16 Asian economies.

5.3.1 Impacts of Exchange Rate Volatility on Foreign Direct Investment and Trade

The world economy appears to enter into turbulent times once again. Fears of 'secular stagnation' in the advanced economies aside, the gradual normalisation of the US monetary policy has also generated fears that there will be a repeat of the 'taper tantrum' that hit the emerging market economies in Asia in 2013. Changes in the US monetary policies have always exerted significant spillover effects on Asian economies, especially pronounced during the period following the global financial crisis.

While ultra-loose monetary policies in the form of quantitative easing resulted in currency appreciations in the region due to massive capital inflows flowing into several Asian economies, the tapering of the same through interest rate hikes generated massive capital outflows, and consequent currency depreciations. Although the severity of the impact was more on fragile emerging economies with weak macroeconomic fundamentals, ebbs and flows of international capital flows sensitive to US monetary policy changes have in general given rise to a variety of policy challenges concerning monetary and exchange rate management in Asia, which in turn has a significant bearing on their overall macroeconomic and financial stability.

Swings in emerging market currencies also matter because of the potential macroeconomic implications they can have on a country's trade and foreign investments. Considering the degree of importance of trade and foreign investments in Asia, one of the most significant challenges for the policy makers in the region is to understand the potential implications of exchange rate volatility on trade flows as well as foreign investments.

---

[19]See Chew (2015).

There is particularly a large extant academic literature that has well documented the importance of FDI as a source of investment, technology and plugging countries into regional value chains helping them with their exports. While there are number of factors that could affect FDI, countries are keenly aware of the need to remain cost competitive if they want to continue to attract FDI inflows. This is especially applicable to the economies in ASEAN. Our future research will thus focus on empirically quantifying the impact of real exchange rate volatility on FDI inflows to ASEAN economies.

Also, given that the world has seen a proliferation of global value chains in the way international trade transactions take place and that ASEAN has been plugged into this network, it naturally becomes important to understand what role exchange rate volatility plays in the region's exports, which will also be a topic of future research.

5.3.2 Multi-country Global Vector Autoregressive Model on Economic Interdependencies

As ASEAN has embarked on the path of deeper regional and global economic integration, macroeconomic risk management has become more pertinent than ever. While ASEAN as a whole has successfully weathered the global financial crisis of 2008–2009, threats of crisis contagion still loom large as the regional bloc is increasingly embedded in the global economy. Such increasing economic interdependencies between different economies invariably mean that the many channels of shock transmission and their potential ramifications in individual ASEAN member states must be thoroughly explored and mitigated for ASEAN to remain resilient in the uncertain global economic landscape.

Recognising the importance of accounting for shock transmission in policy analysis, in 2014 we constructed the Structural Vector Autoregressive (SVAR) model on trade flows to evaluate economic interdependencies amongst eight ASEAN economies and their three key trading partners, namely the Organisation for Economic Cooperation and Development (OECD), China and India, before and after the implementation of the MPAC in 2010.[20] Expanding on this model, our future research looks to examine a larger set of developed and emerging economies under the novel Global Vector Autoregressive (GVAR) framework to empirically quantify and predict the effects of changing macroeconomic conditions at both the national and the global level on the economic performances of various economies, with a particular focus on ASEAN-10.

The GVAR model, initially developed in the aftermath of the 1997–1998 AFC as a tool for credit risk modelling, is particularly suited to the analysis of shock transmissions as well as long-run relations such as purchasing power parity and uncovered interest parity between different economies. It can therefore be used for *ex ante* forecasting of macroeconomic variables in a theoretically coherent and statistically consistent manner. The application of the GVAR framework as a topic of our future research will enrich our existing extensive body of evidence-based quantitative policy analysis that contributes to astute policy formulation in ASEAN.

---

[20] See Tan *et al.* (2015a, Chapter 1, Sec. 1.2.4).

5.3.3 Tracking Firm-Level Productivity, Efficiency and Competitiveness for the 16 Asian Economies

The economic literature has long recognised that it is paramount to not only look at macroeconomic factors to devise policies for improving an economy's competitiveness but also to examine firm-level performances that underpin sectoral and correspondingly export competitiveness. Furthermore, SME development is one of the key policy measures under the AEC as well as the post-2015 AEC agenda.[21] ACI thus seeks to contribute to the policy discourse on SMEs in the region by tracking and benchmarking firm-level productivity, efficiency and competitiveness in ASEAN-10 and six other Asian economies including Japan, South Korea, Taiwan, Hong Kong, China and India.

In particular, as an expansion of our extensive datasets used for conducting competitiveness ranking and simulation analysis at the national level for ASEAN-10 and at the sub-national level for Greater China, India and Indonesia, ACI is looking to build a firm-level database comprising of both hard and soft data, which will be used for constructing an index for tracking productivity and monitoring efficiency for companies across the 16 Asian economies. Our company-level dataset would capture the various dimensions of firms' activities and operations which could potentially be upgraded for an overall improvement of competitiveness *vis-à-vis* similar firms in the same sectors in Asia. Such dimensions are crystallised into four environments, namely: (i) financial and business statistics, (ii) human resource and manpower training, (iii) branding, marketing and internationalisation, and (iv) governance, management and execution manpower. The study would allow us to draw important policy insights as to how policy-makers in the region can nurture and foster competitive enterprises, particularly as the wave of liberalisation is about to sweep in with the conclusion of the TPP, the ongoing negotiation of RCEP and the potential formation of the FTAAP.

REFERENCES

ASEAN Secretariat. 2008. *ASEAN Economic Community Blueprint*. Jakarta: ASEAN Secretariat.

ASEAN Secretariat. 2010. *Master Plan on ASEAN Connectivity*. Jakarta: ASEAN Secretariat. www.asean.org/resources/publications/asean-publications/item/master-plan-on-asean-connectivity-2.

ASEAN Secretariat. 2015. *A Blueprint for Growth — ASEAN Economic Community 2015: Progress and Key Achievements*. Jakarta: ASEAN Secretariat.

Asian Development Bank Institute. 2014. *ASEAN 2030: Toward a Borderless Economic Community*. Tokyo: Asian Development Bank Institute.

Chew, Hui Min. 2015. "Renewed hype over China–Thai canal project: 5 things about the Kra Canal." *The Strait Times*, May 21. www.straitstimes.com/asia/se-asia/renewed-hype-over-china-thai-canal-project-5-things-about-the-kra-canal.

Chia, Yan Min. 2015. "Raising citizens' awareness of Asean benefits a challenge." *The Strait Times*, September 15. www.straitstimes.com/business/economy/raising-citizens-awareness-of-asean-benefits-a-challenge.

---

[21] See ASEAN Secretariat (2015).

Heng, Swee Keat. 2015. "Singapore, a Connected and Trusted Home Base for Companies to Go Global." Opening Remarks at the 2015 Asia Competitiveness Institute Annual Conference, Singapore, November 26–27.

Hu, Albert G. 2013. "ASEAN Economic Community Business Survey." In *The ASEAN Economic Community — A Work in Progress*, eds. S. B. Das, J. Menon, R. Severino and O. L. Shrestha. Singapore: Asian Development Bank and Institute of Southeast Asian Studies.

Intal, Ponciano Jr., Yoshifumi Fukunaga, Fukunari Kimura, Phoumin Han, Philippa Dee, Dionisius Narjoko and Sothea Oum. 2014. *ASEAN Rising: ASEAN and AEC Beyond 2015*. Jakarta: Economic Research Institute for ASEAN and East Asia (ERIA).

Mahbubani, Kishore. 2015. "ASEAN as a Living, Breathing Modern Miracle." *CIRSD Horizons* (2): 136–148.

McKinsey Global Institute. 2014. *Southeast Asia at the crossroads: Three paths to prosperity.*http://www.mckinsey.com/insights/energy_resources_materials/three_paths_to_sustained_economic_growth_in_southeast_asia

Tan, Khee Giap, Sangiita Wei Cher Yoong, Sasidaran Gopalan, and Le Phuong Anh Nguyen. 2015a. *2014 Annual Competitiveness Ranking and Simulation Study for ASEAN-10 and Development Strategies to Enhance Asia Economic Connectivity*. Singapore: World Scientific.

Tan, Khee Giap, Randong Yuan, and Sangiita Wei Cher Yoong. 2015b. *2014 Annual Provincial and Regional Competitiveness Analysis for Greater China Economies: Development Strategies under the New Normal*. Singapore: World Scientific.

Tan, Khee Giap, Le Phuong Anh Nguyen, and Trieu Duong Luu Nguyen. 2016. "Make or Break for Vietnam in the TPP Era?" *The Straits Times*, February 25. www.straitstimes.com/opinion/make-or-break-for-vietnam-in-the-tpp-era

Appendix 1

Policy Brief: Make or Break for Vietnam in the TPP Era?

On 5th October 2015, the United States and 11 Pacific Rim nations including Vietnam concluded the Trans-Pacific Partnership (TPP) negotiation, the largest trade pact since the 1994 Uruguay Round that led to the establishment of the World Trade Organisation (WTO). Together the 12 TPP members make up about 40% of the world's Gross Domestic Product (GDP) and account for around one-third of global trade, and have since signed the TPP agreement on 4th February 2016.

Long before the trade agreement was signed, it had been predicted that Vietnam will stand to achieve the biggest gain among the 12 members of the TPP in terms of GDP, foreign investment and exports. In a widely cited study, the Peterson Institute of International Economics estimated that the TPP could bring about an additional 13.6% GDP growth for Vietnam over 10 years to 2025. Sectors that are forecast to benefit Vietnam the most from TPP include garments and textile, footwear, aquaculture, automotive and timber.

Many pundits have doubted Vietnam's ability to fulfil the high quality agreements of the TPP, and they also cautioned about the risk of the Vietnamese economy being dominated by foreign firms once the country opens door to investment from the TPP members. Although the recent Twelfth National Congress of the Vietnamese Communist Party re-elected the conservative Party Secretary Nguyen Phu Trong, it is widely believed that the economic reforms will continue and political confrontation with China should moderate somewhat. Reformist government leaders and scholars in Vietnam had long anticipated the completion of the TPP deal, hoping it would provide the critical impetus to the current lacklustre economic performance at home.

According to data compiled by the World Bank, Vietnam's real GDP growth has significantly slowed from an average of 7.2% between 1990 and 2007 to 5.7% from 2008 to 2014. Nonetheless, unless inflationary pressure is contained, it will dilute Vietnam's growth momentum further. While Vietnam has enjoyed a renewed surge in foreign direct investment (FDI) inflows in recent years, it is also vital that Vietnam should maintain a stable Vietnamese Dong. If history is any indication, the country's reliance on low labour costs will soon run its course.

Thirty years have passed since Vietnam promulgated its *Doi Moi* (renovation) reform dismantling the central planning system in pursuit of a market economy. Nonetheless, Vietnam has only managed to reduce the scope of the state in the economy without changing the fundamental roles and functions of the government, neither did we see significant improvement on public governance. The results of such half-hearted reform efforts are rampant market as well as state failures, as manifested in the country's multiple ailments: chronic budget deficit, high public debt, proliferation of non-performing loans and an inefficient yet dominant state sector, among others. High-profile default cases of large state-owned corporations such as Vinashin and Vinalines offer just the tip of the much bigger iceberg of poor governance, corporate graft and misallocation of resources in favor of the ailing state sector.

Against such a backdrop, it is widely believed among the reformist government leaders and academics that increased international integration through entering into the so-called "new generation" of "high-standard" free trade agreements (FTAs) is the key driver of domestic reform. Membership in such FTAs usually requires a revamp of national legislation toward levelling the playing field across all sectors, higher standards of intellectual property rights as well as labor conditions. More importantly, such agreements could enhance the rule of law, and thus transparency and predictability, in the member states. Therefore, Vietnam's membership in the TPP could not have been more opportune at a time when Vietnam is struggling to give its fizzling reform momentum a push.

While there has been much talk on what Vietnam could potentially benefit from the TPP, what is lacking in the discourse is what Vietnam has to address in preparation for when TPP fully takes effect about two years from now. In this respect, it is crucial to keep in mind that TPP is not the "holy grail" for Vietnam's structural problems. Globalisation only rewards the stronger and more competitive ones; the weaker ones often risk being trapped in low value-added activities and thus low levels of income. In other words, whether the TPP will be a make or break for Vietnam depends on latter's ability to resolve its most urgent economic malaises before the waves of global competition sweep in.

In that regard, the Asia Competitiveness Institute's (ACI) flagship project on competitiveness analysis and simulation study for Vietnam *vis-à-vis* its ASEAN peers offers an instructive policy roadmap for Vietnamese policy-makers. ACI's analysis shows that Vietnam's Overall Competitiveness score has plunged more negatively, or further from the regional average, between 2009 and 2013 (see Figure 1). The result highlights the urgency of deepening economic reform in the face of intensified competition as a result of the TPP.

A competitiveness ranking in itself is like a beauty contest; it merely identifies who are doing well and who are facing challenges, but stops short of giving more constructive advice on improving the rankings. ACI's Competitiveness Index constructively went one step further by identifying the top weaknesses of individual economies as a guide for impactful policy formulation. Our studies have shown that amongst Vietnam's top 20% weakest indicators out of the 121 indicators used in ACI's competitiveness framework for ASEAN-10, the country remains uncompetitive in a wide array of areas. The most salient and indeed recurring over the decades, in the country's top weaknesses could be summarised under three groups of indicators which include: prevalence of market distortions, corporate governance and transparency and quality of human capital.

Therefore, for Vietnam to successfully ride the wave of the TPP, it has to focus on three urgent policy pillars as the beginning of a deeper, more comprehensive economic reform. If

Figure 1: Overall Competitiveness Scores and Rankings of ASEAN-10, 2000–2013

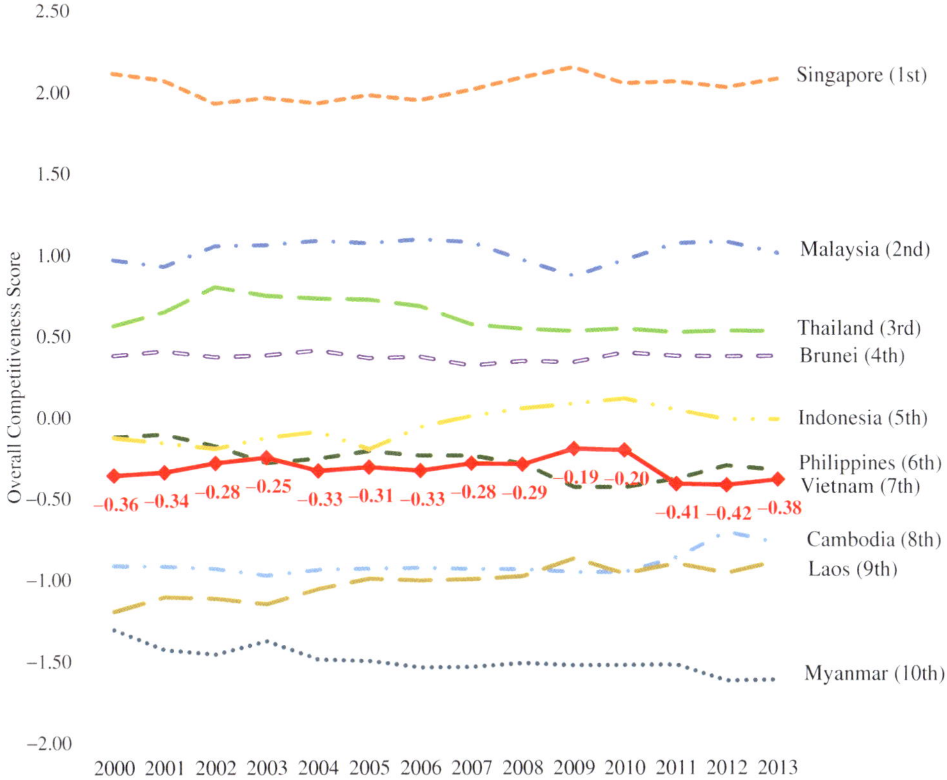

Source: Asia Competitiveness Institute
Note: Value in the parenthesis denotes 2015 competitiveness ranking amongst ASEAN-10

implemented well, these initiatives will provide the much needed revitalisation of Vietnam's sluggish economy.

Eliminating market distortions through deepening SOE reform and levelling the playing field

Despite three decades of economic reforms toward a market economy, market distortions in favour of SOEs remain widespread, which perpetuates unfair competition between the private and the state-owned sector. In particular, SOEs tend to occupy the majority of land allocated for business activity and are often granted a considerable portion of new land provided by the government for business rental. Private firms applying for new industrial land are required to go through complex procedures which typically take two to three years and reportedly involve corruption. Another area where favouritism prevails is access to credit. SOEs are still able to obtain bank loans even when their asset size or profitability does not meet the banks' lending criteria. In terms of market access, while the government usually offers important market opportunities for firms to grow and become competitive through state contracts, eligibility criteria for bidding such contracts are clearly biased toward SOEs. Consequently, the majority of public contracts have been awarded to SOEs.

The promulgation of the third Enterprise Law and the launch of a new round of SOE restructuring in 2005 raised new hopes for private sector development as well as more robust economic growth for Vietnam. Nonetheless, the progress has been notably slow with the state-owned sector still accounting for a significant 40% of annual total investment in Vietnam. According to the latest World Bank Report, as of the first quarter of 2015, only 29 SOEs were equitised out of the target of 289 for the whole year. The key obstacle to the reform process thus far lies in big state-owned groups such as Vinalines, MobiFone and Vinacomin. The amount of state resources held by these groups as well as their impacts on the economy has made it difficult for the government to evaluate and move forward with the privatisation plans. While the pace of the SOE reform has to be quickened, it is important for Vietnam to engage strategic investors with industry expertise in the process.

More fundamental is the elimination of monopoly power and special privileges SOEs have been enjoying which have resulted in market distortions at the expense of private enterprises. To be sure, levelling the playing field between state-owned and private enterprises is one prominent provision under the TPP. While Vietnam has been granted concessions in complying with this particular provision, it is exigent that Vietnam puts in place fundamental market institutions required for local enterprises to develop and be able to cope with the anticipated surge in FDI. To that extent, deepening the SOE reform entails relinquishing the state's direct involvement in the economy to focus on its facilitative role of setting transparent rules and enforcing them while protecting the society from negative externalities. Vietnam therefore must make it its top priority to move toward a rule-based, transparent and competitive economic environment. Streamlining bureaucracy and improving bureaucratic capabilities are crucial in such efforts.

Building strong and transparent corporate governance

Without competitive enterprises, it is hard to envisage how Vietnam could benefit from the TPP, particularly when joining the trade bloc means domestic firms will have to compete with firms from the other 11 members. Nonetheless, corporate governance in Vietnam remains problematic, notwithstanding the rapid development of the legal and regulatory framework for investor protection. For instance, new laws on independent audit was enacted in 2011, and in 2013 the State Securities Commission of Vietnam (SSC) revised Corporate Governance Regulations and a Model Charter that detail good practice for listed companies. Nonetheless, weaknesses remain in terms of poor disclosure, outdated accounting standards, lack of internal control mechanisms and limited accountability of firm management. In fact, current practices in many public enterprises fall far short of what is required in the revised Corporate Governance Regulations, which also highlights the government's difficulties in enforcing the rule of law.

As competitive pressures become more ferocious now that Vietnam is part of the TPP, the need for building strong and transparent corporate governance in accordance with international good practices has become more acute. The roles and functions of the SSC could actually be enhanced as an independent supervisory body to ensure compliance of corporate laws and regulations. Incentives for firms to uphold accountability to investors should be introduced. For example, the rankings of the Corporate Governance Scorecard could be made public for investors to monitor the listed firms as well as for the firms to improve their management. An organisation dedicated to providing training and technical support to firms could be established, ideally with international industry experts on board or acting as advisors.

Upgrading labour skills

The central thrust of Vietnam's *Doi Moi* (renovation) reform has been resource mobilisation from agriculture to manufacturing and services on the back of a large young, literate and low-cost workforce. Nonetheless, such model is not sustainable as competing based on cheap labour is simply a race to the bottom, especially when Vietnam's youth population is shrinking. Only by upgrading labour skills and labour productivity can Vietnam maintain healthy growth and successfully industrialise. After 30 years of reform, firms still find it hard to find skilled workers to fill the jobs, particularly managers, executives and technicians. Foreign corporations have also been complaining about the lack of reliable local suppliers, which is a testament to the wide capability gap between the FDI and the local sector.

Without concrete policies and programmes for improving labour skills, Vietnam will certainly risk being dominated by foreign firms. In fact, many foreign textile companies have announced million-dollar investment plans in Vietnam to take advantage of the TPP's "yarn forward" rule. Upstream electronics manufacturers are expected to follow suit to supply for technology giants such as Samsung, Foxconn, Intel, etc. which are already present in Vietnam. Measures to support and upgrade best-performing local firms, as well as pair them up with multinational corporations, should be put in place. Tax incentives could be introduced to foreign firms engaging in technology transfer with local suppliers. Most importantly, technical education and training must be revamped in alignment with industry needs. Collaboration between universities, industry associations and multinational corporations is thus vital.

Geopolitical conflicts must not derail growth

Geopolitically Vietnam is also at a delicate juncture as its tensions with China have reached new heights since May 2014 when China deployed an oil-drilling rig inside Vietnam's exclusive economic zone near the Paracel and Spratly islands claimed by both countries. The incident sparked a string of anti-China protests, some of which turned into riots targeted not only at Chinese factories but also at Taiwanese, South Korean and Singaporean ones in Binh Duong, Dong Nai and Ha Tinh, eroding investors' trust in the country.

Growing frictions between Vietnam and China have led some experts to predict a military crisis between the two countries. Nonetheless given its current economic woes, Vietnam must not militarise or further escalate the dispute, or else it will get side-tracked from its reform which is of utmost importance. Instead, Vietnam should draw on its rising economic leverage from its TPP membership to facilitate the conclusion of the Code of Conduct (CoC) between ASEAN and China. By emphasising pluralism, Vietnam could play a critical role in strengthening ASEAN's unity and preventing the bloc from splitting and pitting against one another over China. Concurrently, Vietnam can focus on jumpstarting its second, more wide-ranging reform starting with the foregoing urgent policy pillars.

As the third largest foreign investor in Vietnam, Singapore stands to gain from the combination of Vietnam's membership in the TPP and its existing advantages, notably political stability, relatively low business and labour costs. In fact, Singapore, also a TPP member, could become a key strategic partner to Vietnam in its new development trajectory following its accession to the TPP. As more foreign firms seek to relocate their facilities to Vietnam to take advantage of the TPP's preferential tariffs, Singapore could facilitate this process given its extensive experience in the country, thereby enhancing the economic partnership between the two countries. In light of the anti-China riots which went out of control, shocked foreign investors and caused severe damage to Vietnam's image and international reputation, the Vietnamese authorities should reiterate their protection towards all investors and firm commitment to a new wave of reform that would bolster investors' confidence.

While the TPP presents Vietnam with substantial potential economic benefits, it is not the magic bullet for Vietnam's unresolved structural weaknesses. If Vietnam fails to address its shortfalls, the window of opportunity may last at most 10 years or even shorter. Instead of just "renovating" (*Doi Moi*) the economy as during the past 30 years, what Vietnam has to focus on is "revitalising" (*Hoi Sinh*) the economy starting with: eliminating market distortions in favour of SOEs, improving corporate governance and upgrading labour quality. Implementation of these three policies will pave the way for the second, more rigorous round of reform which is imperative for Vietnam right now. Failure to complete these three pressing pillars of reform could exacerbate income disparity which may happen sooner rather than later. It is make or break for Vietnam in the TPP era.

An abridged version of the policy brief was published as an editorial opinion for The Straits Times on 25th February 2016 (see below)

THURSDAY FEBRUARY 25 2016 THE STRAITS TIMES OPINION

Competing based on cheap labour is simply a race to the bottom, especially when Vietnam's youth population is shrinking. Only by upgrading labour skills and labour productivity can Vietnam maintain healthy growth and successfully industrialise. PHOTO: REUTERS

Make or break for Vietnam in the TPP era?

Its economy needs urgent reform as a competitiveness study finds it lagging behind its Asean peers

Tan Khee Giap, Nguyen Le Phuong Anh and Luu Nguyen Trieu Duong

For The Straits Times

Thirty years have passed since Vietnam began its "Doi Moi" (renovation) reform, dismantling the central planning system in pursuit of a market economy. However, the reform was carried out in half-hearted fashion.

The effects of this are rampant market as well as state failures, as manifested in the country's multiple ailments: a chronic budget deficit, high public debt, proliferation of non-performing loans and an inefficient yet dominant state sector.

According to the World Bank, Vietnam's real GDP growth has significantly slowed from an average of 7.2 per cent between 1990 and 2007 to 5.7 per cent from 2008 to 2014.

And compared to its Asean counterparts, its competitiveness has plunged. An analysis by the Asia Competitiveness Institute (ACI) shows that Vietnam's "overall competitiveness" score has plunged more negatively, or further from the regional average, between 2009 and 2013.

ACI's Competitiveness Index found that Vietnam remains uncompetitive in a wide array of areas. The most salient, recurring indicators are the prevalence of market distortions; issues of corporate governance and lack of transparency; and quality of human capital.

Hence, Vietnam's membership in the Trans-Pacific Partnership (TPP) – it was one of 12 signatories on Feb 2 – may seem to have come at an opportune time.

However, the TPP is not the "holy grail" for Vietnam's

provided by the government for business rental. Private firms applying for new industrial land are required to go through complex procedures, which typically take two to three years, and reportedly involve corruption.

Another area where favouritism prevails is access to credit. SOEs are still able to obtain bank loans even when their asset size or profitability does not meet the banks' lending criteria.

In terms of market access, while the government usually offers important opportunities for firms to grow and become competitive through state contracts, the eligibility criteria for bidding are clearly biased toward SOEs.

The promulgation of the third Enterprise Law and the launch of a new round of SOE restructuring in 2005 raised new hopes for private sector development, as well as more robust economic growth for Vietnam. But progress has been notably slow, with the state-owned sector still accounting for a significant 40 per cent of annual total investment in Vietnam.

According to the latest World Bank Report, as of the first quarter of last year, only 29 SOEs were equitised, out of the target of 289 for the whole year. The key obstacle to the reform process lies in big state-owned groups such as Vinalines, MobiFone and Vinacomin. The amount of state resources held by these groups, as well as their impact on the economy, has made it difficult for the

Given its economic woes, Vietnam must not militarise or further escalate the dispute, or

government to evaluate and move forward with the privatisation plans.

To be sure, levelling the playing field between state-owned and private enterprises is one prominent provision under the TPP. While Vietnam has been granted concessions in complying with this provision, it is vital that Vietnam puts in place fundamental market institutions required for local enterprises to develop and be able to cope with the anticipated surge in foreign direct investment.

To that extent, deepening the SOE reform entails relinquishing the state's direct involvement in the economy to focus on its facilitative role of setting transparent rules and enforcing them, while protecting society from negative externalities.

Vietnam, therefore, must make it the top priority to move towards a rule-based, transparent and competitive economic environment through streamlining bureaucracy and strengthening capacity building.

Building transparent corporate governance

Without competitive enterprises, it is hard to envisage how Vietnam could benefit from the TPP. Yet, corporate governance remains problematic even amid the rapid development of the legal and regulatory framework for investor protection.

For instance, new laws on independent audits were enacted in 2011, and in 2013, the State Securities Commission of Vietnam (SSC) revised Corporate Governance Regulations and a Model Charter that detail good practice for listed companies. But weaknesses remain in terms of poor disclosure, outdated accounting standards, lack of internal control mechanisms and

An organisation dedicated to providing training and technical support to firms could be established, ideally with international industry experts on board or acting as advisers.

Upgrading labour skills

The central thrust of Doi Moi reform has been resource mobilisation from agriculture to manufacturing and services, on the back of a large young, literate and low-cost workforce.

However, such a model is not sustainable, as competing based on cheap labour is simply a race to the bottom, especially when Vietnam's youth population is shrinking. Only by upgrading labour skills and labour productivity can Vietnam maintain healthy growth and successfully industrialise.

After 30 years of reform, firms still find it hard to find skilled workers to fill jobs, particularly managers, executives and technicians. Foreign corporations have also been complaining about the lack of reliable local suppliers.

In fact, many foreign textile companies have announced million-dollar investment plans in Vietnam to take advantage of the TPP's "yarn forward" rule. Upstream electronics manufacturers are expected to follow suit to supply technology giants, such as Samsung, Foxconn and Intel, which are already present in Vietnam.

Measures to support and upgrade best-performing local firms, as well as pair them with multinational corporations, should be put in place. Tax incentives could be introduced to foreign firms engaging in technology transfer with local suppliers.

Most importantly, technical education and training must be revamped in alignment with industry needs. Collaboration between universities, industry associations and multinational corporations is vital.

GEOPOLITICAL CONFLICTS MUST NOT DERAIL GROWTH

Geopolitically, Vietnam is also at a delicate juncture as tensions with China have reached new levels since May 2014's sovereignty dispute near the Paracel and Spratly islands.

Given its economic woes, Vietnam must not militarise or further escalate the dispute, or else it will get side-tracked from reform. Instead, Vietnam should draw on its rising economic leverage from the TPP membership to facilitate the conclusion of the Code of Conduct between Asean and China.

By emphasising pluralism, Vietnam could play a critical role in strengthening Asean's unity and preventing the bloc from splitting

Appendix 2

Competitiveness Analysis of ASEAN-10 Economies, 2000–2012 — List of Indicators

No.	Indicator	Unit	Reversed Indicator
1	MACROECONOMIC STABILITY		
1.1	***Economic Vibrancy***		
1.1.01	Gross Domestic Product (GDP)	US Dollars, Constant 2000 Prices	
1.1.02	Real GDP Growth	Percentage Change Per Annum	
1.1.03	GDP Per Capita	US Dollars, Constant 2000 Prices	
1.1.04	Output Agriculture (Value Added)	US Dollars, Constant 2005 Prices	
1.1.05	Output Industry (Value Added)	US Dollars, Constant 2005 Prices	
1.1.06	Output Services (Value Added)	US Dollars, Constant 2005 Prices	
1.1.07	Gross Domestic Savings	Percentage of GDP	
1.1.08	Gross Fixed Capital Formation	Percentage of GDP	
1.1.09	Inflation	Percentage Change Per Annum	*R*
1.1.10	Money Supply Growth (M2)	Percentage Change Per Annum	*R*
1.2	***Openness to Trade and Services***		
1.2.01	Current Account Balance	Percentage of GDP	
1.2.02	Exports of Goods and Services	US Dollars, Current Market Prices	
1.2.03	Imports of Goods and Services	US Dollars, Current Market Prices	
1.2.04	Openness to Trade	Ratio	
1.2.05	International Tourism Receipts	US Dollars, Current Market Prices	
1.3	***Attractiveness to Foreign Investors***		
1.3.01	Foreign Direct Investment Inflows	US Dollars, Current Market Prices	
1.3.02	Foreign Owned Companies	Survey Data	
1.3.03	Corporate Income Tax	Percentage of Profit	*R*
1.3.04	Lack of Discrimination Against Investors	Survey Data	
1.3.05	Levelness of Playing Field	Survey Data	
1.3.06	Transparency of Processes for Investors	Survey Data	
1.3.07	Business Impact of Rules on FDI	Survey Data	

(*Continued*)

Appendix 2 (*Continued*)

No.	Indicator	Unit	Reversed Indicator
2	**GOVERNMENT AND INSTITUTIONAL SETTING**		
2.1	***Government Policies and Fiscal Sustainability***		
2.1.01	Government Revenue (Including Grants)	US Dollars, Constant 2000 Prices	
2.1.02	Tax Revenue	US Dollars, Constant 2000 Prices	
2.1.03	Tax Revenue/Government Revenue	Percentage	
2.1.04	Government Consumption Expenditure	Percentage of GDP	
2.1.05	External Debt	US Dollars, Current Market Prices	*R*
2.1.06	International Reserves	US Dollars, Current Market Prices	
2.1.07	Net Foreign Assets	US Dollars, Current Market Prices	
2.1.08	International Monetary Fund Credit	US Dollars, Current Market Prices	*R*
2.1.09	Budget Balance	Percentage of GDP	
2.1.10	Tax Burden	Percentage of GDP	*R*
2.1.11	Money Supply (M2)	Percentage of GDP	
2.1.12	Deposit Interest Rate	Percentage	
2.1.13	Lending Rate	Percentage	*R*
2.1.14	Interest Rate Spread	Percentage	*R*
2.1.15	Exchange Rate Appreciation	Annual Average Percentage Change (Local Currency Unit per US Dollar)	*R*
2.2	***Institutions, Governance and Leadership***		
2.2.01	Quality of Government Policies	Survey Data	
2.2.02	Government Effectiveness	Survey Data	
2.2.03	Government Environmental Protection Policies	Percentage of Total Territorial Area	
2.2.04	Quality of Central Bank/Monetary Authority	Survey Data	
2.2.05	Government Stability	Survey Data	
2.2.06	Government Responsiveness	Survey Data	
2.2.07	Quality of Political Leadership	Survey Data	*R*
2.2.08	Political System Risk	Survey Data	*R*
2.2.09	Legislature/Parliament (Functioning of Government System)	Survey Data	
2.2.10	Strength of Auditing and Reporting Standards	Survey Data	
2.2.11	Institutional Sophistication	Survey Data	
2.2.12	Bureaucracy/Red Tape	Survey Data	
2.2.13	Lack of Corruption	Survey Data	
2.2.14	Lack of Cronyism	Survey Data	
2.2.15	Lack of Market Distortion due to State-Ownership	Survey Data	

(*Continued*)

Appendix 2 (*Continued*)

No.	Indicator	Unit	Reversed Indicator
2.3	***Competition, Regulatory Standards and Rule of Law***		
2.3.01	Corporate Governance	Survey Data	
2.3.02	Quality of Regulatory Environment	Survey Data	
2.3.03	Transparency of Government	Survey Data	
2.3.04	Monopolies and Cartel	Survey Data	
2.3.05	Policy Towards Private Enterprise Rating	Survey Data	
2.3.06	Ease of Starting Business	Number	*R*
2.3.07	Barriers to Trade	Survey Data	
2.3.08	Intellectual Property Rights Protection	Survey Data	
2.3.09	Regional Competitiveness	Survey Data	
2.3.10	Quality of Judiciary	Survey Data	
2.3.11	Effectiveness of Legal System	Survey Data	
2.3.12	Integrity of Legal System	Survey Data	
2.3.13	Quality of Police	Survey Data	
2.3.14	Political Change	Survey Data	
3	**FINANCIAL, BUSINESSES AND MANPOWER CONDITIONS**		
3.1	***Financial Deepening and Business Efficiency***		
3.1.01	Claims on Private Sector	Annual Growth as Percentage of Broad Money	
3.1.02	Domestic Credit to Private Sector	Percentage of GDP	
3.1.03	Quality of Banking and Financial System	Survey Data	
3.1.04	Domestic Credit Provided by the Financial Sector	Percentage of GDP	
3.1.05	Listed Domestic Companies	Number	
3.1.06	Market Capitalisation	US Dollars, Current Market Prices	
3.1.07	Total Value of Stocks Traded	Percentage of GDP	
3.1.08	Sovereign Debt Risk	Number	*R*
3.1.09	Quality of Stock Market Regulatory Authority	Survey Data	
3.2	***Labour Market Flexibility***		
3.2.01	Labour Force	Number	
3.2.02	Unemployment Rate	Percentage	*R*
3.2.03	Cost of Production Labour	US Dollars, Current Market Prices	*R*
3.2.04	Availability of Scientists and Engineers	Survey Data	
3.2.05	Cost of Management Staff	US Dollars, Current Market Prices	*R*
3.2.06	High Quality Management Staff	Survey Data	
3.2.07	Extent of Staff Training	Survey Data	

(*Continued*)

Appendix 2 (*Continued*)

No.	Indicator	Unit	Reversed Indicator
3.2.08	Labour Market Rating	Survey Data	
3.3	***Productivity Performance***		
3.3.01	Agriculture, Value Added Per Worker	US Dollars, Constant 2005 Prices	
3.3.02	Industry, Value Added Per Worker	US Dollars, Constant 2005 Prices	
3.3.03	Services, Value Added Per Worker	US Dollars, Constant 2005 Prices	
4	**QUALITY OF LIFE AND INFRASTRUCTURE DEVELOPMENT**		
4.1	***Physical Infrastructure***		
4.1.01	Population	Number	
4.1.02	Local Road System	Survey Data	
4.1.03	Rail Facilities	Survey Data	
4.1.04	Airport Facilities	Survey Data	
4.1.05	Port Facilities	Survey Data	
4.1.06	Electric Power	Survey Data	
4.1.07	Access to Water	Percentage of Population	
4.1.08	Access to Sanitation	Percentage of Population	
4.2	***Technological Infrastructure***		
4.2.01	Telecommunications System	Percentage of GDP	
4.2.02	Telephone Main Lines	Number Per 100 Persons	
4.2.03	Internet Subscribers	Number Per 100 Persons	
4.2.04	Cellular Mobile Telephone Subscribers	Number Per 100 Persons	
4.2.05	Networked Readiness Index	Survey Data	
4.2.06	Personal Computers	Number Per 100 Persons	
4.2.07	Information and Communication Technology (ICT) Expenditure	Percentage of GDP	
4.3	***Standard of Living, Education and Social Stability***		
4.3.01	Literacy Rate	Percentage	
4.3.02	Primary Schooling	Percentage	
4.3.03	Secondary Schooling	Percentage	
4.3.04	University and Post-Graduate Education	Percentage	
4.3.05	Human Development Index	Composite Statistic	
4.3.06	Health Expenditure	Percentage of GDP	
4.3.07	Urban Population	Percentage of Total Population	
4.3.08	Urban Population Growth	Percentage Per Annum	
4.3.09	Public Security and Safety	Survey Data	

(*Continued*)

Appendix 2 (*Continued*)

No.	Indicator	Unit	Reversed Indicator
4.3.10	Freedom of Press	Survey Data	*R*
4.3.11	Freedom of Speech	Survey Data	
4.3.12	Access to Print Media	Number	
4.3.13	Freedom of Religion	Survey Data	
4.3.14	Religious Unrest	Survey Data	
4.3.15	Labour Unrest	Survey Data	
4.3.16	Social Unrest	Survey Data	
4.3.17	Labour Activism	Survey Data	*R*
4.3.18	Daily Organic Water Pollutant Emissions	Kilogram Per 100 Persons	*R*
4.3.19	Air Quality	Micrograms Per Cubic Meter	*R*
4.3.20	Traffic Congestion	Number of Vehicles Per Km of Road	*R*

Appendix 3

Competitiveness Analysis — The Algorithm

A step-by-step description of the ranking process is described below for N economies, M practical indicators and C environments, with each environment comprising S sub-environments.

(1) Compute the mean value of practical indicator j ($j = 1, \ldots, M$),

$$\bar{X}_j = \frac{1}{N}\sum_{i=1}^{N} X_{ij},$$

where X_{ij} represents the value that economy i ($i = 1, \ldots, N$) takes for practical indicator j.

(2) For each practical indicator j ($j = 1, \ldots, M$), calculate its standard deviation (SD),

$$\mathrm{SD}_j = \sqrt{\frac{1}{N}\sum_{i=1}^{N} (X_{ij} - \bar{X}_j)^2}.$$

(3) Compute the standardised value of indicator (SVI) that each economy i ($i = 1, \ldots, N$) takes under each of the practical indicators j ($j =1, \ldots, M$),

$$\mathrm{SVI}_{ij} = \frac{X_{ij} - \bar{X}_j}{SD_j}.$$

(4) Compute the 'ranked' standardised value of indicator (RSVI) that each economy i ($i = 1, \ldots, N$) takes under each of the practical indicators j ($j = 1, \ldots, M$):

$$\mathrm{RSVI}_{ij} = \begin{cases} \mathrm{SVI}_{ij}, & \text{if a higher value is better} \\ -\mathrm{SVI}_{ij}, & \text{if a lower value is better.} \end{cases}$$

(5) For each of the practical indicators j ($j = 1, \ldots, M$), a ranking can be obtained for economies: economies with a higher value of RSVI for indicator j are ranked ahead of those with a lower value.

(6) For each economy i ($i = 1, \ldots, N$), calculate the RSVI for each sub-environment k ($k = 1, \ldots, S$) belonging to environment l ($l = 1, \ldots, C$),

$$\text{Raw\_RSVI}_{i,lk} = \frac{1}{y_{lk}} \sum_{p=1}^{y_{lk}} \text{RSVI}_{i,jlk,p},$$

$$\text{Mean\_RSVI}_{lk} = \frac{1}{N} \sum_{i=1}^{N} \text{Raw\_RSVI}_{i,lk},$$

$$\text{SD\_RSVI}_{lk} = \sqrt{\frac{1}{N} \sum_{i=1}^{N} (\text{Raw\_RSVI}_{i,lk} - \text{Mean\_RSVI}_{lk})^2},$$

$$\text{RSVI}_{i,lk} = \frac{\text{Raw\_RSVI}_{i,lk} - \text{Mean\_RSVI}_{lk}}{\text{SD\_RSVI}_{lk}},$$

where y_{lk} is the total number of practical indicators under sub-environment k of environment l and ($\text{RSVI}_{i,jlk,1}$, …, $\text{RSVI}_{i,jlk,ylk}$) are the RSVIs for economy i that make up sub-environment k of environment l.

(7) For each economy i ($i = 1, \ldots, N$), calculate the RSVI for each environment l ($l = 1, \ldots, C$),

$$\text{Raw\_RSVI}_{i,l} = \frac{1}{S_l} \sum_{k=1}^{S_l} \text{RSVI}_{i,lk},$$

$$\text{Mean\_RSVI}_{l} = \frac{1}{N} \sum_{i=1}^{N} \text{Raw\_RSVI}_{i,l},$$

$$\text{SD\_RSVI}_{l} = \sqrt{\frac{1}{N} \sum_{i=1}^{N} (\text{Raw\_RSVI}_{i,l} - \text{Mean\_RSVI}_{l})^2},$$

$$\text{RSVI}_{i,l} = \frac{\text{Raw\_RSVI}_{i,l} - \text{Mean\_RSVI}_{l}}{\text{SD\_RSVI}_{l}},$$

where ($\text{RSVI}_{i,l1}$, …, $\text{RSVI}_{i,lS}$) are the RSVIs for the S sub-environments under each environment l.

(8) Overall score of economy i ($i = 1, \ldots, N$),

$$\text{Raw\_}R_i = \frac{1}{C} \sum_{l=1}^{C} \text{RSVI}_{i,l},$$

$$\text{Mean\_}R = \frac{1}{N} \sum_{i=1}^{N} \text{Raw\_}R_i,$$

$$\text{SD}\_R = \sqrt{\frac{1}{N}\sum_{i=1}^{N}(\text{Raw}\_R_i - \text{Mean}\_R)^2},$$

$$R_i = \frac{\text{Raw}\_R_i - \text{Mean}\_R}{\text{SD}\_R}.$$

Economies with a higher R_i are ranked ahead of those with lower value of R_i, and the economy with the highest R_i is the most competitive economy.

Step (5) provides the ranking of each economy for each individual practical indicator. To achieve this ranking, Step (4) adjusts the value of the SVIs so that a higher value will lead to a better ranking in terms of 'competitiveness.' Depending on the nature of the indicator in question, a higher or lower value may reflect a more 'competitive' economy. Take, for instance, the practical sub-indicators 'Gross Domestic Product (GDP)' and 'Unemployment Rate.' A higher GDP but a lower 'unemployment rate' suggests better economic performance which makes an economy more 'competitive.' In most cases where a higher value is better (e.g., GDP), the SVIs of regions are considered, and those with a higher SVI value will have a better ranking. However, for indicators where the inverse is true (e.g., Unemployment Rate), the 'negative SVI' values are compared between economies and a lower SVI value will lead to a better ranking. Step (4) thus seeks to make all standardised values of all practical indicators consistent for ranking purposes.

Step (6) determines the sub-environment rankings of each economy. The average RSVI of all the indicators in the sub-environment are calculated, re-standardised across all the economies and compared to other economies. Economies with a higher average RSVI rank better in the sub-environment.

To arrive at the ranking for each environment, the RSVIs of the sub-environments are averaged and re-standardised across all economies as detailed in Step (7). Finally, Step (8) requires the RSVI values of each environment to be averaged and re-standardised across all economies to determine the overall ranking of the economy. Economies with a higher RSVI are ranked ahead of those with a lower RSVI.

Although the number of sub-environments and indicators may vary for each environment, the aggregate score for each environment is given an equal weight which is 25% of the ASEAN-10 Overall Competitiveness Index. This method is repeated and applied consistently across all economies to ensure precision of the ranking. Mathematically, this can be illustrated as follows:

ASEAN-10 Overall Competitiveness Index:

= 25% × (Macroeconomic Stability)
+25% × (Government and Institutional Setting)
+25% × (Financial, Businesses and Manpower Conditions)
+25% × (Quality of Life and Infrastructure Development).